Springer Series on

Atoms + Plasmas 18

Editor: I.I. Sobel'man

Springer-Verlag Berlin Heidelberg GmbH

Springer Series on

Atoms+Plasmas

Editors: G. Ecker P. Lambropoulos I.I. Sobel'man H. Walther
Managing Editor: H.K.V. Lotsch

1 **Polarized Electrons**
2nd Edition
By J. Kessler

2 **Multiphoton Processes**
Editors: P. Lambropoulos and S.J. Smith

3 **Atomic Many-Body Theory**
2nd Edition
By I. Lindgren and J. Morrison

4 **Elementary Processes in Hydrogen-Helium Plasmas**
Cross Sections and Reaction Rate Coefficients
By R.K. Janev, W.D. Langer, K. Evans, Jr. and D.E. Post, Jr.

5 **Pulsed Electrical Discharge in Vacuum**
By G.A. Mesyats and D.I. Proskurovsky

6 **Atomic and Molecular Spectroscopy** 2nd Edition
Basic Aspects and Practical Applications
By S. Svanberg

7 **Interference of Atomic States**
By E.B. Alexandrov, M.P. Chaika and G.I. Khvostenko

8 **Plasma Physics** 2nd Edition
Basic Theory with Fusion Applications
By K. Nishikawa and M. Wakatani

9 **Plasma Spectroscopy**
The Influence of Microwave and Laser Fields
By E. Oks

10 **Film Deposition by Plasma Techniques**
By M. Konuma

11 **Resonance Phenomena in Electron-Atom Collisions**
By V.I. Lengyel, V.T. Navrotsky and E.P. Sabad

12 **Atomic Spectra and Radiative Transitions** 2nd Edition
By I.I. Sobel'man

13 **Multiphoton Processes in Atoms**
By N.B. Delone and V.P. Krainov

14 **Atoms in Plasmas**
By V.S. Lisitsa

15 **Excitation of Atoms and Broadening of Spectral Lines**
By I.I. Sobel'man, L. Vainshtein and E. Yukov

16 **Reference Data on Multicharged Ions**
By V.G. Pal'chikov and V. Shevelko

17 **Lectures on Non-linear Plasma Kinetics**
By V.N. Tsytovich

18 **Atoms and Their Spectroscopic Properties**
By V.P. Shevelko

V.P. Shevelko

Atoms and Their Spectroscopic Properties

With 69 Figures and 74 Tables

Dr. Viatcheslav P. Shevelko
P.N. Lebedev Physics Institute
Optical Division
Russian Academy of Sciences
117924 Moscow, Russia

Series Editors:

Professor Dr. Günter Ecker
Ruhr-Universität Bochum, Fakultät für Physik und Astronomie,
Lehrstuhl für Theoretische Physik I, Universitätsstrasse 150, D-44801 Bochum, Germany

Professor Peter Lambropoulos, Ph.D.
Max-Planck-Institut für Quantenoptik, D-85748 Garching, Germany, and
Foundation for Research and Technology – Hellas (FO.R.T.H.),
Institute of Electronic Structure & Laser (IESL),
University of Crete, PO Box 1527, Heraklion, Crete 71110, Greece

Professor Igor I. Sobel'man
Lebedev Physics Institute, Optical Division, Russian Academy of Sciences,
Leninsky Prospekt 53, 117924 Moscow, Russia

Professor Dr. Herbert Walther
Sektion Physik der Universität München, Am Coulombwall 1,
D-85748 Garching/München, Germany

Managing Editor: Dr.-Ing. Helmut K.V. Lotsch
Springer-Verlag, Tiergartenstrasse 17, D-69121 Heidelberg, Germany

ISSN 0177-6495

CIP-data applied for

Die Deutsche Bibliothek – CIP-Einheitsaufnahme
Sevelko, Vjačeslav P.:
Atoms and their spectroscopic properties: with 74 tables/V.
P. Shevelko.
(Springer series on atoms + plasmas; 18)
DOI 10.1007/978-3-662-03434-7
NE: GT

Originally published by Springer-Verlag Berlin Heidelberg New York in 1997
MyCopy version of the original edition 1997

Cover design: Design & Production GmbH, Heidelberg
Typesetting: Asco Trade Typesetting Ltd., Hong Kong

SPIN: 10009991 54/3144/SPS – 5 4 3 2 1 0 – Printed on acid-free paper
www.springer.com/mycopy

Preface

Atomic spectroscopy and the physics of atomic collisions have many applications in investigation of confined plasmas, thermonuclear fusion, laser-produced plasmas, the upper planetary atmosphere, plasma diagnostics, and many others.

The aim of this reference book is to give brief information on atomic radiative characteristics and elementary processes occurring in astrophysical and laboratory plasmas. These topics include energy levels, transition probabilities, oscillator strengths, electric multipole polarizabilities, photoionization, excitation, ionization and charge transfer cross sections.

This monograph deals mainly with the interaction of neutral atoms with photons, electrons and ions. Some elementary atomic processes involving negative and positive ions are discussed.

The book can be conditionally divided into two main parts. The first one contains information on the energy-level transition probabilities, atomic polarizabilities, fine and hyperfine structure, angular momenta coupling schemes and selection rules. The second part comprises data on collisional characteristics such as cross sections and the corresponding Maxwellian rate coefficients for different elementary processes: photoionization, excitation, single and multi-electron ionization and electron capture. In the last chapter, recent data on collisions involving H^- ions are considered.

The monograph does not contain a detailed description of complicated theoretical approaches and formulas. It presents data in a dense form using figures, tables and simple analytical formulas, which allows one to estimate the atomic characteristics without resorting to computers. This may be of a special interest to experimentalists working in some fields of atomic physics. As well as the pure atomic data, a brief physical background of specific problems is presented.

I am very grateful to Professor H.-J. Kluge (Gesellschaft für Schwerionenforschung, Darmstadt, Germany) and Professor Hiro Tawara (National Institute for Fusion Science, Nagoya, Japan) for their hospitality during my stays at GSI (October 1994 to February 1995) and NIFS (March 1995 to August 1995), where the major part of this book was written.

It is a great pleasure to thank my colleagues for their help during the writing of this book, especially I.L. Beigman, H.F. Beyer, B.N. Chichkov, V.S. Lebedev, V.G. Pal'chikov, L.P. Presnyakov, E. Salzborn, I.Yu. Tolstikhina, O.I. Tolstikhin, D.B. Uskov, L.A. Vainshtein and E.A. Yukov.

Special thanks are addressed to M.A. Evteeva and N.V. Kozulina for expertly typing the manuscript and to T.A. Shergina for preparing the figures.

Moscow-Darmstadt-Nagoya
June 1996

V.P. Shevelko

Contents

Glossary of Terms

Units

The system of atomic units (a.u.) is used: $e^2 = m = \hbar = 1$.

Length (Bohr radius)	$a_0 = 0.529\ 177\ 249(24) \times 10^{-8}$ cm
Energy	$E_0 = e^2/a_0 = 27.211\ 3961(81)$ eV = 2Ry
Rydberg	$1\ \mathrm{Ry} = me^4/2\hbar^2 = 13.605\ 6981(40)$ eV $= 109\ 737.315\ 34(13)\ \mathrm{cm}^{-1}$
Time	$\tau_0 = \hbar^3/me^4 = a_0/v_0$ $= 2.418\ 88433(11) \times 10^{-17}$ s
Velocity	$v_0 = e^2\hbar = 2.187\ 691\ 417(98) \times 10^8$ cm/s
Cross section	$\pi a_0^2 = 0.879\ 735\ 6696(80) \times 10^{-16}\ \mathrm{cm}^2$
Fine-structure constant	$\alpha = e^2/\hbar c = 1/137.035\ 9895(61)$
Velocity of light	$c = 1/\alpha = 137.035\ 9895(61)$ a.u. $= 2.997\ 92458 \times 10^{10}$ cm/s

The values of the fundamental physical constants are given in a report of the CODATA Task Group on Fundamental Constants, CODATA Bulletin No. 63, E.R. Cohen, B.N. Taylor: Rev. Mod. Phys. **59**, 1121 (1987).

List of Symbols

A	Autoionization transition probability
[A]	Ions of the isoelectronic sequence of an atom A or A-like ions
E	Incident particle energy
E_{cm}	Center-of-mass energy
$E\varkappa$	Electric $2^\varkappa$-pole transition
f	Oscillator strength
I	Binding energy; ionization potential
l	Orbital quantum number
M	Nuclear mass
$M\varkappa$	Magnetic $2^\varkappa$-pole transition
m	Electron mass
N	Total number of atomic electrons
n	Principal quantum number
q	Number of equivalent electrons

T	Electron or ion temperature
v	Relative velocity
$\langle v\sigma \rangle$	Maxwellian rate coefficient
X_z	Ion with a charge $z-1$: $X_z = X^{(z-1)+}$
W	Radiative transition probability
Z	Nuclear charge
z	Spectroscopic symbol: $z = Z - N + 1$
$\beta_\varkappa$	Electric $2^\varkappa$-pole polarizability
ΔE	Transition energy, energy shift
γ	Hyperpolarizability
$\varkappa$	Multiplicity
$\varkappa_r$	Radiative recombination rate coefficient
λ	Wavelength
μ	Reduced mass
σ	Cross section
σ_+	Net (gross) ionization cross section

Special mathematical functions used in the book can be found in *Handbook of Mathematical Functions*, ed. by N. Abramowitz, I.A. Stegun (Constable, London 1970).

Introduction

The development of fundamental and applied research in quantum electronics, astrophysics, plasma physics, physics of the upper atmosphere and quantum chemistry requires the knowledge of radiative and collisional characteristics of neutral atoms colliding with photons and atomic particles. By the latter we mean electrons, ions, atoms and molecules. Such characteristics as energy levels, transition probabilities, cross sections and rate coefficients of different elementary processes, occurring in laboratory and astrophysical plasmas, are of high interest in investigations of plasma-kinetic problems, various diagnostic purposes, development of new laboratory ion and atom sources, radiative losses, plasma spectroscopy, injection of energetic neutral atomic beams for plasma heating, plasma modeling and others.

The present book is a comprehensive collection of radiative and collisional characteristics of neutral and weekly ionized atoms. The material covers a broad range of elementary processes: excitation, ionization, recombination, charge transfer etc., and is presented in a brief form giving, if possible, the scaling laws for different atomic characteristics, universal figures and tables. Besides quite a large number of tables, figures and simple formulas, the book comprises a short physical description of the atomic characteristics, i.e., their dependence on nuclear charge, transition energy, number of electrons, relative velocity of colliding particles, and others.

The book is presented in a way similar to the monograph *Reference Data on Multicharged Ions* by *V.G. Pal'chikov* and *V.P. Shevelko* (Springer Ser. At. Plasm., Vol 16 (1995)). First, information about spectroscopic characteristics of neutral and ionized atoms (Lamb shifts, atomic polarizabilities, fine and hyperfine structure, coupling schemes and selections rules) are given and then the properties of the experimental and theoretical cross sections and corresponding rate coefficients of elementary processes are considered.

Some data (i.e., electron affinities, fine and hyperfine energy intervals and isotope shifts) are taken from the well-known monograph *Reference Data on Atoms, Molecules and Ions* by *A.A. Radzig* and *B.M. Smirnov* (Springer Ser. Chem. Phys., Vol. 13 (1985)). The main difference between the present book and that by *A.A. Radzig* and *B.M. Smirnov* relates to the inclusion the collisional characteristics such as cross sections and corresponding rate coefficients.

A significant volume of the recommended data for effective cross sections and rate coefficients is taken from the issues of the journal *Atomic and Plasma-Material Interaction Data for Fusion* edited by *R.K. Janev* (International

Atomic Energy Agency, Vienna). The data are regularly contributed and critically selected by experts in atomic physics who, besides generating a substantial amount of the new spectroscopic and collisional information, also stimulate a general scientific interest in the atomic physics community for the studies of these data.

The book is attempted in a way that the reader who knows quantum mechanics at a certain level can understand and use the main information of this book. It is written for specialists interested in astrophysics, atomic spectroscopy, controlled fusion, gas dynamics, chemical kinetics, isotope separated and physics of atomic collisions.

1 Atomic Structure and Spectra

The spectroscopic notations and characteristics of atoms and ions including binding energies, electron affinities, fine and hyperfine structure intervals, the Lamb shift and others are given in this Chapter. The radial atomic wave functions expressed in a closed analytical form and their properties are also considered.

1.1 Classification of Spectral Lines

1.1.1 Notations

The *spectroscopic symbol* z of an atom or ion is defined by relation

$$z = Z - N + 1\,, \tag{1.1.1}$$

where Z is the nuclear charge and N is the total number of atomic electrons. The spectroscopic symbol z coincides with the Coulomb charge of the core, consisting of $N - 1$ electrons and the nucleus, at large distances

$$U_c(r) \approx -(Z - N + 1)/r = -z/r\,, \quad r \to \infty\,. \tag{1.1.2}$$

For positive ions $z > 1$, for neutrals $z = 1$.

Ions with the spectroscopic symbol z are also designated as

$$X_z = X^{(z-1)+}\,,$$

i.e., the difference between the spectroscopic symbol and the ion charge is unity.

In atomic spectroscopy, the roman notations for z are also used. For example, the neutral Fe atom is written as Fe I, the Fe^{25+} ion as Fe XXVI, etc. The spectroscopic symbol is important quantity used as a scaling factor for many atomic characteristics such as wavelengths, transition probabilities, effective cross sections and rate coefficients.

Ions with a given number of electrons N arranged in an increasing order of the nuclear charge Z belong to the *isoelectronic sequence* of the corresponding atom A and are called A-like ions or [A] ions. For example, for $N = 1$, one has hydrogen-like ions [H], for $N = 2$ He-like ions [He], etc.

Table 1.1. Notations for electric and magnetic $2^\varkappa$-pole transitions

Notation	Transition	Example
E1	Electric dipole	$2\,^1P_1 - 1\,^1S_0$ in [He]
M1	Magnetic dipole	$2\,^3S_1 - 1\,^1S_0$ in [He]
E1M1	Two-photon electro-magnetic	$2\,^3P_0 - 1\,^1S_0$ in [He]
E1M2	Two-photon electro-magnetic	–
E2M1	Two-photon electro-magnetic	–
E2	Electric quadrupole	$2p_{3/2} - 2p_{1/2}$ in [H]
M2	Magnetic quadrupole	$2\,^3P_2 - 1\,^1S_0$ in [He]
2E1	Two-photon electric dipole	$2\,^1S_0 - 1\,^1S_0$ in [He]
2E2	Two-photon electric quadrupole	–
2M1	Two-photon magnetic dipole	–
2M2	Two-photon magnetic quadrupole	–

The spectral line corresponds to radiation of an atom which makes a transition from the excited level to the lower one. The energy levels are usually described in the *LS*-coupling scheme (sect. 1.2) and are designated in the form:

$$^{2S+1}L_J\,,$$

where L and S are the orbital and spin momenta, respectively, and J is the total angularm momentum of an atom which, according to the general rules of quantum mechanics of momenta addition can take the values

$$|L - S| \leqslant J \leqslant L + S\,.$$

Transitions with a change of spin $|\Delta S| = 1$ are called *intercombination* transitions. Transitions which are not allowed by the selection rules (Sect. 2.2) are called *forbidden* transitions. Notations used for Electric ($\mathrm{E}\varkappa$) and Magnetic ($\mathrm{M}\varkappa$) $2^\varkappa$-pole transitions are given in Table 1.1.

1.1.2 Spectral Series of Hydrogen Atom

The spectrum of the hydrogen atom consists of clearly distinguished series of lines with the wavelengths λ satisfying the formulas [1.1]:

$$\frac{1}{\lambda} = R_\mathrm{H}\left(1 - \frac{1}{n^2}\right), \quad n = 2, 3, 4, \ldots \quad \text{Lyman series}$$

$$\frac{1}{\lambda} = R_\mathrm{H}\left(\frac{1}{2^2} - \frac{1}{n^2}\right), \quad n = 3, 4, 5, \ldots \quad \text{Balmer series}$$

$$\frac{1}{\lambda} = R_\mathrm{H}\left(\frac{1}{3^2} - \frac{1}{n^2}\right), \quad n = 4, 5, 6, \ldots \quad \text{Ritz-Paschen series}$$

$$\frac{1}{\lambda} = R_\mathrm{H}\left(\frac{1}{4^2} - \frac{1}{n^2}\right), \quad n = 5, 6, 7, \ldots \quad \text{Brackett series}$$

$$\frac{1}{\lambda} = R_{\mathrm{H}}\left(\frac{1}{5^2} - \frac{1}{n^2}\right), \quad n = 6, 7, 8, \ldots \quad \text{Pfund series} ,$$

where R_{H} is the Rydberg constant for the hydrogen atom: $R_{\mathrm{H}} = \mathrm{Ry}(1 + m/m_{\mathrm{p}})^{-1} = 109\,677.583\,4043(24)\ \mathrm{cm}^{-1}$.

The Lyman-series lines are denoted by $L_\alpha, L_\beta, L_\gamma, \ldots$, in order of decreasing wavelength (Fig. 1.1); the Balmer-series lines by H_α, H_β, H_γ, and so on. The line L_α with $\lambda = 1215.674$ Å is the resonance line in hydrogen, i.e., it corresponds to the transition from the first excited level to the ground state.

In the Balmer series, the lines of main interest lie in the visible and near-ultraviolet region:

H_α	6562.80 Å ,	H_ε	3970.07 Å ,
H_β	4861.33 Å ,	H_ζ	3889.05 Å ,
H_γ	4340.48 Å ,	H_η	3835.38 Å ,
H_δ	4101.73 Å ,	H_θ	3797.90 Å .

The notations from X-ray spectroscopy are also frequently used (Table 1.2).

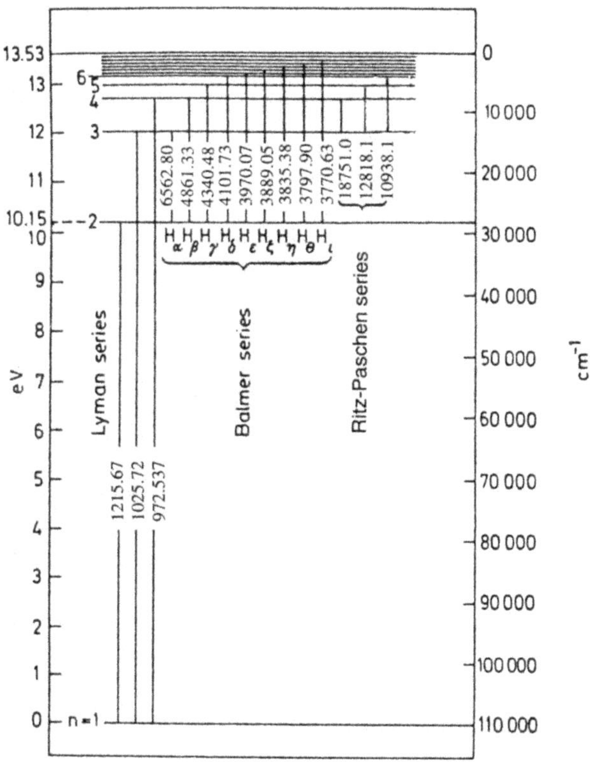

Fig. 1.1. Energy-level diagram for the hydrogen atom; wavelengths are given in Å

Table 1.2. X-ray spectroscopic notations for atomic shells

Shell	Atomic notation	Shell	Atomic notation
K	$1s_{1/2}$	N_6	$4f_{5/2}$
L_1	$2s_{1/2}$	N_7	$4f_{7/2}$
L_2	$2p_{1/2}$	O_1	$5s_{1/2}$
L_3	$2p_{3/2}$	O_2	$5p_{1/2}$
M_1	$3s_{1/2}$	O_3	$5p_{3/2}$
M_2	$3p_{1/2}$	O_4	$5d_{3/2}$
M_3	$3p_{3/2}$	O_5	$5d_{5/2}$
M_4	$3d_{1/2}$	O_6	$5f_{5/2}$
M_5	$3d_{3/2}$	O_7	$5f_{7/2}$
N_1	$4s_{1/2}$	P_1	$6s_{1/2}$
N_2	$4p_{1/2}$	P_2	$6p_{1/2}$
N_3	$4p_{3/2}$	P_3	$6p_{3/2}$
N_4	$4d_{3/2}$	P_4	$6d_{3/2}$
N_5	$4d_{5/2}$	Q_1	$7s_{1/2}$

1.1.3 Spectra of Alkali Elements

The electron structure of alkali atoms (Li, Na, K, Rb, Cs and Fr) is similar to the hydrogen atom, i.e., one (optical) electron outside the atomic core with filled shells. For small principal quantum numbers n and orbital quantum numbers l, the energy levels E_{nl} of alkali atoms strongly differ from those of hydrogen and lie below the hydrogenic states because of the screening effects caused by the atomic core [1.1]. For high nl states, the energy levels in alkali atoms are well described by the hydrogenic formula

$$E_{nl} = -\mathrm{Ry}/n_*^2\,, \quad n_* = n - \Delta\,,$$

where n_* is the *effective* principal quantum number and Δ is the *quantum defect*; the energies E_{nl} are counted from the atomic ionization limit. Obviously, $n_* \equiv n$ for H-like atoms. The dependence of the quantum defect Δ on l is shown in Fig. 1.2. It is seen that Δ is the largest for Cs atoms when $l < 3$.

In general, the principal quantum number n_* is defined as

$$n_* = [E_{nl}/(z^2\mathrm{Ry})]^{-1/2}\,. \tag{1.1.3}$$

Similar to hydrogen, the spectra of alkali atoms have a number of distinguished series:

$ns\ S\text{–}n'p\ P$	principal series
$np\ P\text{–}n's\ S$	sharp series
$np\ P\text{–}n'd\ D$	diffuse series
$nd\ D\text{–}n'f\ F$	fundamental series .

The spectra of other elements are more complicated; their general features are described in [1.1].

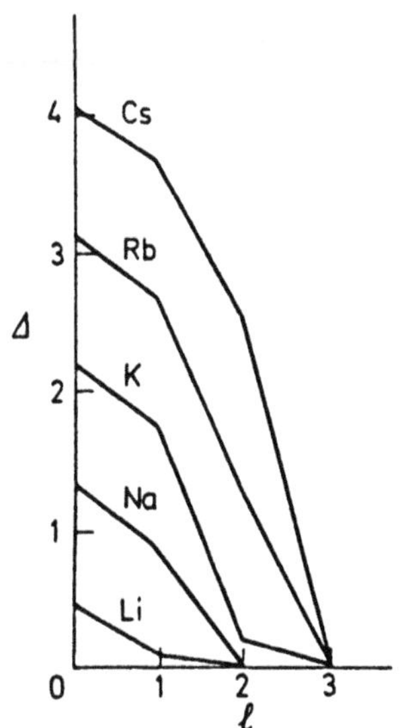

Fig. 1.2. Dependence of the quantum defect Δ for Na, K, Rb and Cs atoms on the orbital quantum number l [1.1]

1.2 Coupling Schemes

The *coupling scheme* (or *vector coupling*) is the term describing the interaction of electrons in atoms, ions and molecules. It shows the scheme of the vector addition of the orbital (l_i) and spin (s_i) momenta of the atomic electrons into the total momentum J.

In the zero-order approximation, the energy of the atomic system is defined by the electrostatic interaction V_{es} of electrons with the nucleus and electrons themselves. In this approximation, the energy level is described by the configuration $n_1 l_1, n_2 l_2, \ldots, n_N$, which is g-times degenerate over all possible projections m_{li} and m_{si}:

$$g = 2(2l_1 + 1)(2l_2 + 2)\ldots(2l_N + 1) .$$

The non-central part of the electrostatic interaction V_{nc} and magnetic spin–orbit interaction V_{so} cause the splitting of the atomic level into sublevels the relative position of which can be described by the scheme of vector coupling of momenta l_i and s_i.

The types of coupling scheme can be easily illustrated by the case of two non-equivalent electrons with momenta l_1, s_1 and l_2, s_2:

$$LS\text{-coupling:}\quad \mathbf{l}_1 + \mathbf{l}_2 = \mathbf{L}\,, \quad \mathbf{s}_1 + \mathbf{s}_2 = \mathbf{S}\,, \quad \mathbf{L} + \mathbf{S} = \mathbf{J}\,;$$

$$LK\text{-coupling:}\quad \mathbf{l}_1 + \mathbf{l}_2 = \mathbf{L}\,, \quad \mathbf{L} + \mathbf{s}_1 = \mathbf{K}\,, \quad \mathbf{K} + \mathbf{s}_2 = \mathbf{J}\,;$$

$$jK\text{-coupling:}\quad \mathbf{l}_1 + \mathbf{s}_1 = \mathbf{j}_1\,, \quad \mathbf{j}_1 + \mathbf{l}_2 = \mathbf{K}\,, \quad \mathbf{K} + \mathbf{s}_2 = \mathbf{J}\,;$$

$$jj\text{-coupling:}\quad \mathbf{l}_1 + \mathbf{s}_1 = \mathbf{j}_1\,, \quad \mathbf{l}_2 + \mathbf{s}_2 = \mathbf{j}_2\,, \quad \mathbf{j}_1 + \mathbf{j}_2 = \mathbf{J}\,.$$

In the case of equivalent electrons nl^q, due to the Pauli principle, the *LS*- and *jj*-couplings are only possible because all electrons participate symmetrically.

Each type of coupling scheme characterizes relative values of different types of interactions. The *LS*-coupling (also called *Russel-Saunders coupling*) is used when the electrostatic interaction V_{es} is much larger than the spin-orbit one:

$$V_{es} \gg V_{so} .$$

The *LS*-coupling scheme is used for not very heavy neutral atoms and ions with a small charge in the not highly excited states.

With increasing nuclear charge Z ($Z \to \infty$), the opposite situation occurs

$$V_{es} \ll V_{so} ,$$

and the *jj*-coupling scheme is used, which is of particular interest for multicharged ions.

In the case $V_{es} \approx V_{so}$, the states are described by the so-called *intermediate coupling scheme*. Figure 1.3 shows a smooth transformation from *LS*- to the *jj*-coupling with increasing nuclear charge Z in the case of Be-like ions. In this book the *LS*-coupling scheme is mainly used.

The jK-coupling is usually called jl-coupling; jl-coupling is used when the spin–orbit interaction of the electrons of the atomic core is larger than the electrostatic interaction of these electrons with the excited electron. The jl-coupling appears in the spectra of excited states of noble-gas atoms (Ne, Ar, Kr, Xe and Rn, e.g., configuration $np^5n's$) and of some other atoms when one of the electrons is in average at a large distance from the atomic core.

The *LS*-coupling scheme is most often used. The designation for terms in

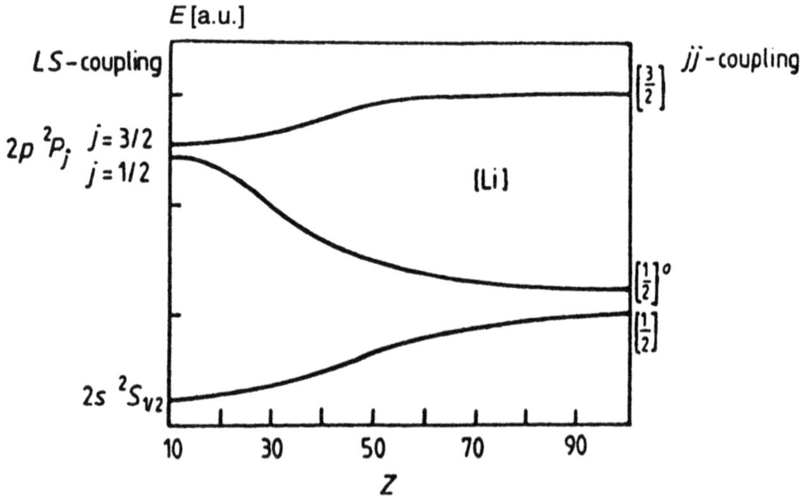

Fig. 1.3. Transformation from *LS*-coupling to *jj*-coupling with increasing nuclear charge Z for low-lying energy levels in Be-like ions [1.2]

LS-coupling is ${}^{2S+1}L_J$, where S, L and J are the spin, orbital and total momenta, respectively. For example, for the two-electron configuration $npn'p$, one has the singlet ($S = 0$) and the triplet ($S = 1$) states:

$$ {}^1S_0\,,\quad {}^1P_1\,,\quad {}^1D_2\,,\quad {}^3S_1\,,\quad {}^3P_{0,1,2}\,,\quad {}^3D_{1,2,3}\,. $$

The designation of terms in other coupling schemes is more complicated. For the same configuration $npn'p$ one has:

LK-coupling, level designation $L[K]_J$:

$$ S\left[\frac{1}{2}\right]_{0,1},\quad P\left[\frac{1}{2}\right]_{0,1},\quad P\left[\frac{3}{2}\right]_{1,2},\quad D\left[\frac{3}{2}\right]_{1,2},\quad D\left[\frac{5}{2}\right]_{2,3}; $$

jK-coupling, level designation $j[K]_J$:

$$ \frac{1}{2}\left[\frac{1}{2}\right]_{0,1},\quad \frac{3}{2}\left[\frac{1}{2}\right]_{0,1},\quad \frac{1}{2}\left[\frac{3}{2}\right]_{1,2},\quad \frac{3}{2}\left[\frac{3}{2}\right]_{1,2},\quad \frac{3}{2}\left[\frac{5}{2}\right]_{2,3}; $$

jj-coupling, level designation $[j_1 j_2]_J$:

$$ \left[\frac{1}{2}\frac{1}{2}\right]_{0,1},\quad \left[\frac{1}{2}\frac{3}{2}\right]_{1,2},\quad \left[\frac{3}{2}\frac{1}{2}\right]_{1,2},\quad \left[\frac{3}{2}\frac{3}{2}\right]_{0,1,2,3}. $$

The number of sublevels is the same for a given total momentum J for all types of the coupling schemes.

In many cases, the use of a “pure” scheme is not possible and it is necessary to use the intermediate coupling scheme [1.1].

1.3 Ionization Potentials and Binding Energies

The ground-state atomic levels are occupied according to the Pauli principle. If the electron configuration is known, the term of the ground state can be determined by the *Hund's rule* [1.1].

For ionization of outer- or inner-shell electrons, it is necessary to assign a certain energy to an atom. In the case of ionization of outer electrons from the ground atomic state, this energy is called the *ionization energy* or *ionization potential* (I). The energy required for ionization of an inner-shell electron or of an electron from excited state is called the *binding energy*.

Configurations of two outer electron shells of neutral atoms from H ($Z = 1$) up to No ($Z = 102$) in their ground states are given in Table 1.3 together with the ground-state terms and ionization potentials. The binding energies of different subshells in neutrals are given in Table 1.4. Calculated ionization potentials for all states of ionization of all elements up to the nuclear charge number $Z = 103$ are given in [1.5].

Table 1.3. Electron configuration and ionization potentials I of atoms [1.1, 3]

Element	Electron configuration	Ground-state term	I [eV]	Element	Electron configuration	Ground state term	I [eV]
1 H	$1s$	$^2S_{1/2}$	13.598	52 Te	$5s^2 5p^4$	$^3P_2$	9.009
2 He	$1s^2$	$^1S_0$	24.587	53 I	$5s^2 5p^5$	$^2P_{3/2}$	10.451
3 Li	$1s^2 2s$	$^2S_{1/2}$	5.392	54 Xe	$5s^2 5p^6$	$^1S_0$	12.130
4 Be	$1s^2 2s^2$	$^1S_0$	9.322	55 Cs	$5p^6 6s$	$^2S_{1/2}$	3.894
5 B	$2s^2 2p$	$^2P_{1/2}$	8.298	56 Ba	$5p^6 6s^2$	$^1S_0$	5.212
6 C	$2s^2 2p^2$	$^3P_0$	11.260	57 La	$5d6s^2$	$^2D_{3/2}$	5.577
7 N	$2s^2 2p^3$	$^4S_{3/2}$	14.534	58 Ce	$4f5d6s^2$	$^1G_4$	5.47
8 O	$2s^2 2p^4$	$^3P_2$	13.618	59 Pr	$4f^3 6s^2$	$^4I_{9/2}$	5.42
9 F	$2s^2 2p^5$	$^2P_{3/2}$	17.422	60 Nd	$4f^4 6s^2$	$^5I_4$	5.49
10 Ne	$2s^2 2p^6$	$^1S_0$	21.564	61 Pm	$4f^5 6s^2$	$^6H_{5/2}$	5.55
11 Na	$2p^6 3s$	$^2S_{1/2}$	5.139	62 Sm	$4f^6 6s^2$	$^7F_0$	5.63
12 Mg	$2p^6 3s^2$	$^1S_0$	7.646	63 Eu	$4f^7 6s^2$	$^8S_{7/2}$	5.67
13 Al	$3s^2 3p$	$^2P_{1/2}$	5.986	64 Gd	$4f^7 5d6s^2$	$^9D_2$	6.14
14 Si	$3s^2 3p^2$	$^3P_0$	8.151	65 Tb	$4f^9 6s^2$	$^6H_{15/2}$	5.85
15 P	$3s^2 3p^3$	$^4S_{3/2}$	10.486	66 Dy	$4f^{10} 6s^2$	$^5I_8$	5.93
16 S	$3s^2 3p^4$	$^3P_2$	10.360	67 Ho	$4f^{11} 6s^2$	$^4I_{15/2}$	6.02
17 Cl	$3s^2 3p^5$	$^2P_{3/2}$	12.967	68 Er	$4f^{12} 6s^2$	$^3H_6$	6.10
18 Ar	$3s^2 3p^6$	$^1S_0$	15.759	69 Tm	$4f^{13} 6s^2$	$^2F_{7/2}$	6.18
19 K	$3p^6 4s$	$^2S_{1/2}$	4.341	70 Yb	$4f^{14} 6s^2$	$^1S_0$	6.254
20 Ca	$3p^6 4s^2$	$^1S_0$	6.113	71 Lu	$5d6s^2$	$^2D_{3/2}$	5.426
21 Sc	$3d4s^2$	$^2D_{3/2}$	6.54	72 Hf	$5d^2 6s^2$	$^3F_2$	7.0
22 Ti	$3d^2 4s^2$	$^3F_2$	6.82	73 Ta	$5d^3 6s^2$	$^4F_{3/2}$	7.89
23 V	$3d^3 4s^2$	$^4F_{3/2}$	6.74	74 W	$5d^4 6s^2$	$^5D_0$	7.98
24 Cr	$3d^5 4s$	$^7S_3$	6.766	75 Re	$5d^5 6s^2$	$^5S_{5/2}$	7.88
25 Mn	$3d^5 4s^2$	$^6S_{5/2}$	7.435	76 Os	$5d^6 6s^2$	$^5D_4$	8.7
26 Fe	$3d^6 4s^2$	$^5D_4$	7.870	77 Ir	$5d^7 6s^2$	$^4F_{9/2}$	9.1
27 Co	$3d^7 4s^2$	$^4F_{9/2}$	7.86	78 Pt	$5d^9 6s$	$^3D_3$	9.0
28 Ni	$3d^8 4s^2$	$^3F_4$	7.635	79 Au	$5d^{10} 6s$	$^2S_{1/2}$	9.225
29 Cu	$3d^{10} 4s$	$^2S_{1/2}$	7.726	80 Hg	$5d^{10} 6s^2$	$^1S_0$	10.437
30 Zn	$3d^{10} 4s^2$	$^1S_0$	9.394	81 Tl	$6s^2 6p$	$^2P_{1/2}$	6.108
31 Ga	$4s^2 4p$	$^2P_{1/2}$	5.999	82 Pb	$6s^2 6p^2$	$^3P_0$	7.416
32 Ge	$4s^2 4p^2$	$^3P_0$	7.899	83 Bi	$6s^2 6p^3$	$^4S_{3/2}$	7.289
33 As	$4s^2 4p^3$	$^4S_{3/2}$	9.81	84 Po	$6s^2 6p^4$	$^3P_2$	8.42
34 Se	$4s^2 4p^4$	$^3P_2$	9.752	85 At	$6s^2 6p^5$	$^2P_{3/2}$	9.5
35 Br	$4s^2 4p^5$	$^2P_{3/2}$	11.814	86 Rn	$6s^2 6p^6$	$^1S_0$	10.748
36 Kr	$4s^2 4p^6$	$^1S_0$	13.999	87 Fr	$6p^6 7s$	$^2S_{1/2}$	4.0
37 Rb	$4p^6 5s$	$^2S_{1/2}$	4.177	88 Ra	$6p^6 7s^2$	$^1S_0$	5.279
38 Sr	$4p^6 5s^2$	$^1S_0$	5.695	89 Ac	$6d7s^2$	$^2D_{3/2}$	5.170
39 Y	$4d5s^2$	$^2D_{3/2}$	6.38	90 Th	$6d^2 7s^2$	$^3F_2$	6.080
40 Zr	$4d^2 5s^2$	$^3F_2$	6.84	91 Pa	$5f^2 6d7s^2$	$^4K_{11/2}$	5.890
41 Nb	$4d^4 5s$	$^6D_{1/2}$	6.88	92 U	$5f^3 6d7s^2$	$^5L_6$	6.050
42 Mo	$4d^5 5s$	$^7S_3$	7.099	93 Np	$5f^4 6d7s^2$	$^6L_{11/2}$	6.190
43 Tc	$4d^5 5s^2$	$^6S_{5/2}$	7.28	94 Pu	$5f^6 7s^2$	$^7F_0$	6.062
44 Ru	$4d^7 5s$	$^5F_5$	7.37	95 Am	$5f^7 7s^2$	$^8S_{7/2}$	5.993
45 Rh	$4d^8 5s$	$^4F_{9/2}$	7.46	96 Cm	$5f^7 6d7s^2$	$^9D_2$	6.021
46 Pd	$4p^6 4d^{10}$	$^1S_0$	8.34	97 Bk	$5f^9 7s^2$	$^6H_{15/2}$	6.229
47 Ag	$4d^{10} 5s$	$^2S_{1/2}$	7.576	98 Cf	$5f^{10} 7s^2$	$^5I_8$	6.298
48 Cd	$4d^{10} 5s^2$	$^1S_0$	8.993	99 Es	$5f^{11} 7s^2$	$^4I_{15/2}$	6.422
49 In	$5s^2 5p$	$^2P_{1/2}$	5.786	100 Fm	$5f^{12} 7s^2$	$^3H_6$	6.500
50 Sn	$5s^2 5p^2$	$^3P_0$	7.344	101 Md	$5f^{13} 7s^2$	$^2F_{7/2}$	6.580
51 Sb	$5s^2 5p^3$	$^4S_{3/2}$	8.641	102 No	$5f^{14} 7s^2$	$^1S_0$	6.650

Table 1.4. Semiempirical binding energies (in eV) of subshells in neutral atoms [1.4]. Notations for subshells are given in Table 1.2

Z	K	L_1	L_2	L_3	M_1	M_2	M_3	M_4	M_5	N_1	N_2	N_3	N_4	N_5	N_6	N_7	O_1	O_2	O_3	O_4	O_5	O_6	O_7	P_1	P_2	P_3	P_4	Q_1
1 H	13.60																											
2 He	24.59																											
3 Li	58	5.392																										
4 Be	115	9.322																										
5 B	192	12.93	8.298																									
6 C	288	16.59	11.26																									
7 N	403	20.33	14.53																									
8 O	538	28.48	13.62																									
9 F	694	37.85	17.42																									
10 Ne	870.1	48.87	21.66	21.56																								
11 Na	1075	66	34	34	5.139																							
12 Mg	1308	92	54	54	7.646																							
13 Al	1564	121	77	77	10.62	5.986																						
14 Si	1844	154	104	104	13.46	8.151																						
15 P	2148	191	135	134	16.15	10.49																						
16 S	2476	232	170	168	20.20	10.36																						
17 Cl	2829	277	208	206	24.54	12.97																						
18 Ar	3206.3	326.5	250.6	248.5	29.24	15.94	15.76																					
19 K	3610	381	299	296	37	19	18.7			4.341																		
20 Ca	4041	441	353	349	46	28	28			6.113																		
21 Sc	4494	503	408	403	55	33	33	8		6.540																		
22 Ti	4970	567	465	459	64	39	38	8		6.820																		
23 V	5470	633	525	518	72	44	43	8		6.740																		
24 Cr	5995	702	589	580	80	49	48	8.25		6.765																		
25 Mn	6544	775	656	645	89	55	53	9		7.434																		
26 Fe	7117	851	726	713	98	61	59	9		7.870																		
27 Co	7715	931	800	785	107	68	66	9		7.864																		
28 Ni	8338	1015	877	860	117	75	73	10	10	7.635																		
29 Cu	8986	1103	958	938	127	82	80	11	10.4	7.726																		
30 Zn	9663	1198	1047	1024	141	94	91	12	11.2	9.394																		
31 Ga	10371	1302	1146	1119	162	111	107	21	20	11	6.00																	
32 Ge	11107	1413	1251	1220	184	130	125	33	32	14.3	7.90																	
33 As	11871	1531	1362	1327	208	151	145	46	45	17	9.81																	
34 Se	12662	1656	1479	1439	234	173	166	61	60	20.15	9.75																	
35 Br	13481	1787	1602	1556	262	197	189	77	76	23.80	11.85																	

Table 1.4 (*Cont.*)

Z	K	L_1	L_2	L_3	M_1	M_2	M_3	M_4	M_5	N_1	N_2	N_3	N_4	N_5	N_6	N_7	O_1	O_2	O_3	O_4	O_5	O_6	O_7	P_1	P_2	P_3	P_4	Q_1
36 Kr	14327	1927	1731	1678	292	222	214	95.0	93.8	27.51	14.67	14.00																
37 Rb	15203	2068	1867	1807	325	251	242	116	114	32	16	15.3					4.18											
38 Sr	16108	2219	2010	1943	361	283	273	139	137	40	23	22					5.69											
39 Y	17041	2375	2158	2083	397	315	304	163	161	48	30	29	6.38				6.48											
40 Zr	18002	2536	2311	2227	434	348	335	187	185	56	35	33	8.61				6.84											
41 Nb	18990	2702	2469	2375	472	382	367	212	209	62	40	38	7.17				6.88											
42 Mo	20006	2872	2632	2527	511	416	399	237	234	68	45	42	8.56				7.10											
43 Tc	21050	3048	2800	2683	551	451	432	263	259	74	49	45	8.6				7.28											
44 Ru	22123	3230	2973	2844	592	488	466	290	286	81	53	49	8.50				7.37											
45 Rh	23225	3418	3152	3010	634	526	501	318	313	87	58	53	9.56				7.46											
46 Pd	24357	3611	3337	3180	677	565	537	347	342	93	57	8.87	8.34															
47 Ag	25520	3812	3530	3357	724	608	577	379	373	101	69	63	11	10			7.58											
48 Cd	26715	4022	3732	3542	775	655	621	415	408	112	78	71	14	13			8.99											
49 In	27944	4242	3943	3735	830	707	669	455	447	126	90	82	21	20			10	5.79										
50 Sn	29204	4469	4160	3933	888	761	719	497	489	141	102	93	29	28			12	7.34										
51 Sb	30496	4703	4385	4137	949	817	771	542	533	157	114	104	38	37			15	8.64										
52 Te	31820	4945	4618	4347	1012	876	825	589	578	174	127	117	48	46			17.84	9.01										
53 I	33176	5195	4858	4563	1078	937	881	638	626	193	141	131	58	56			20.61	10.45										
54 Xe	34565	5452	5106	4785	1149	1001	939	689	676	213	157	147	69.5	67.5			23.40	13.44	12.13									
55 Cs	35987	5717	5362	5014	1220	1068	1000	742	728	233	174	164	81	79			25	14	12.3					3.89				
56 Ba	37442	5991	5626	5249	1293	1138	1063	797	782	254	193	181	94	92			31	18	16					5.21				
57 La	38928	6269	5894	5486	1365	1207	1124	851	834	273	210	196	105	103			36	22	19	5.75				5.58				
58 Ce	40446	6552	6167	5726	1437	1275	1184	903	885	291	225	209	114	111	6		39	25	22	6				5.65				
59 Pr	41995	6839	6444	5968	1509	1342	1244	954	934	307	238	220	121	117	6		41	27	24					5.42				
60 Nd	43575	7432	6727	6213	1580	1408	1303	1005	983	321	250	230	126	122	6		42	28	25					5.49				
61 Pm	45188	7432	7017	6464	1653	1476	1362	1057	1032	335	261	240	131	127	6		43	28	25					5.55				
62 Sm	46837	7740	7315	6720	1728	1546	1422	1110	1083	349	273	251	137	132	6		44	29	25					5.63				
63 Eu	48522	8056	7621	6981	1805	1618	1848	1164	1135	364	286	262	143	137	6		45	30	26					5.68				
64 Gd	50243	8380	7935	7247	1884	1692	1547	1220	1189	380	300	273	150	143	6		46	31	27	6				6.16				
65 Tb	51999	8711	8256	7518	1965	1768	1612	1277	1243	398	315	285	157	150	6		48	32	28	6				5.85				
66 Dy	53792	9050	8585	7794	2048	1846	1678	1335	1298	416	331	[illegible]	[illegible]															
67 Ho	55622	9398	8922	8075	2133	1926	1746	1395	1354	434	348	310	172	164	6		52	34	29					6.02				
68 Er	57489	9754	9267	8361	2220	2008	1815	1456	1412	452	365	323	181	172	6		54	35	30					6.10				
69 Tm	59393	10118	9620	8651	2309	2092	1885	1518	1471	471	382	336	190	181	7		56	36	30					6.18				
70 Yb	61335	10490	9981	8946	2401	2178	1956	1580	1531	400	399	349	200	190	8	7	58	37	31					6.25				

71 Lu	63310	10876	10355	9250	2499	2270	2032	1647	1596	514	420	366	213	202	13	12	62	39	32	6.6				7.0				
72 Hf	65350	11275	10742	9564	2604	2369	2113	1720	1665	542	444	386	229	217	21	20	68	43	35	7.0				7.5				
73 Ta	67419	11684	11139	9884	2712	2472	2197	1796	1737	570	469	407	245	232	30	28	74	47	38	8.3				7.9				
74 W	69529	12103	11546	10209	2823	2577	2283	1874	1811	599	495	428	261	248	38	36	80	51	41	9.0				8.0				
75 Re	71681	12532	11963	10540	2937	2686	2371	1953	1887	629	522	450	278	264	47	45	86	56	45	9.6				7.9				
76 Os	73876	12972	12390	10876	3054	2797	2461	2035	1964	660	551	473	295	280	56	54	92	61	49	9.6				8.5				
77 Tr	76115	13422	12828	11219	3175	2912	2554	2119	2044	693	581	497	314	298	67	64	99	66	53	9.6				9.1				
78 Pt	78399	13883	13277	11567	3300	3030	2649	2206	2126	727	612	522	335	318	78	75	106	71	57	9.6				9.0				
79 Au	80729	14356	13738	11923	3430	3153	2748	2295	2210	764	645	548	357	339	91	87	114	76	61	12.5	11.1			9.23				
80 Hg	83108	14845	14214	12288	3567	3283	2852	2390	2300	806	683	579	382	363	107	103	125	85	68	14	12			10.4				
81 Tl	85536	15350	14704	12662	3710	3420	2961	2490	2394	852	726	615	411	391	127	123	139	98	79	21	19			8	6.11			
82 Pb	88011	15867	15206	13041	3857	3560	3072	2592	2490	899	769	651	441	419	148	144	153	111	90	27	25			10	7.42			
83 Bi	90534	16396	15719	13426	4007	3704	3185	2696	2588	946	813	687	472	448	170	165	167	125	101	34	32			12	7.29			
84 Po	93106	16937	16244	13816	4161	3852	3301	2802	2687	994	858	724	503	478	193	187	181	139	112	41	38			15	8.43			
85 At	95729	17490	16782	14212	4320	4005	3420	2910	2788	1044	904	761	535	508	217	211	196	153	123	48	44			19	11	9.3		
86 Rn	98404	18055	17334	14615	4483	4162	3542	3019	2890	1096	951	798	567	538	242	235	212	167	134	55	51			24	14	10.7		
87 Fr	101134	18637	17903	15028	4652	4324	3666	3134	2998	1153	1003	839	603	572	268	260	231	183	147	65	61			33	17	13		4.0
88 Ra	103919	19237	18488	15449	4827	4491	3793	3254	3111	1214	1060	884	642	609	296	287	253	201	161	77	73			46	28	23		5.28
89 Ac	106759	19850	19086	15874	5005	4661	3921	3374	3223	1274	1116	928	680	645	322	313	274	218	174	88	83			56	39	33	5.7	6.3
90 Th	109654	20475	19696	16303	5185	4833	4049	3494	3335	1333	1171	970	717	679	347	338	293	233	185	97	91			64	48	41	6	6
91 Pa	112604	21112	20318	16735	5368	5008	4178	3613	3446	1390	1225	1011	752	712	372	362	312	248	195	104	97	6		70	54	46	6	6
92 U	115611	21762	20953	17171	5553	5187	4308	3733	3557	1446	1278	1050	785	743	396	386	329	261	203	110	101	6		74	57	48	6.1	6
93 Np	118676	22427	21602	17612	5742	5370	4440	3854	3669	1504	1331	1089	819	774	421	410	346	274	211	116	106	6		78	61	51	6	6
94 Pu	121800	23109	22267	18059	5936	5557	4574	3977	3783	1563	1384	1128	853	805	446	434	363	287	219	122	111	6		83	65	54		6
95 Am	124984	23808	22949	18512	6135	5748	4710	4102	3898	1623	1439	1167	888	837	471	458	380	301	227	128	116	6		87	69	57		6.0
96 Cm	128229	24524	23648	18971	6339	5944	4848	4230	4016	1684	1495	1207	923	869	497	484	397	315	235	134	121	6		92	73	60	6	6
97 Bk	131536	25257	24361	19435	6548	6145	4989	4360	4136	1746	1553	1248	959	902	525	517	414	330	243	141	126	6		97	78	64	6	6
98 Cf	135906	26007	25097	19905	6762	6351	5132	4493	4258	1809	1613	1289	995	935	554	539	431	345	252	149	131	6		103	83	68	6	6
99 Es	139340	26774	25847	20380	6982	6562	5278	4628	4382	1873	1675	1331	1032	968	584	569	448	361	261	157	137	6		108	88	72	6	6
100 Fm	142839	27559	26614	20860	7208	6779	5426	4766	4508	1939	1739	1373	1069	1002	615	599	465	377	270	166	143	6		114	93	76	6	6
101 Md	146404	28361	27399	21345	7440	7002	5577	4906	4636	2006	1805	1416	1107	1036	647	630	482	393	279	175	149	7		120	99	81		6
102 No	150036	29181	28202	21835	7678	7231	5731	5048	4766	2074	1873	1459	1146	1071	680	662	499	410	288	185	155	8	7	126	105	86		6
103 Lw	153736	30023	29027	22332	7925	7469	5891	5196	4901	2147	1946	1506	1188	1109	716	697	519	430	300	198	163	13	12	134	113	93	7	7
104 Ku	157500	30887	29874	22837	8182	7716	6057	5350	5042	2225	2024	1557	1233	1151	755	735	542	454	315	213	173	21	20	144	123	102	8	8
105	161340	31770	30740	23346	8446	7970	6227	5507	5185	2304	2105	1609	1279	1193	795	774	565	479	330	229	183	29	27	154	133	111	9	8
106	165250	32670	31630	23860	8717	8231	6401	5667	5330	1385	2189	1662	1326	1236	836	814	588	505	345	246	193	37	35	164	143	120	9	8
107	169240	33600	32530	24380	8995	8499	6579	5830	5478	2468	2276	1716	1374	1279	878	855	611	529	360	263	203	46	44	174	153	129	10	8
108	173290	34540	33460	24900	9280	8770	6760	5996	5628	2553	2366	1771	1422	1323	921	897	634	554	375	281	213	56	54	184	163	138	10	9

1.4 Electron Affinity

Some neutral atoms and molecules can capture one electron and create negative ions in the stable states. The binding energy of an outer electron of the negative ion in its ground state is called the *electron affinity*. The quantities of electron affinities taken from [1.6–7] and recent publications are given in Table 1.5.

1.5 Fine and Hyperfine Structure

In the case of neutral hydrogen or hydrogen-like ion, the relativistic effects such as the velocity dependence of the electron mass and the spin–orbit interaction lead to the splitting of the nl level into two components: $j = l + 1/2$ and $j = l - 1/2$. This splitting is called *fine* or *multiplet splitting*. Due to the relativistic effects mentioned, each of the components is also shifted by

$$\Delta E_{nlj} = -\left(\frac{1}{j+1/2} - \frac{3}{4n}\right)\frac{\alpha^2 Z^4}{n^3}\mathrm{Ry} < 0\,, \quad j = l \pm 1/2\,, \tag{1.5.1}$$

where α is the dimensionless constant called the *fine-structure constant*. Lines arising from transitions between the fine structure components (transitions nlj–$n'l'j'$) are called *multiplet* with the selection rule $\Delta j = j' - j = 0,\ \pm 1$ (Sect. 2.2).

The scale of splitting and shifting is of the same order of magnitude $\propto \alpha^2$ (Fig. 1.4). According to (1.5.1), the splitting between two components $j = l \pm 1/2$ is given by

$$\delta E = \frac{\alpha^2 Z^4}{n^3 l(l+1)}\mathrm{Ry}\,. \tag{1.5.2}$$

For the $2p$ state, one has from (1.5.1, 2) (in units of $\alpha^2 Z^4$Ry):

$$\Delta E_{2p1/2} = -\frac{1}{8}\frac{5}{8} = -\frac{5}{64}; \quad \Delta E_{2p3/2} = -\frac{1}{8}\frac{1}{8} = -\frac{1}{64}; \quad \delta E = \frac{4}{64} = \frac{1}{16}.$$

For the hydrogen atom ($Z = 1$) in the $n = 2$, 3 and 4 states, the splittings of the levels with $j = 1/2$ and $3/2$ are, respectively, $\delta E = 0.36, 0.12$ and $0.044\ \mathrm{cm}^{-1}$.

In the case of a multielectron system, the spin–orbit interaction plays a dominant role and leads to the dependence of the energy term on the orbital L, spin S and total angular momentum J.

If $L \geqslant S$, the atomic term splits into $2S + 1$ different components. The quantity $2S + 1$, determining the number of components is called the multiplicity of the term. If $L < S$ the term splits into $2L + 1$ components.

Table 1.5. Electron affinities I

Nuclear charge Z	Ion, term	Configuration	I [eV]
1	$D^-(^1S)$	$1s^2$	0.754593
1	$H^-(^1S)$	$1s^2$	0.754202
2	$He^-(^4P)$	$1s2s2p$	0.077
3	$Li^-(^1S)$	$1s^22s^2$	0.6180
	$Li^-(^3P)$	$2s2p^2$	0.050
4	$Be^-(^4P)$	$2s2p^2$	0.291
	$Be^-(^4S)$	$1s^22p^3$	0.295
5	$B^-(^3P)$	$2s^22p^2$	0.277
	$B^-(^1P)$	$2s^22p^2$	0.104
6	$C^-(^4S^0)$	$2s^22p^3$	1.2629
	$C^-(^2D)$	$2s^22p^3$	0.035
7	$N^-(^3P)$	$2s^22p^4$	0.2–0.7
8	$O^-(^2P)$	$2s^22p^5$	1.4611103
9	$F^-(^1S)$	$2s^22p^6$	3.401190
11	$Na^-(^1S)$	$3s^2$	0.547926
13	$Al^-(^3P_0)$	$3p^2$	0.441
	$Al^-(^1D_2)$	$3p^2$	0.33
14	$Si^-(^4S)$	$3p^3$	1.389
	$Si^-(^2D)$	$3p^3$	0.526
	$Si^-(^2P_2)$	$3p^3$	0.034
15	$P^-(^3P)$	$3p^4$	0.7465
16	$S^-(^2P)$	$3p^5$	2.077104
17	$Cl^-(^1S)$	$3p^6$	3.61269
19	$K^-(^1S)$	$4s^2$	0.50147
20	$Ca^-(^2P_{1/2})$	$4s^24p$	0.0175–0.0246
	$Ca^-(^2P_{3/2})$	$4s^24p$	0.0197
21	$Se^-(^1D)$	$3d4s^24p$	0.188
	$Se^-(^3D)$	$3d4s^24p$	0.04
22	$Ti^-(^4F)$	$3d^34s^2$	0.079
23	$V^-(^5D)$	$3d^44s^2$	0.525
24	$Cr^-(^6S)$	$3d^54s^2$	0.666
26	$Fe^-(^4F)$	$3d^74s^2$	0.151
27	$Co^-(^3F)$	$3d^84s^2$	0.662
28	$Ni^-(^2D)$	$3d^94s^2$	1.156
29	$Cu^-(^1S)$	$3d^{10}4s^2$	1.235
31	$Ga^-(^3P)$	$4p^2$	0.3
32	$Ge^-(^4S)$	$4p^3$	1.233
33	$As^-(^3P)$	$4p^4$	0.81
34	$Se^-(^2P)$	$4p^5$	2.020670
35	$Br^-(^1S)$	$4p^6$	3.363590
37	$Rb^-(^1S)$	$5s^2$	0.48592
38	$Sr^-(^2P_{1/2})$	$5s^25p$	0.054
	$Sr^-(^2P_{3/2})$	$5s^25p$	0.029
39	$Y^-(^1D)$	$4d5s^25p$	0.307
	$Y^-(^3D)$	$4d5s^25p$	0.16
40	$Zr^-(^4F)$	$4d^35s^2$	0.426
41	$Nb^-(^5D)$	$4d^45s^2$	0.893
42	$Mo^-(^6S)$	$4d^55s^2$	0.746
43	$Tc^-(^5D)$	$4d^65s^2$	0.55
44	$Ru^-(^4F)$	$4d^75s^2$	1.05

Table 1.5 (*Cont.*)

Nuclear charge Z	Ion, term	Configuration	I [eV]
45	$Rh^-(^3F)$	$4d^85s^2$	1.137
46	$Pd^-(^2D)$	$4d^95s^2$	0.557
47	$Ag^-(^1S)$	$4d^44d^{10}$	1.302
49	$In^-(^3P)$	$5p^2$	0.30
50	$Sn^-(^4S)$	$5p^3$	1.112
51	$Sb^-(^3P)$	$5p^4$	1.07
52	$Te^-(^2P)$	$5p^5$	1.9708
53	$I^-(^1S)$	$5p^6$	3.0591
55	$Cs^-(^1S)$	$6s^2$	0.4716
57	$La^-(^3F)$	$5d^26s^2$	0.5
73	$Ta^-(^5D)$	$5d^46s^2$	0.322
74	$W^-(^6S)$	$5d^56s^2$	0.815
75	$Re^-(^5D)$	$5d^66s^2$	0.15
76	$Os^-(^4F)$	$5d^76s^2$	1.2
77	$Ir^-(^3F)$	$5d^86s^2$	1.565
78	$Pt^-(^2D)$	$5d^96s^2$	2.128
79	$Au^-(^1S)$	$5d^{10}6s^2$	2.30863
80	$Hg^-(^2P)$	$6s^26p$	1.54–1.8
81	$Tl^-(^3P)$	$6p^2$	0.2
82	$Pb^-(^4S)$	$6p^3$	0.364
83	$Bi^-(^3P)$	$6p^4$	0.946
84	$Po^-(^2P)$	$6p^5$	1.9
85	$At^-(^1S)$	$6p^6$	2.8
87	$Fr^-(^1S)$	$6p^67s^2$	0.46
92	$U^-(J = 13/2)$	$5f^36d7s^27p$	0.175

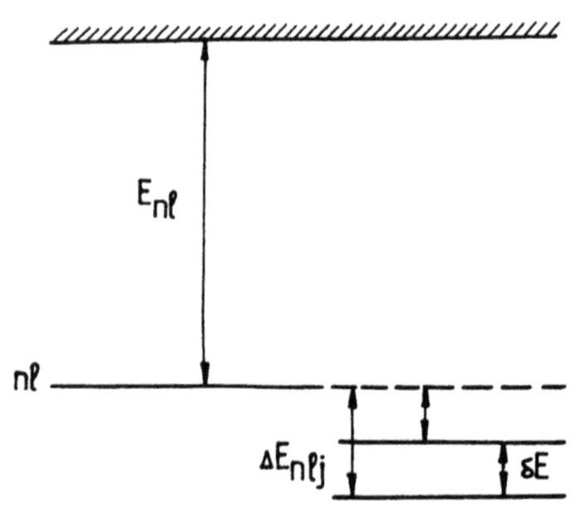

Fig. 1.4. Fine structure of the nl level in H-like atoms; E_{nl} is the binding energy

In the *LS*-coupling scheme (Sect. 1.2), the energy splitting between neighbour levels is given by the *Lande interval rule* [1.1]

$$\Delta E_{j,j-1} = E_j - E_{j-1} = A(LS)J\,, \tag{1.5.3}$$

i.e., is proportional to J. The *multiplet splitting constant* $A(LS)$ depends only on the quantum numbers L and S.

If $A > 0$ the multiplet is called *normal*. In this case, the component with the smallest possible $J = |L - S|$ has the lowest energy. If $A < 0$, the multiplet is called *inverted*. The component with the largest possible $J = L + S$ has the lowest energy.

For highly excited atomic states $n \gg 1$, the splitting ΔE may be approximated by [1.6]:

$$\Delta E_{j,j-1} = B/n_*^3 , \quad n_* = n - \Delta , \tag{1.5.4}$$

where B is a constant, n_* is the effective principal quantum number and Δ is the quantum defect (1.1.3). The quantities B, Δ and the range Δn of the change of the principal quantum number n for alkali atoms are listed in Table 1.6.

The interaction of nuclei having a non-zero magnetic moment ($\mu \neq 0$) or a non-zero quadrupole moment ($Q \neq 0$) with atomic electrons leads to a splitting of a level with the angular momentum J into components. This splitting is called the *hyperfine splitting*. Each component is characterized by the set of quantum numbers $JIFM_F$, where $\mathbf{J} = \mathbf{L} + \mathbf{S}$ is the total (angular and spin) momentum of the atomic electrons, I is the nuclear spin, $\mathbf{F} = \mathbf{J} + \mathbf{I}$ is the total angular momentum of an atom and M_F is the projection of F on the quantum axis.

The total splitting of the level J is given by [1.1]

$$\begin{aligned} \Delta E_F &= \frac{1}{2}AC + B\frac{(3C/4)(C+1) - I(I+1)J(J+1)}{2I(2I-1)J(2J-1)} \\ C &= F(F+1) - J(J+1) - I(I+1) , \end{aligned} \tag{1.5.5}$$

where A and B are the *hyperfine splitting constants*: A is the magnetic dipole interaction constant and B is the electric quadrupole interaction constant. In general, the determination of the constants A and B is quite complicated. In the

Table 1.6. Constant B (in cm^{-1}), quantum defect Δ and the value Δn for the experimental fine doublet splittings in alkali atoms [1.6]; $\Delta n = n - n_0$, with n_0 the principal quantum number of the ground state

Term	Parameters in (1.5.5)	Atom				
		Li	Na	K	Rb	Cs
$2p^0_{1/2,3/2}$	B	–	179	680	2870	7140
	Δ	0.047	0.855	1.712	2.647	3.5699
	Δn	7–10	10–40	9–21	13–68	10–80
$^2D_{3/2,5/2}$	B	0.97	−3.07	−39	350	2006
	Δ	0.002	0.0155	0.277	1.347	2.471
	Δn	7–10	7–30	10–36	10–40	10–50
$^2F^0_{5/2,7/2}$	B	0.48	0.47	–	–	−32.5
	Δ	−0.00008	0.00145	0.010	0.016	0.0335
	Δn	7–10	11–17	7–14	7–12	10–30
$^2G_{7/2,9/2}$	B	0.3	–	–	–	–
	Δ	≈ 0	0.00044	–	–	0.0070
	Δn	8–9	10–20	–	–	25–35

case of H-like ions with the orbital quantum number 1, one has in the non-relativistic approximation:

$$A(l=0) = \frac{8g_I}{3}\left(\frac{m}{m_p}\right)\frac{\alpha^2 Z^3}{n^3}\,\mathrm{Ry}\,, \tag{1.5.6}$$

$$A(l \neq 0) = \frac{g_I}{(l+1/2)j(j+1)}\left(\frac{\mathrm{m}}{\mathrm{m_p}}\right)\frac{\alpha^2 Z^3}{n^3}\,\mathrm{Ry}\,, \tag{1.5.7}$$

$$B = e^2 Q \frac{2j-1}{2j+2}\langle r^{-3}\rangle = \frac{Q a_0^{-2}(2j-1)}{j(j+1)(l+1)(l+1/2)l}\frac{Z^3}{n^3}\,\mathrm{Ry}\,, \tag{1.5.8}$$

where m_p is the proton mass and Q is the nuclear quadrupole moment; Q has the dimension of an area and is a measure of the deviation of the charge distribution from spherical symmetry. The factor g_I is a dimensionless quantity called the *gyromagnetic ratio* or *nuclear g-factor* and is defined by

$$g_I = \mu/(\mu_N I)\,, \tag{1.5.9}$$

where μ is the nuclear *magnetic dipole moment*, I is the nuclear spin and $\mu_N = e\hbar/(2m_p)$ is the *nuclear magneton*. According to (1.5.5), the energy difference $\Delta E_{FF'}$ between two components F and $F' = F - 1$ is given by

$$\Delta E_{FF'} = AF + \frac{3FB(F^2 + 1/2 - I(I+1) - J(J+1)]}{2I(2I-1)J(2J-1)}\,,$$

$$F' = F - 1\,; \quad I, J \geqslant 1 \tag{1.5.10}$$

the constant $B = 0$ if $I \leqslant 1/2$ or $J = 1/2$.

Experimental values of the hyperfine splitting and parameters (1.5.10) for some stable isotopes of atoms are given in Table 1.7.

1.6 Isotope Shift

The energy levels of two isotopes of an atom are shifted relative to each other. For H-like systems, the Schrödinger equation for the stationary states has the solution:

$$E = -\frac{1}{2}\frac{Z^2\mu e^4}{n^2\hbar^2} = -\frac{Z^2\mu}{mn^2}\,\mathrm{Ry}\,, \tag{1.6.1}$$

where $\mu = mM/(m+M)$ is the reduced mass. For example, the isotope shift of the n levels in deuterium relative to hydrogen levels is

$$\Delta E = -\frac{\mathrm{Ry}}{n^2}\left(\frac{2m_p}{m+2m_p} - \frac{m_p}{m+m_p}\right) \approx -\frac{\mathrm{Ry}}{2n^2}\frac{m}{m_p}\,, \tag{1.6.2}$$

i.e., the lines in deuterium spectra are shifted towards shorter wavelengths.

The isotope shift is related to the finite mass and finite size of the atomic nucleus, and the mass effect is opposite in sign to the volume effect. For light

Table 1.7. Hyperfine splitting (1.5.10) and its parameters of low-lying atomic energy levels for stable isotopes [1.6]

Nuclear charge Z	Isotope, ground term	Nuclear spin I	Electronic term	$F-F'$	$\Delta E_{FF'}$ [MHz] A, B	$\Delta E_{FF'}$ [10^{-3} cm^{-1}]
1	$^1H(^2S_{1/2})$	1/2	$1\,^2S_{1/2}$	(1 − 0)	1420.40575	47.3796
			$2\,^2S_{1/2}$	(1 − 0)	177.5568	5.92266
	$^2H(^2S_{1/2})$	1	$1\,^2S_{1/2}$	(3/2 − 1/2)	327.38435	10.9204
			$2\,^2S_{1/2}$	(3/2 − 1/2)	40.9244	1.36509
	$^3H(2S_{1/2})$	1/2	$1\,^2S_{1/2}$	(1 − 0)	1516.70147	50.5917
2	$^3He(^1S_0)$	1/2	$2\,^3S_1$	(3/2 − 1/2)	6739.701	224.812
	$^3He^+(^2S_{1/2})$		$1\,^2S_{1/2}$	(1 − 0)	8665.6499	289.055
3	$^6Li(^2S_{1/2})$	1	$2\,^2S_{1/2}$	(3/2 − 1/2)	228.20526	7.61211
			$2\,^2P_{1/2}$	–	$A = 17.37$	–
			$2\,^2P_{3/2}$	–	$A = -1.16$	–
					$B = -0.1$	
	$^7Li(^2S_{1/2})$	3/2	$2\,^2S_{1/2}$	(2 − 1)	803.50409	26.80203
			$2\,^2P_{1/2}$	–	$A = 45.9$	–
			$2\,^2P_{3/2}$	–	$A = -3.06$	–
					$B = -0.2$	–
4	$^9Be(^1S_0)$	3/2	$2\,^3P_1$	(5/2 − 3/2)	354.44	11.823
				(3/2 − 1/2)	202.95	6.7697
			$2\,^3P_2$	(7/2 − 5/2)	435.48	14.526
				(5/2 − 3/2)	312.02	10.408
				(3/2 − 1/2)	187.62	6.2583
5	$^{10}B(^2P_{1/2})$	3	$2\,^2P_{1/2}$	(7/2 − 5/2)	429.05	14.312
	$^{11}B(^2P_{1/2})$	3/2	$2\,^2P_{1/2}$	(2 − 1)	732.15	24.422
			$2\,^2P_{1/2}$	(3 − 2)	222.7	7.428
				(2 − 1)	144.0	4.803
				(1 − 0)	71	2.37
6	$^{13}C(^3P_0)$	1/2	$2\,^3P_1$	(3/2 − 1/2)	4.3	0.14
			$2\,^3P_2$	(5/2 − 3/2)	372.6	12.43
7	$^{14}N(^4S_{3/2})$	1	$2\,^4S_{3/2}$	–	$A = 10.45093$	–
					$B = 1.3$	–
8	$^{17}O(^3P_2)$	5/2	$2\,^3P_2$	–	$A = -219.6$	–
			$2\,^3P_1$	–	$A = 4.7$	–
9	$^{19}F(^2P_{3/2})$	1/2	$2\,^2P_{3/2}$	(2 − 1)	4020	134
			$2\,^2P_{3/2}$	(1 − 0)	10250	342
10	$^{21}Ne(^1S_0)$	3/2	$3\,^3P_2$	(7/2 − 5/2)	1034.5	34.51
				(5/2 − 3/2)	599.4	19.99
				(3/2 − 1/2)	303.9	10.14
11	$^{23}Na(^2S_{1/2})$	3/2	$3\,^2S_{1/2}$	(2 − 1)	1771.62613	59.09513
			$3\,^2P_{1/2}$	–	$A = 94.3$	–
			$3\,^2P_{3/2}$	–	$A = 18.7$	–
					$B = 2.9$	–
12	$^{25}Mg(^1S_0)$	5/2	$3\,^3P_1$	(7/2 − 5/2)	516.1	17.22
				(5/2 − 3/2)	350.0	11.7
			$3\,^3P_2$	(9/2 − 7/2)	567.3	18.92
				(7/2 − 5/2)	452.3	15.09
				(5/2 − 3/2)	329.0	10.97
				(3/2 − 1/2)	199.8	6.66

Table 1.7 (*Cont.*)

Nuclear charge Z	Isotope, ground term	Nuclear spin I	Electronic term	$F-F'$	$\Delta E_{FF'}$ [MHz] A, B	$\Delta E_{FF'}$ [10^{-3} cm^{-1}]
13	$^{27}Al(^2P_{1/2})$	5/2	$3\,^2P_{1/2}$	(3 – 2)	1506.1	50.24
			$3\,^2P_{3/2}$	(4 – 3)	392	13.1
				(3 – 2)	274	9.14
			$4\,^2S_{1/2}$	–	$A = 420$	–
15	$^{31}P(^4S_{3/2})$	1/2	$3\,^4S_{3/2}$	–	$A = 55.06$	–
17	$^{35}Cl(^2P_{3/2})$	3/2	$3\,^2P_{3/2}$	(3 – 2)	670.0135	22.349
				(2 – 1)	355.2210	11.849
				(1 – 0)	150.1736	5.009
			$3\,^2P_{1/2}$	(2 – 1)	2074.38	69.19
	$^{37}Cl(^2P_{3/2})$	3/2	$3\,^2P_{3/2}$	(3 – 2)	555.3043	18.523
				(2 – 1)	298.1277	9.944
				(1 – 0)	127.4408	4.251
			$3\,^2P_{1/2}$	(2 – 1)	1726.7	57.60
19	$^{39}K(^2S_{1/2})$	3/2	$4\,^2S_{1/2}$	(2 – 1)	461.71972	15.40132
			$4\,^2P_{1/2}$	–	$A = 27.8$	–
			$4\,^2P_{3/2}$	–	$A = 6.1$	–
					$B = 2.8$	–
	$^{40}K(^2S_{1/2})$	4	$4\,^2S_{1/2}$	–	$A = -285.73$	–
			$4\,^2P_{1/2}$	–	$A = -34.5$	–
			$4\,^2P_{3/2}$	–	$A = -7.5$	–
					$B = -3$	–
	$^{41}K(^2S_{1/2})$	3/2	$4\,^2S_{1/2}$	(2 – 1)	254.01387	8.47300
			$4\,^2P_{1/2}$	–	$A = 15.2$	–
			$4\,^2P_{3/2}$	–	$A = 3.4$	–
					$B = 3.3$	–
21	$^{45}Sc(^2D_{3/2})$	7/2	$3\,^2D_{3/2}$	(5 – 4)	1329	44.3
				(4 – 3)	1085.8	36.22
			$3\,^2D_{5/2}$	(6 – 5)	635.0	21.18
				(5 – 4)	543.8	18.14
				(4 – 3)	444.7	14.83
			$4\,^4F_{3/2}$	–	$A = 158.5$	–
					$B = -5.2$	–
			$4\,^4F_{5/2}$	–	$A = 154.0$	–
					$B = -6.5$	–
			$4\,^4F_{7/2}$	–	$A = 250.0$	–
					$B = -9.1$	–
			$4\,^4F_{9/2}$	–	$A = 286.0$	–
					$B = -15$	–
	$^{47}Ti(^3F_2)$	5/2	$3\,^3F_2$	–	$A = -85.703$	–
					$B = 25.70$	–
	$^{49}Ti(^3F_2)$	7/2	$3\,^3F_2$	–	$A = -85.726$	–
					$B = 21.07$	–
23	$^{51}V(^4F_{3/2})$	7/2	$3\,^4F_{3/2}$	–	$A = 560.07$	–
					$B = 3.98$	–
24	$^{59}Cr(^7S_3)$	3/2	$3\,^7S_3$	(9/2 – 7/2)	371.7	12.40
				(7/2 – 5/2)	289.09	9.643
				(5/2 – 3/2)	206.50	6.888

Table 1.7 (*Cont.*)

Nuclear charge Z	Isotope, ground term	Nuclear spin I	Electronic term	$F-F'$	$\Delta E_{FF'}$ [MHz] A, B	$\Delta E_{FF'}$ [10^{-3} cm^{-1}]
25	^{55}Mn($^6S_{5/2}$)	5/2	$3\,^6S_{5/2}$	–	$A=-72.4208$	–
					$B=-0.018$	–
			$4\,^6D_{9/2}$	–	$A=510.3$	–
					$B=132.2$	–
			$4\,^6D_{7/2}$	–	$A=458.9$	–
					$B=21.7$	–
			$4\,^6D_{5/2}$	–	$A=436.7$	–
					$B=-46.8$	–
			$4\,^6D_{3/2}$	–	$A=469.4$	–
					$B=-65.1$	–
			$4\,^6D_{1/2}$	–	$A=882.1$	–
26	^{57}Fe($^5D_4$)	1/2	$3\,^5D_4$	–	$A=38.08$	–
			$4\,^5F_5$	–	$A=87.25$	–
			$4\,^5F_4$	–	$A=78.43$	–
			$4\,^5F_3$	–	$A=69.63$	–
			$4\,^5F_2$	–	$A=55.99$	–
27	^{59}Co($^4F_{9/2}$)	7/2	$3\,^4F_{9/2}$	(8 – 7)	3655	121.9
				(7 – 6)	3169.4	105.7
				(6 – 5)	2695	89.9
				(5 – 4)	2230.6	74.40
				(4 – 3)	1774.5	59.19
28	^{61}Ni($^3F_4$)	3/2	$3\,^3F_4$	–	$A=-215.04$	–
					$B=-56.9$	–
29	^{53}Cu($^2S_{1/2}$)	3/2	$4\,^2S_{1/2}$	(2 – 1)	11733.8174	391.398
	^{65}Cu($^2S_{1/2}$)	3/2	$4\,^2S_{1/2}$	(2 – 1)	12568.780	419.250
30	^{67}Zn($^1S_0$)	5/2	$4\,^3P_2$	(9/2 – 7/2)	2418.1	80.66
				(7/2 – 5/2)	1855.7	61.90
				(5/2 – 3/2)	1312.1	43.77
				(3/2 – 1/2)	781.9	26.08
31	^{69}Ga($^2P_{1/2}$)	3/2	$4\,^2P_{1/2}$	(2 – 1)	2677.987	89.328
			$4\,^2P_{3/2}$	–	$A=190.794$	–
					$B=62.522$	–
			$5\,^2S_{1/2}$	(2 – 1)	2140	71.3
	^{71}Ga($^2P_{1/2}$)	3/2	$4\,^2P_{1/2}$	(2 – 1)	3402.69	113.50
			$4\,^2P_{3/2}$	(3 – 2)	766.696	25.574
				(2 – 1)	445.470	14.859
				(1 – 0)	203.043	6.773
			$5\,^2S_{1/2}$	(2 – 1)	2720	90.6
32	^{78}Ge($^3P_0$)	9/2	$4\,^3P_1$	–	$A=15.55$	–
					$B=-54.57$	–
			$4\,^3P_2$	–	$A=-64.427$	–
					$B=111.8$	–
33	^{75}As($^4S_{3/2}$)	3/2	$4\,^4S_{3/2}$	–	$A=-66.20$	–
					$B=-0.53$	–
				(3 – 2)	819.45	27.33
				(2 – 1)	595.12	19.85
36	^{83}Kr($^1S_0$)	9/2	$5p[1/2]_1$	–	$A=-143.0$	–

Table 1.7 (*Cont.*)

Nuclear charge Z	Isotope, ground term	Nuclear spin I	Electronic term	$F-F'$	$\Delta E_{FF'}$ [MHz] A, B	$\Delta E_{FF'}$ [10^{-3} cm^{-1}]
37	$^{85}Rb(^2S_{1/2})$	5/2	$5\,^2S_{1/2}$	(3 – 2)	3035.732	101.261
			$5\,^2P_{1/2}$	–	$A = 120.7$	–
			$5\,^2P_{3/2}$	–	$A = 25.0$	–
					$B = 26.0$	–
			$4\,^2D_{5/2}$	–	$A = -5$	–
			$4\,^2D_{3/2}$	–	$A = 7$	–
			$6\,^2S_{1/2}$	–	$A = 239$	–
	$^{87}Rb(^2S_{1/2})$	3/2	$5\,^2S_{1/2}$	(2 – 1)	6834.6826	227.98
			$5\,^2P_{1/2}$	–	$A = 406$	–
			$5\,^2P_{3/2}$	–	$A = 84.9$	–
					$B = 12.6$	–
			$4\,^2D_{5/2}$	–	$A = -17$	–
			$4\,^2D_{3/2}$	–	$A = 25$	–
			$6\,^2S_{1/2}$	–	$A = 810$	–
39	$^{89}Y(^2D_{3/2})$	1/2	$4\,^2D_{3/2}$	–	$A = -57.2$	–
			$4\,^2D_{5/2}$	–	$A = -28.7$	–
47	$^{107}Ag(^2S_{1/2})$	1/2	$5\,^2P_{3/2}$	–	$A = -32$	–
49	$^{119}In(P_{1/2})$	9/2	$5\,^2P_{1/2}$	(5 – 4)	11385	379.8
			$6\,^2S_{1/2}$	(5 – 4)	8410	281
	$^{115}In(^2P_{1/2})$	9/2	$5\,^2P_{1/2}$	(5 – 4)	11410	380.6
			$5\,^2P_{3/2}$	(5 – 4)	242.165	–
			$6\,^2S_{1/2}$	(5.4)	8430	281
51	$^{123}Sb(^4S_{3/2})$	7/2	$5\,^4S_{3/2}$	(5 – 4)	815.6	27.20
				(4 – 3)	648.5	21.63
				(3 – 2)	484.0	16.1
52	$^{125}Te(^3P_2)$	1/2	$5\,^3P_2$	–	$A = -1010.3$	–
			$5\,^3P_1$	–	$A = 782.5$	–
			$5\,^1D_2$	–	$A = -2887.0$	–
53	$^{127}(^2P_{3/2})$	5/2	$5\,^2P_{3/2}$	(4 – 3)	4226.17	140.97
				(3 – 2)	1965.9	65.58
				(2 – 1)	737.49	24.60
54	$^{129}Xe(^1S_0)$	1/2	$6\,^3P_2$	(5/2 – 3/2)	5961.258	198.85
	$^{131}Xe(^1S_0)$	3/2	$6\,^3P_2$	(7/2 – 5/2)	2693.623	89.850
				(5/2 – 3/2)	1608.348	53.649
				(3/2 – 1/2)	838.764	27.978
55	$^{133}Cs(^2S_{1/2})$	7/2	$6\,^2S_{1/2}$	(4 – 3)	9192.63177	306.63342
			$6\,^2P_{1/2}$	–	$A = 292$	–
			$6\,^3P_{3/2}$	–	$A = 50.3$	–
					$B = -0.4$	–
			$5\,^2D_{3/2}$	–	$A = 16.3$	–
			$5\,^2D_{5/2}$	–	$A = -22$	–
			$7\,^3S_{1/2}$	–	$A = 550$	–
56	$^{135}Ba(^1S_0)$	3/2	$5\,^3D_1$	–	$A = 470$	–
					$B = 12$	–
			$5\,^3D_2$	–	$A = 371$	–
					$B = 18$	–
			$5\,^3D_3$	–	$A = 408$	–
					$B = 20$	–

Table 1.7 (*Cont.*)

Nuclear charge Z	Isotope, ground term	Nuclear spin I	Electronic term	$F - F'$	$\Delta E_{FF'}$ [MHz] A, B	$\Delta E_{FF'}$ [10^{-3} cm^{-1}]
			$5\,^1D_2$	–	$A = -73.4$	–
					$B = 38.7$	–
			$5\,^1D_2$	–	$A = -73.4$	–
					$B = 38.7$	–
	^{137}Ba($^1S_0$)	3/2	$5\,^3D_1$	–	$A = -520$	–
					$B = 17$	–
			$5\,^3D_2$	–	$A = 414$	–
					$B = 27$	–
			$5\,^3D_3$	–	$A = 455$	–
					$B = 40$	–
			$5\,^1D_2$		$A = -82.2$	–
					$B = 59.6$	–
57	^{139}La($^2D_{3/2}$)	7/2	$5\,^2D_{3/2}$	(5 – 4)	737.97	24.62
				(4 – 3)	551.98	18.41
				(3 – 2)	391.6	13.06
			$5\,^2D_{5/2}$	(6 – 5)	1120.90	37.39
				(5 – 4)	912.79	30.45
				(4 – 3)	716.29	23.89
				(3 – 2)	529.1	17.65
			$6\,^4F_{3/2}$	(5 – 4)	2390.6	79.74
				(4 – 3)	1925.5	64.23
			$6\,^4F_{5/2}$	(6 – 5)	1808.9	60.34
				(5 – 4)	1503.2	50.14
				(4 – 3)	1199.8	40.02
			$6\,^4P_{1/2}$	(4 – 3)	9840.6	328.2
			$6\,^4P_{3/2}$	(4 – 3)	3707.8	123.68
			$6\,^4P_{5/2}$	(4 – 3)	3216.5	107.3
59	^{141}Pr($^4I_{9/2}$)	5/2	$4\,^4I_{9/2}$	–	$A = 926.209$	–
					$B = -11.88$	–
			$4\,^4I_{11/2}$	–	$A = 730.393$	–
					$B = -11.88$	–
			$4\,^4I_{13/2}$	–	$A = 613.240$	–
					$B = -12.85$	–
			$4\,^4I_{15/2}$	–	$A = 541.575$	–
					$B = -14.56$	–
60	^{143}Nd($^5I_4$)	7/2	$4\,^5I_4$	(15/2 – 13/2)	1418	47.3
				(13/2 – 11/2)	1257.5	41.95
				(11/2 – 9/2)	1084.7	36.18
				(9/2 – 7/2)	901.5	30.07
				(7/2 – 5/2)	710	23.7
			$4\,^5I_5$	–	$A = -153.68$	–
					$B = 115.7$	–
	^{145}Nd($^5I_4$)	7/2	$4\,^5I_4$	–	$A = -121.63$	–
					$B = 64.6$	–
			$4\,^5I_5$	–	$A = -95.53$	–
					$B = 61.0$	–

Table 1.7 (*Cont.*)

Nuclear charge Z	Isotope, ground term	Nuclear spin I	Electronic term	$F - F'$	$\Delta E_{FF'}$ [MHz] A, B	$\Delta E_{FF'}$ [10^{-3} cm^{-1}]
62	$^{147}Sm(^7F_0)$	7/1	$4\,^7F_1$	–	$A = -33.44$	–
					$B = -58.62$	–
			$4\,^7F_2$	–	$A = -41.184$	–
					$B = -62.23$	–
			$4\,^7F_3$	–	$A = -50.240$	–
					$B = -33.68$	–
	$^{149}Sm(^7F_0)$	7/2	$4\,^7F_1$	–	$A = -27.611$	–
					$B = 16.962$	–
			$4\,^7F_2$	–	$A = -33.95$	–
					$B = 17.99$	–
			$4\,^7F_3$	–	$A = -41.418$	–
					$B = 9.75$	–
63	$^{151}Eu(^8S_{7/2})$	5/2	$4\,^8S_{7/2}$	(6 – 5)	120.67	4.025
				(5 – 4)	100.29	3.345
				(4 – 3)	80.05	2.67
	$^{153}Eu(^8S_{7/2})$	5/2	$4\,^8S_{7/2}$	(6 – 5)	54.04	1.803
				(5 – 4)	44.00	1.47
				(4 – 3)	35.00	1.17
64	$^{155}Gd(^9D_2)$	3/2	$5\,^9D_2$	–	$A = 36.575$	–
					$B = 179.4$	–
			$5\,^9D_3$	–	$A = 4.92$	–
					$B = -406.67$	–
			$5\,^9D_4$	–	$A = -6.86$	–
					$B = -352.8$	–
	$^{157}Gd(^9D_2)$	3/2	$5\,^9D_2$	–	$A = 47.96$	–
					$B = 191.2$	–
			$5\,^9D_3$	–	$A = 6.45$	–
					$B = -433.2$	–
			$5\,^9D_4$	–	$A = -9.00$	–
					$B = -375.9$	–
65	$^{159}Tb(^6H_{15/2})$	3/2	$4\,^6H_{15/2}$	–	$A = 673.75$	–
					$B = 1449.3$	–
			$4\,^6H_{18/2}$	–	$A = 682.91$	–
					$B = 1167.5$	–
			$5\,^8G_{13/2}$	–	$A = 532.20$	–
					$B = 928.9$	–
66	$^{161}Dy(^5I_8)$		$4\,^5I_8$	–	$A = -116.232$	–
					$B = 1091.57$	–
	$^{163}Dy(^5I_8)$		$4\,^5I_8$	–	$A = -162.7543$	–
					$B = 1152.86$	–
			$4\,^5I_7$	–	$A = 177.53$	–
					$B = 1066.4$	–
67	$^{165}Ho(^4I_{15/2})$	7/2	$4\,^4I_{15/2}$	(9 – 8)	7184.8	239.7
				(8 – 7)	6540.8	218.2
				(7 – 6)	5842.4	194.9
				(6 – 5)	5096.3	170.0
				(5 – 4)	4309.3	143.7

Table 1.7 (*Cont.*)

Nuclear charge Z	Isotope, ground term	Nuclear spin I	Elec-tronic term	$F-F'$	$\Delta E_{FF'}$ [MHz] A, B	$\Delta E_{FF'}$ [10^{-3} cm^{-1}]
68	$^{167}Er(^3H_6)$	7/2	$4\,^3H_6$	–	$A = -120.486$	–
					$B = -4552.96$	–
69	$^{169}Tm(^2F_{7/2})$	1/2	$4\,^2F_{7/2}$	(4 – 3)	1496.5507	49.920
				–	$A = -374.13766$	–
71	$^{175}Lu(^2D_{3/2})$	7/2	$5\,^2D_{3/2}$	(5 – 4)	2051.2201	68.421
				(4 – 3)	345.497	11.524
				(3 – 2)	496.578	16.564
			$5\,^2D_{5/2}$	(6 – 5)	1837.570	61.295
				(5 – 4)	800.343	26.70
				(4 – 3)	161.815	5.398
				(3 – 2)	157.73	5.26
				(2 – 1)	238.058	7.941
	$^{176}Lu(^2D_{3/2})$	7	$5\,^2D_{3/2}$	–	$A = 137.99$	–
					$B = 2131$	–
			$5\,^2D_{5/2}$	–	$A = 104.0$	–
					$B = 2624$	–
72	$^{177}Hf(^3F_2)$	7/2	$5\,^3F_2$	(11/2 – 9/2)	991.792	33.08
				(9/2 – 7/2)	477.008	15.91
				(7/2 – 5/2)	162.887	5.433
				(5/2 – 3/2)	4.864	0.16
	$^{179}Hf(^3F_2)$	9/2		(13/2 – 11/2)	82.132	2.74
				(11/2 – 9/2)	392.848	13.104
				(9/2 – 7/2)	541.9104	18.076
				(7/2 – 5/2)	558.672	18.635
73	$^{181}Ta(^4F_{3/2})$	7/2	$5\,^4F_{3/2}$	–	$A = 509.08$	–
					$B = -1012.24$	–
			$5\,^4F_{5/2}$	–	$A = 313.47$	–
					$B = -834.8$	–
			$5\,^4F_{7/2}$	–	$A = 264.41$	–
					$B = -787.5$	–
			$5\,^4F_{9/2}$	–	$A = 256.62$	–
					$B = -650.4$	–
			$5\,^4P_{1/2}$	–	$A = 884.17$	–
			$5\,^4P_{3/2}$	–	$A = 379$	–
					$B = -1350$	–
74	$^{183}W(^5D_0)$	1/2	$5\,^5D_1$	–	$A = 29.12$	–
			$6\,^7S_3$	–	$A = 505.6$	–
			$5\,^5D_2$	–	$A = 56.3$	–
			$5\,^5D_3$	–	$A = 78.0$	–
			$5\,^5D_4$	–	$A = 88.3$	–
75	$^{185}Re(^6S_{5/2})$	5/2	$5\,^6S_{5/2}$	–	$A = -56.596$	–
					$B = 29.635$	–
			$5\,^4P_{5/2}$	–	$A = 880.44$	–
					$B = 1618.5$	–
	$^{187}Re(^6S_{5/2})$	5/2	$5\,^6S_{5/2}$	–	$A = -57.149$	–
					$B = 28.05$	–

Table 1.7 (*Cont.*)

Nuclear charge Z	Isotope, ground term	Nuclear spin I	Elec-tronic term	$F - F'$	$\Delta E_{FF'}$ [MHz] A, B	$\Delta E_{FF'}$ [10^{-3} cm^{-1}]
			$5\,^4P_{5/2}$	–	$A = 889.24$	–
					$B = 1531.7$	–
			$6\,^6D_{9/2}$	–	$A = 2600$	–
					$B = 2000$	–
77	^{191}Ir($^4F_{9/2}$)	3/2	$5\,^4F_{9/2}$	(6 – 5)	659.265	21.991
				(5 – 4)	189.440	6.319
				(4 – 3)	84.050	2.804
	^{198}Ir($^4F_{9/2}$)	3/2	$5\,^4F_{9/2}$	(6 – 5)	660.090	22.018
				(5 – 4)	224.478	7.4888
				(4 – 3)	33.535	1.119
78	^{195}Pt($^3D_3$)	1/2	$5\,^3D_3$	–	$A = 5702.6$	–
			$5\,^3D_2$	–	$A = -2609.6$	–
			$6\,^3F_4$	(9/2 – 7/2)	3820.56	127.4
79	^{197}Au($^2S_{1/2}$)	3/2	$6\,^2S_{1/2}$	(2 – 1)	6099.320	203.452
			$5\,^2D_{5/2}$	–	$A = 80.24$	–
					$B = 1049.8$	–
			$5\,^2D_{3/2}$	–	$A = 199.842$	–
					$B = 911.077$	–
80	^{199}Hg($^1S_0$)	1/2	$6\,^3P_2$	–	$A = 9066.45$	–
	^{201}Hg($^1S_0$)	3/2	$6\,^3P_2$	(7/2 – 5/2)	11382.629	379.68
				(5/2 – 3/2)	8629.522	287.85
				(3/2 – 1/2)	5377.49	179.37
			$6\,^3D_3$	–	$A = -2450$	–
					$B = 60$	–
81	^{203}Tl($^2P_{1/2}$)	1/2	$6\,^2P_{1/2}$	(1 – 0)	21105.45	704.0026
			$6\,^2P_{3/2}$	(2 – 1)	524.0599	17.4808
	^{205}Tl($^2P_{1/2}$)	1/2	$6\,^2P_{1/2}$	(1.0)	21310.83	710.8534
			$6\,^2P_{3/2}$	(2 – 1)	530.0765	17.6815
82	^{207}Pb($^3P_0$)	1/2	$6\,^1D_2$	(5/2 – 3/2)	1524.5	50.85
83	^{209}Bi($^4S_{3/2}$)	9/2	$^4S_{3/2}$	(6 – 5)	2884.67	96.22
				(5 – 4)	2171.42	72.43
				(4 – 3)	1584.50	52.85
			$6\,^2P_{3/2}$	(6 – 5)	3598.65	120.04
				(5 – 4)	2251.04	75.09
				(4 – 3)	1311.9	43.76
92	^{92}U($^5L_6^0$)	7/2	$6\,^5L_6^0$	–	$A = -60.57$	–
					$B = 4104.1$	
			$6\,^5K_5^0$	–	$A = -68.35$	
				–	$B = 40.1$	
93	^{237}Np($^6L_{11/2}$)	5/2	$5\,^6L_{11/2}$	–	$A = 778$	–
					$B = 645$	–

elements, the volume effect is quite small while for heavy atoms with a nuclear change $Z \geqslant 50$ it becomes dominant.

The isotope shift ΔE (the nuclear-size correction) of the energy levels depends on the ion charge Z and the principal n and orbital 1 quantum numbers. For the lowers states in H-like ions one has (see [1.12]):

$$\Delta E(1s_{1/2}) = \frac{4}{3}\mathrm{Ry}\, Z^2[1 + 0.50(\alpha Z)^2]\left[\frac{Z\langle r^2\rangle^{1/2}}{a_0}\right]^{2\gamma} \quad (1.6.3a)$$

$$\Delta E(2s_{1/2}) = \frac{1}{6}\mathrm{Ry}\, Z^2[1 + 1.38(\alpha Z)^2]\left[\frac{Z\langle r^2\rangle^{1/2}}{a_0}\right]^{2\gamma} \quad (1.6.3b)$$

$$\Delta E(2p_{1/2}) = \frac{1}{32}\mathrm{Ry}\, Z^2(\alpha Z)^2\left[\frac{Z\langle r^2\rangle^{1/2}}{a_0}\right]^{2\gamma} \quad (1.6.3c)$$

$$\Delta E(2p_{3/2}) = 0\,, \quad (1.6.3d)$$

where $\gamma = \sqrt{1-(\alpha Z)^2}$, a_0 is the Bohr radius and $\langle r^2\rangle^{1/2}$ is the root-mean-square radius of the nucleus.

The isotope shift ΔE of the level is negative ($\Delta E < 0$) if the level of the heavier isotope lies deeper than the ionization limit, i.e., its distance from the ionization limit is larger than that of the light isotope. In Table 1.8, the experimental values of the isotope shift in atoms are given for isotopes with different mass numbers A.

The isotope shift of a specific spectral line is given by

$$\Delta E = \Delta E_1 - \Delta E_0\,, \quad (1.6.4)$$

where ΔE_1 and ΔE_0 are the isotope shifts of the upper and lower levels. The value ΔE is considered positive ($\Delta E > 0$) if the wavelength λ_{h} of the heavier isotope (the mass number A_{h}) is shifted towards shorter wavelength and, respectively, the line λ_{l} of the light isotope (the mass number A_{l}) is shifted towards longer wavelengths, i.e., $\lambda_{\mathrm{l}} > \lambda_{\mathrm{h}}$. The experimental isotope shifts in atomic resonance lines are given in Table 1.9.

1.7 Lamb Shift

In the theory of atomic spectra for neutral and weakly ionized atoms, relativistic effects are considered as small corrections in contrary to the case of multicharged ions when relativistic effects strongly depend on the ion charge [1.8].

In relativistic theory, the Dirac equation for an electron in the Coulomb field is used [1.1]. For H-like systems with charge $Z \ll 137$, the corresponding Dirac energy is given by

$$E_{\mathrm{D}} \approx E_0 + E_1 + \cdots\,, \quad (1.7.1)$$

$$E_0 = -\frac{Z^2}{n^2}\mathrm{Ry}\,,\quad E_1 = -\left(\frac{1}{j+1/2} - \frac{3}{4n}\right)\frac{\alpha^2 Z^4}{n^3}\mathrm{Ry}\,,\quad j = l \pm 1/2\,,$$

Table 1.8. Experimental isotope shifts ΔE of atomic terms [1.6]

Nuclear charge Z	Atom	Term	Mass numbers A_1–A_2	ΔE [cm^{-1}]
1	H	$1S_{1/2}$	1–2	−29.84284
			1–3	−39.77218
		$2P^0_{1/2}$	1–2	−7.46090
			1–3	−9.94326
		$2S_{1/2}$	1–2	−7.46085
			1–3	−9.94322
		$2P^0_{3/2}$	1–2	−7.46080
			1–3	−9.94313
		$3P^0_{1/2}$	1–2	−3.31595
			1–3	−4.41922
		$3S_{1/2}$	1–2	−3.31594
			1–3	−4.41921
		$3P^0_{3/2}$	1–2	−3.31592
			1–3	−4.41918
		$3D_{3/2}$	1–2	−3.31592
			1–3	−4.41918
		$3D_{5/2}$	1–2	−3.31591
			1–3	−4.41917
2	He	$1\,^1S_0$	3–4	−10.5
		$2\,^3S_1$	3–4	−1.86
		$2\,^1S_0$	3–4	−1.56
		$2\,^3P^0$	3–4	−0.68
		$2\,^1P^0_1$	3–4	−1.68
		$3\,^3S_1$	3–4	−0.7
		$3\,^1S_0$	3–4	−0.64
		$3\,^3P^0$	3–4	−0.41
		$3\,^3D$	3–4	−0.54
		$3\,^1D$	3–4	−0.57
		$3\,^1P^0_1$	3–4	−0.71
3	Li	$2\,^2S_{1/2}$	6–7	−0.603
		$2\,^2P^0$	6–7	−0.251
		$3\,^2P^0$	6–7	−0.120
12	Mg	$3\,^1S_0$	24–25	−0.115
		$3\,^3P^0$	24–25	−0.032
		$3\,^1P^0_1$	24–25	−0.068
		$4\,^3S_1$	24–25	−0.045
19	K	$4\,^2S_{1/2}$	39–41	−0.0220
		$4\,^2P^0$	39–41	−0.0142
29	Cu	$4\,^2P^0$	63–65	−0.018
		$3d^9 4s^2\,^2D_{5/2}$	63–65	−0.085
		$3d^9 4s^2\,^2D_{3/2}$	63–65	−0.074
37	Rb	$5\,^2S_{1/2}$	85–87	−0.0053
		$5\,^2P^0$	85–87	−0.0027
		$6\,^2P^0$	85–87	−0.0013

Table 1.9. Experimental isotope shifts ΔE in atomic lines [1.6]

Nuclear charge Z	Atom	Transition	Wave-length [Å]	A_1–A_2	ΔE [10^{-3} cm^{-1}]
1	H	$1\,^2S_{1/2}$–$2\,^2P^0$	1215.7	1–2	2.238×10^4
				1–3	2.983×10^4
				2–3	7.477×10^3
2	He	$1\,^1S_0$–$2\,^1P^0$	584.3	3–4	8.8×10^3
3	Li	$2\,^2S_{1/2}$–$2\,^2P^0$	6708	6–7	351.3
5	B	$2p\,^2P^0$–$3s\,^2S_{1/2}$	2497	10–11	−170
6	C	$2p^2\,^1S_0$–$2p3s\,^1P_1^0$	2478.6	12–13	−160
7	N	$3s\,^2P_{3/2}$–$3p\,^2P_{3/2}^0$	8629.2	14–15	70
		$3s\,^4P_{5/2}$–$3p\,^4P_{5/2}^0$	8216.3	14–15	−60
8	O	$3s\,^3S_1$–$3p\,^3P_1$	8446.8	16–18	140
10	Ne	$3s'[1/2]_1^0$–$3p[5/2]_1$	7173.9	20–22	70
		$3s[3/2]_2^0$–$3p[1/2]_1$	7032.4	20–22	50
11	Na	$3s\,^2S_{1/2}$–$4p\,^2P_{1/2}^0$	3303.0	23–24	24
12	Mg	$3s^2\,^1S_0$–$3s3p\,^1P_1^0$	2852.1	24–25	50
		$3s^2\,^1S_0$–$3s3p\,^1P_1^0$		24–26	60
18	Ar	$4s[3/2]_2^0$–$4p'[3/2]_1$	7147.0	36–40	20
		$4s'[1/2]_1^0$–$5p[1/2]_0$	4510.7	36–40	50
19	K	$4s\,^2S_{1/2}$–$4p\,^2P_{1/2}^0$	7699.0	39–40	4.19
		$4s\,^2S_{1/2}$–$4p\,^2P_{1/2}^0$		39–41	7.85
		$4s\,^2S_{1/2}$–$4p\,^2P_{3/2}^0$	7664.9	39–40	4.22
				39–41	7.88
20	Ca	$4s^2\,^1S_0$–$4s4p\,^3P_1^0$	4226.7	40–42	13.0
				40–43	20.4
				40–44	25.7
				40–48	50.4
		$4s^2\,^1S_0$–$4s4p\,^3P_1^1$	6572.8	40–41	9.37
				40–42	17.0
				40–43	26.1
				40–44	33.2
				40–48	64.1
29	Cu	$4s\,^2S_{1/2}$–$4p\,^2P_{1/2}^0$	3274.0	63–65	20
30	Zn	$4s^2\,^1S_0$–$4s4p\,^1P_1^0$	2138.6	$\Delta A = 2$ (mean value)	16
31	Ga	$4p\,^2P_{1/2}^0$–$5s\,^2S_{1/2}$	4033.0	69–71	−1.1
		$4p\,^2P_{3/2}^0$–$5s\,^2S_{1/2}$	4172.1	69–71	−1.3
36	Kr	$5s[3/2]_1^0$–$5p[5/2]_2$	8776.7	82–84	2
37	Rb	$5s\,^2S_{1/2}$–$5p\,^2P_{1/2}^0$	7947.6	85–87	2.6
		$5s\,^2S_{1/2}$–$5p\,^2P_{3/2}^0$	7800.2	85–87	2.6
38	Sr	$5s^2\,^1S_0$–$5s5p\,^1P_1^0$	4607.3	84–88	9.0
				86–88	4.2
				87–88	1.5
		$5s^2\,^1S_0$–$5s6p\,^1P_1^0$	2931.8	84–88	15.9
				86–88	7.2
				87–88	3.2
		$5s^2\,^1S_0$–$5s6p\,^3P_1^0$	6892.6	84–88	14.1
				86–88	6.7
				87–88	3.2

Table 1.9 (*Cont.*)

Nuclear charge Z	Atom	Transition	Wave-length [Å]	A_1–A_2	ΔE [10^{-3} cm^{-1}]
40	Zr	$4d^3 5s\,{}^5F_5$–$4d^3 5p\,{}^5G_6^0$	4687.8	90–92	−12
				92–94	−7
				94–96	−5
47	Ag	$4d^{10} 5s\,{}^2S_{1/2}$–$5p\,{}^2P_{1/2}^0$	3382.9	107–109	−15
		$4d^{10} 5s\,{}^2S_{1/2}$–$5p\,{}^2P_{3/2}^0$	3280.7	107–109	−15
48	Cd	$5s^2\,{}^1S_0$–$5s5p\,{}^3P_1^0$	3261.0	$\Delta A = 2$ (mean value)	−15
49	In	$5p\,{}^2P_{1/2}^0$–$6s\,{}^2S_{1/2}$	4101.8	113–115	8.6
		$5p\,{}^2P_{3/2}^0$–$6s\,{}^2S_{1/2}$	4511.3	113–115	8.5
54	Xe	$6s[3/2]_2^0$–$6p[3/2]_2$	8231.6	134–136	−3
55	Cs	$6s\,{}^2S_{1/2}$–$6p\,{}^2P_{3/2}^0$	8521.1	133–134	1.2
				133–135	1.2
70	Yb	$6s^2\,{}^1S_0$–$6s6p\,{}^1P_1^0$	3988.0	174–176	17
				174–172	18
71	Lu	$5p6s^2\,{}^2D_{3/2}$–$5d6s6p\,{}^4F_{3/2}^0$	5736.5	175–176	−13.1
		$5d6s^2\,{}^2D_{5/2}$–$5d6s6p\,{}^4F_{5/2}^0$	6055.0	175–176	−13.9
		$5p6s^2\,{}^2D_{5/2}$–$5d6s6p\,{}^4F_{7/2}^0$	5421.9	175–176	−13.6
79	Au	$6s\,{}^2S_{1/2}$–$6p\,{}^2P_{1/2}^0$	2675.9	195–197	−97
80	Hg	$6s^2\,{}^1S_0$–$6s6p\,{}^3P_1$	2536.5	198–199	−9
				198–200	−160
				199–201	−210
				200–202	−180
				202–204	−160
82	Pb	$6p^2\,{}^3P_0$–$6p7s\,{}^3P_1^0$	2833.1	207–208	−47
				206–208	−75
				204–208	−140
				202–208	−207

where α is the fine-structure constant, E_0 is the non-relativistic part and E_1 is the fine structure splitting (Sect. 1.5); it is seen that $E_1 \propto (\alpha Z)^2 E_0$.

In the Dirac theory, the levels with different $l = j \pm 1/2$ have equal energies (1.7.1). However, the degeneration of levels with orbital quantum number l in the Dirac formula is removed due to the interaction of the electron with the radiation field described by Quantum ElectroDynamics (QED) and causes the splitting between $2s_{1/2}$ and $2p_{1/2}$ levels (Fig. 1.5). The radiative corrections are of the order of $\Delta E_R \propto \alpha(\ln\alpha)E_1 \ll E_1$.

The main contribution to the total Lamb shift ΔE_L is given by a sum of so-called electron *self-energy*, *vacuum-polarization* and nuclear-finite size corrections [1.9]. The Lamb shift is of the order of

$$\Delta E_L \approx \frac{\alpha^3 Z^4 \mathrm{Ry}}{n^3} F(\alpha Z),$$

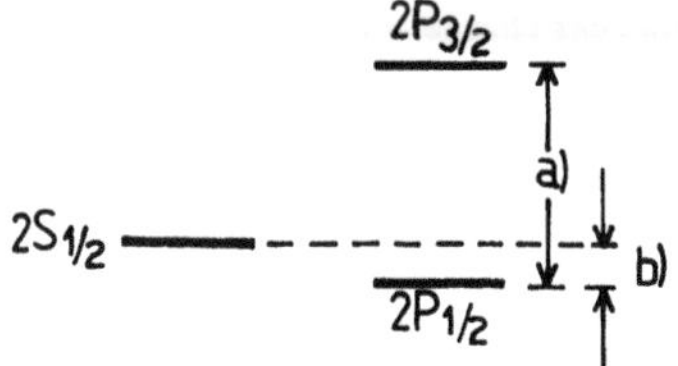

Fig. 1.5. Low-lying levels in H-like ions: (a) fine structure; (b) Lamb shift

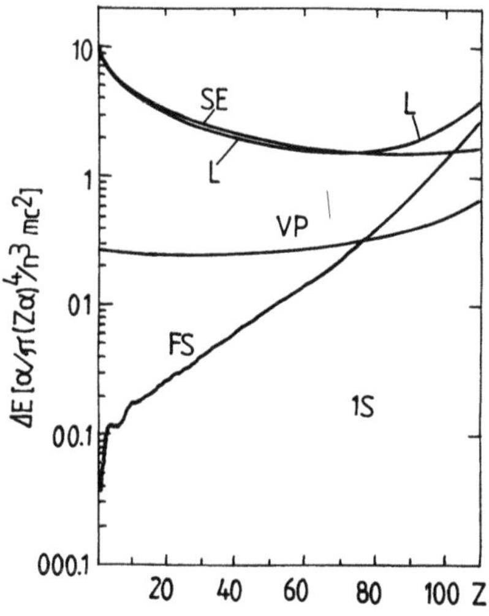

Fig. 1.6. Lamb shift of the 1s state in H-like ions as a function of the nuclear charge Z [1.9]: *SE* electron self-energy contribution; *VP* vacuum polarization; *FS* nucleus finite-size correction; *L* the total Lamb shift

where $F(\alpha Z)$ is a dimensionless slowly varying function describing the sum of different corrections; $F(\alpha Z) \sim 1$ for states with $l \neq 0$, and $F(\alpha Z) \sim \ln \alpha(\alpha Z)^{-1}$ for the s-states. Therefore, the Lamb shift is largest for the ground state and strongly increases ($\sim Z^4$) with the nuclear charge Z. For ions with $Z \geqslant 50$ the standard QED perturbation theory (expansion in terms of $(\alpha Z)^n$) is not valid and non-perturbation calculations are required (see, e.g., [1.10, 11]). The calculated contributions of different corrections to the total Lamb shift for the 1s state in H-like ions are shown in Fig. 1.6. Detailed analysis of difficulties arising in measurements and QED calculations of the Lamb shift in neutral hydrogen is given in the works [1.12, 13] and in highly charged ions in the monograph [1.14]. Experimental and theoretical values of the 1s Lamb shift and Lamb $2s_{1/2}$–$2p_{1/2}$ splitting in H-like ions are given in Tables 1.10, 11.

Table 1.10. The 1*s* Lamb shift (in eV) in H-like ions vs the nuclear charge *Z* (from [1.14])

Z	Experiment	Uncertainty	Theory
1	3.380006×10^{-5}	2.9×10^{-10}	3.38011×10^{-5}
	3.380024×10^{-5}	2.5×10^{-10}	
	3.380011×10^{-5}	2.07×10^{-10}	
	3.379999×10^{-5}	1.9×10^{-10}	
2			4.45426×10^{-4}
3			1.97829×10^{-3}
4			5.64215×10^{-3}
5			1.26377×10^{-2}
6			2.43257×10^{-2}
7			4.21795×10^{-2}
8			6.77822×10^{-2}
9			0.102759
10			0.148942
16	1.015	0.35	0.762000
17	0.85	0.10	0.938310
	0.84	0.12	
	0.90	0.10	
	0.825	0.12	
	0.807	0.15	
18	1.0	0.5	1.14140
	1.145	0.016	
20			1.63461
22	2.12890	0.6	2.26020
26	4.13	0.7	3.97246
	3.4	0.6	
	4.07	0.7	
	4.29	0.6	
28	5.07	0.10	5.09575
30			6.42983
36	11.95	0.50	11.8579
40			16.8743
50			36.0174
54	54.0	10.0	47.1016
60			68.3649
66	110.0	38.0	97.4020
70			122.831
79	212	15	205.300
	202.3	7.9	
80			217.840
82	290	75	244.745
83	215	100	259.870
90			401.833
92	520	130	458.490
	470	16	463.4
	429	63	464.6
	508	98	465.5
100			808.998
110			1853.57

Table 1.11. The $2s_{1/2}$–$2p_{1/2}$ Lamb splitting (in eV) in H-like ions vs the nuclear charge Z [1.14]

Z	Experiment	Uncertainty	Theory
1	4.374897×10^{-6}	3.7×10^{-11}	4.37500×10^{-6}
	4.374923×10^{-6}	7.9×10^{-12}	
	4.374872×10^{-6}	5.0×10^{-11}	
	4.374864×10^{-6}	3.7×10^{-11}	
2	5.80752×10^{-5}	6.6×10^{-10}	5.8074×10^{-5}
	5.80731×10^{-5}	5.0×10^{-9}	
3	2.59575×10^{-4}	8.7×10^{-8}	2.5940×10^{-4}
6	3.22624×10^{-3}	3.3×10^{-5}	3.2340×10^{-3}
8	9.06539×10^{-3}	6.2×10^{-5}	9.0823×10^{-3}
	9.16299×10^{-3}	3.1×10^{-5}	
	9.10674×10^{-3}	4.5×10^{-5}	
9	0.013809	1.4×10^{-4}	0.01383
15	0.08325	0.00083	0.08376
	0.08343	0.00029	
	0.08349	0.00012	
	0.08342	0.00016	
16	0.10397	0.00099	0.10494
	0.10449	0.00026	
17	0.12899	9.9×10^{-4}	0.12964
	0.129	9.1×10^{-4}	
18	0.15716	2.5×10^{-3}	0.15817
	0.15670	1.6×10^{-3}	
92			75.94 [3.140]
			75.296 [3.5]

1.8 Radial Analytical Wave Functions

In this Section, the analytical expressions for the radial atomic wave functions are given, which are used in atomic spectroscopy for the calculation of the matrix elements, the central effective field of the atomic core and other characteristics.

1.8.1 Hydrogen-like Wave Functions

The Schrödinger equation for the radial part $P_{nl}(r)$ of a particle wave function

$$\Psi_{nlm}(r) = Y_{lm}(\theta, \varphi) P_{nl}(r)/r \tag{1.8.1}$$

in the Coulomb field $-Ze^2/r$ is written in the form

$$\left[\frac{\mathrm{d}^2}{\mathrm{d}r^2} - \frac{l(l+1)}{r^2} + \frac{2\mu}{\hbar^2}\left(E + \frac{Ze^2}{r}\right)\right] P_{nl}(r) = 0\,, \tag{1.8.2}$$

where nlm are the quantum numbers, $Y_{lm}(\theta, \varphi)$ are the spherical angular functions and μ is the reduced mass of colliding particles. If a particle is in the bound state ($E < 0$) then (1.8.2) has finite solutions at certain discrete values (so-called *eigenvalues*) of energies:

$$E = -\frac{1}{2}\frac{Z^2}{n^2}\frac{\mu e^4}{\hbar^2}. \qquad (1.8.3)$$

The radial wave functions $P_{nl}(r)$ for $Z = 1$ have the form:

$$P_{nl}(r) = \frac{(2r)^{l+1}}{n^{l+2}(2l+1)!}\left(\frac{(n+1)!}{(n-l-1)!}\right)^{1/2} e^{-r/n} F(-n+l+1, 2l+2; 2r/n)$$

$$= -\frac{1}{n}\left(\frac{(n-l-1)!}{[(n+1)!]^3}\right)^{1/2} e^{-r/n}(2r/n)^{l+1} L_{n+l}^{2l+1}(2r/n)$$

$$= \left(\frac{(n+1)!}{(n-l-1)!}\right)^{1/2} \frac{M_{n,l+1/2}(2r/n)}{n(2l+1)!}, \qquad (1.8.4)$$

where $F(m, n; x)$ is the confluent hypergeometric function, $L_n^m(x)$ is the generalized Laguerre polynomial and $M_{m,n}(x)$ is the Whittaker function.

The functions $P_{nl}(r)$ are orthonormalized

$$\int_0^\infty P_{nl}(r)P_{n'l}(r)\,\mathrm{d}r = \delta_{nn'}, \qquad (1.8.5)$$

$$P_{nl}(0) = P_{nl}(\infty) = 0. \qquad (1.8.6)$$

The function $P_{nl}(r)$ has $n - l - 1$ nodes and the asymptotics

$$P_{nl}(r) \approx \begin{cases} \dfrac{(2r)^{l+1}}{n^{2l+1}(2l+1)!}\left(\dfrac{(n+l)!}{(n-l-1)!}\right)^{1/2}, & r \to 0 \\[2ex] (-1)^{n-l-1}\dfrac{(2r/n)^n e^{-r/n}}{n^{n+1}[(n+l)!(n-l-1)!]^{1/2}}, & r \to \infty. \end{cases} \qquad (1.8.7)$$

The wave function for the neutral hydrogen $P_{nl}^{\mathrm{H}}(r)$ ($Z = 1$) is related with that of H-like ions, $P_{nl}^{Z}(r)$, by

$$P_{nl}^{Z}(r) = Z^{3/2} P_{nl}^{\mathrm{H}}(Zr), \qquad (1.8.8)$$

where $P_{nl}^{\mathrm{H}}(r)$ is given by (1.8.4–7).

The first few functions $P_{nl}^{\mathrm{H}}(r)$ for hydrogen are:

$$P_{10}(r) = 2r\mathrm{e}^{-r}\,,\quad P_{20}(r) = \frac{r}{\sqrt{2}}(1 - r/2)\mathrm{e}^{-r/2}\,,\quad P_{21}(r) = \frac{r^2}{2\sqrt{6}}\mathrm{e}^{-r/2}\,,$$

$$P_{30}(r) = \frac{2r}{3\sqrt{3}}(1 - 2r/3 + 2r^2/27)\mathrm{e}^{-r/3}\,, \tag{1.8.9}$$

$$P_{31}(r) = \frac{8r^2}{27\sqrt{6}}(1 - r/6)\mathrm{e}^{-r/3}\,,\quad P_{32}(r) = \frac{4r^3}{81\sqrt{30}}\mathrm{e}^{-r/3}\,.$$

The mean values of the quantities r^k with the functions (1.8.8) are

$$\langle r^k\rangle = \int_0^\infty P_{nl}^2(r) r^k\,\mathrm{d}r\,,$$

$$\langle r\rangle = \frac{1}{2}[3n^2 - l(l+1)]\frac{a_0}{Z}\,,$$

$$\langle r^2\rangle = \frac{n^2}{2}[5n^2 + 1 - 3l(l+1)]\frac{a_0^2}{Z^2}\,, \tag{1.8.10}$$

$$\langle r^3\rangle = \frac{n^2}{8}[35n^2(n^2-1) - 30n^2(l+2)(l-1) + 3(l+2)(l+1)l(l-1)]\frac{a_0^3}{Z^3}\,,$$

$$\langle r^{-1}\rangle = \frac{1}{n^2}\frac{Z}{a_0}\,,$$

$$\langle r^{-2}\rangle = \frac{1}{n^3(l+1/2)}\frac{Z^2}{a_0^2}\,,$$

$$\langle r^{-3}\rangle = \frac{1}{n^3 l(l+1)(l+1/2)}\frac{Z^3}{a_0^3}\,,$$

where a_0 is the Bohr radius.

The radial wave function $P_{kl}(r)$ of the continuous spectra ($E > 0$) has the form:

$$P_{kl}(r) = \frac{C_{kl} r}{(2l+1)!}(2kr)^l \mathrm{e}^{-\mathrm{i}kr} F(\mathrm{i}/k + l + 1, 2l + 2; 2\mathrm{i}kr)\,,\quad k^2 = (E/\mathrm{Ry})^2\,.$$

The functions $P_{kl}(r)$ are real. The constant C_{kl} depends on the way of normalization. If one normalizes on the momentum k, then

$$\int_0^\infty P_{kl}(r) P_{k'l}(r)\,\mathrm{d}r = \pi\delta(k' - k)\,,$$

$$C_{kl} = \begin{cases} \left(\dfrac{4k}{1 - \mathrm{e}^{-2\pi/k}}\right)^{1/2} \prod\limits_{s=1}^{l} \sqrt{s^2 + k^{-2}}\,, & l \neq 0\,, \\[2ex] \left(\dfrac{4k}{1 - \mathrm{e}^{-2\pi/k}}\right)^{1/2}\,, & l = 0\,. \end{cases}$$

When $k \to 0$, one has

$$C_{kl} \approx 2k^{-l+1/2}, \quad P_{kl}(r) \approx \sqrt{2r}\, J_{2l+1}(8r),$$

where $J_n(x)$ is the Bessel function.

The function $P_{kl}(r)$ has the asymptotics

$$P_{kl}(r) \approx \begin{cases} \sqrt{2/\pi}\dfrac{(kr)^{l+1}}{(2l+1)!!}, & r \to 0, \\ \sqrt{2/\pi}\sin[kr + k^{-1}\ln(2kr) - \pi l/2 + \delta_l], & r \to \infty, \end{cases}$$

$$\delta_l = \arg[\Gamma(l + 1 - \mathrm{i}/k)],$$

where $\Gamma(x)$ is the Gamma function.

The wave function $P_{\varepsilon l}(r)$ normalized to the energy, $\varepsilon = E/\mathrm{Ry}$, is

$$P_{\varepsilon l}(r) = (\pi/2k)^{1/2} P_{kl}(r); \quad \int_0^\infty P_{\varepsilon l}(r) P_{\varepsilon' l}(r)\,\mathrm{d}r = \pi\delta(\varepsilon' - \varepsilon). \tag{1.8.11}$$

In the vicinity of discrete and continuous spectra one has

$$P_{nl}(r) = (2Z^2/\pi n^3)^{1/2} P_{\varepsilon l}(r); \quad n \to \infty, \quad \varepsilon \to 0. \tag{1.8.12}$$

The expansion of the plane wave on the partial waves has the form

$$\mathrm{e}^{\mathrm{i}\mathbf{kr}} = \sum_{l=0}^{\infty} \mathrm{i}^l (2l+1) P_l(\cos\theta) j_l(kr) = \sum_{\mu,\varkappa} 4\pi^{\varkappa}\, Y_{\varkappa\mu}(\mathbf{k})\, Y^*_{\varkappa\mu}(\mathbf{r}) j_l(kr), \tag{1.8.13}$$

where θ is the angle between vectors k and r, $P_l(x)$ is the Legendre polynomial and $j_l(x)$ is the spherical Bessel function related to the usual one by

$$j_\varkappa(x) = (\pi/2x)^{1/2} J_{\varkappa - 1/2}(x). \tag{1.8.14}$$

The function $j_\varkappa(x)$ has the asymptotics

$$j_\varkappa(x) \approx \begin{cases} 1 - x^2/6 + \cdots, & \varkappa = 0, \quad x \to 0, \\ \dfrac{x^\varkappa}{(2\varkappa+1)!!}, & \varkappa \neq 0, \quad x \to 0 \\ x^{-1}\sin(x - \pi\varkappa/2), & x \to \infty \end{cases} \tag{1.8.15}$$

The angular functions $Y_{lm}(\theta, \varphi)$ can be expressed in terms of the associated Legendre polynomials $P_l^m(x)$:

$$Y_{lm}(\theta, \varphi) = \Theta_{lm}(\theta)\Phi_m(\varphi),$$

$$\Theta_{lm} = (-1)^m \frac{(2l+1)(l-m)!}{2(l+m)!} P_l^m(\cos\theta), \quad \Phi_m = \frac{e^{im\varphi}}{\sqrt{2\pi}}. \tag{1.8.16}$$

Here, it is assumed that $m > 0$. For $m < 0$, $\Theta_{l,-|m|} = (-1)^m \Theta_{l,|m|}$. The functions Y_{lm} are orthonormalized

$$\int_0^{2\pi}\int_0^{\pi} Y^*_{l'm'} Y_{lm} \sin\theta\,\mathrm{d}\theta\,\delta\varphi = \delta_{ll'}\delta_{mm'}. \tag{1.8.17}$$

When $l = 0, 1, 2$, the expressions for the functions Θ_{lm} are

$$\Theta_{00} = \frac{1}{\sqrt{2}}, \quad \Theta_{10} = \left(\frac{3}{2}\right)^{1/2} \cos\theta, \quad \Theta_{1,\pm1} = \mp\left(\frac{3}{4}\right)^{1/2} \sin\theta,$$

$$\Theta_{20} = \left(\frac{5}{2}\right)^{1/2} \left(\frac{3}{2}\cos^2\theta - \frac{1}{2}\right), \quad \Theta_{2,\pm1} = \mp(15/4)^{1/2} \cos\theta \sin\theta,$$

$$\Theta_{2,\pm2} = \frac{1}{5}\sqrt{15}\sin^2\theta. \tag{1.8.18}$$

1.8.2 The Slater Wave Functions and Potential

For atoms having more than one electron, the Schrödinger equation (1.8.2) has no analytical solutions, and to solve it one has to use approximations. As the starting point for calculations, the central-field approximation is used when each atomic electron is assumed to move in a certain effective central symmetric field $U_c(r)$ of the atomic core created by the nucleus and the other electrons. In this case, the radial Schrödinger equation (1.8.2) for the wave function $P_{nl}(r)$ can be written in the form:

$$\left[\frac{1}{2}\frac{d^2}{dr^2} - \frac{l(l+1)}{2r^2} - U_c(r) + E_{nl}\right] P_{nl}(r) = 0, \tag{1.8.19}$$

where E is the known energy value (experimentally or theoretically).

There are several approximations for the effective central potential of the core $U_c(r)$ [1.15]. Here, we consider the Slater potential which can be obtained using the Staler analytical wave functions.

The central field $U_c(r)$ of the core with a charge distribution $P_c(r)$ is given by

$$U_c(r) = -\frac{\zeta_c(r)}{r}, \qquad \zeta_c(r) = Z - \int_0^r \rho_c(r')\,dr' - r\int_r^\infty \frac{\rho_c(r')}{r'}\,dr',$$
$$\rho_c(r) = \sum_\gamma N_\gamma P_\gamma^2(r), \tag{1.8.20}$$

where $\zeta_c(r)$ is the effective charge of the core, and

$$\zeta_c(0) = Z, \quad \zeta_c(\infty) = z = Z - \int_0^\infty \rho_c(r)\,dr. \tag{1.8.21}$$

Here, z is the spectroscopic symbol and Z is the nuclear charge. Summation in (1.8.20) is over all γ shells of the core, N_γ being the number of electrons in the shell γ. If the nodeless Slater functions $P(r)$ are used

$$P_\gamma(r) = \left(\frac{(2b)^{2a+1}}{\Gamma(2a+1)}\right)^{1/2} r^a e^{-br}, \tag{1.8.22}$$

the effective charge $\zeta_c(r)$ can be written in a closed analytical form

$$\zeta_c(r) = Z - \sum_\gamma N_\gamma \left(1 - e^{-2br} \sum_{k=\mu}^{2\mu-1} \frac{2\mu - k}{2\mu} \frac{(2br)^k}{k!}\right)_\gamma . \tag{1.8.23}$$

To find the effective field of the atomic core $U_c(r)$ one has to know the parameters b, μ and N for each shell of the core. The values for b, μ and N can be calculated by the Slater method. For the shell $\gamma = nl$, one has

$$\begin{aligned} n &= 1 \quad 2 \quad 3 \quad 4 \quad\ \ 5 \quad 6 \quad \ldots, \\ a_\gamma &= 1 \quad 2 \quad 3 \quad 3.5 \quad 4 \quad 4.5 \quad \ldots, \end{aligned} \tag{1.8.24}$$

$$b_\gamma = (Z - s_\gamma)/\mu_\gamma , \tag{1.8.25}$$

where s_γ is the screening constant

$$s_\gamma = \sum_{\gamma' \leqslant \gamma} s_{\gamma\gamma'}(N_{\gamma'} - \delta_{\gamma\gamma'}) . \tag{1.8.26}$$

According to the Slater method, one has for s_γ:

$$\begin{aligned} &\text{if}\quad \gamma' > \gamma, \quad \text{then } s_{\gamma\gamma'} = 0 , \\ &\gamma' = \gamma, \quad s_{\gamma\gamma} = 0.35 ; \quad \text{for } \gamma = 1s , \quad s_{\gamma\gamma} = 0.30 , \\ &\gamma' < \gamma, \quad \gamma = nsp , \quad s_{\gamma\gamma'} = 0.85 , \quad \text{if } n' = n - 1 , \\ &\gamma' < \gamma, \quad \gamma = nsp , \quad s_{\gamma\gamma'} = 1 , \quad \text{if } n' \leqslant n - 2 , \\ &\gamma' < \gamma, \quad \gamma = ndf , \quad s_{\gamma\gamma'} = 1 . \end{aligned}$$

The parameters s, μ and N for the closed atomic shells are given in Table 1.12.

For atomic shells which are less than half-filled, it is better to use another method: a equals to the nearest semi-integer number to $z/\sqrt{I}$, $b = aI/z$, where I is the binding energy of the shell in Ry units.

The Slater analytical functions (1.8.22) and the Slater potential (1.8.23) are used in the semiempirical method for calculating the radial wave functions $P_{nl}(r)$ of the optical electron in the ATOM code [1.16].

Table 1.12. The Slater parameters s_γ, μ_γ and N_γ

Shell γ	s_γ	μ_γ	N_γ
$1s^2$	0.30	1	2
$2(sp)^8$	4.15	2	8
$3(sp)^8$	11.25	3	8
$3d^{10}$	21.15	3	10
$4(sp)^8$	27.75	3.5	8
$4d^{10}$	39.15	3.5	10
$4(sp)^8$ (without $4f$-shell)	45.75	4	8
$4(df)^{24}$	44.05	3.5	24
$5(sp)^8$ (with $4f$-shell)	57.65	4	8
$5d^{10}$ (with $4f$-shell)	71.15	4	10

1.8.3 Bates-Damgaard Approximation

If a behavior of the wave function $P_{nl}(r)$ of an optical electron at large distances is the most important (e.g., calculation of the oscillator strengths in the length gauge; Sect. 2.3) one can use the Bates-Damgaard approximation [1.17] when the core effective potential in (1.8.17) is changed to its Coulomb asymptotic:

$$U_c(r) \approx -z/r\,. \qquad (1.8.27)$$

Here, z in the spectroscopic symbol of an atom. Then the solution of (1.8.17) can be obtained in a closed analytical form:

$$P_{nl}(r) = \left(\frac{2zr}{n_*}\right)^{n_*} \exp(-zr/n_*) \sum_{k=0}^{k_{max}} a_k/r^k\,, \qquad (1.8.28)$$

$$P_{nl}(0) = P_{nl}(\infty) = 0\,,$$

$$k_{max} = n_* + 1\,, \quad n_* = \sqrt{|E|/Z^2\mathrm{Ry}}\,, \qquad (1.8.29)$$

where n_* is the effective quantum number (1.1.3). The coefficients a_k satisfy the recurrent relations:

$$a_k = a_{k-1}\frac{n_*[l(l+1) - (n_* - k)(n_* - k + 1)]}{2kz}\,, \qquad (1.8.30)$$

$$a_0 = \frac{1}{n_*}\left(\frac{z}{\Gamma(n+l+1)\Gamma(n-1)}\right)^{1/2}. \qquad (1.8.31)$$

The radial wave functions (1.8.28) are used for the calculation of dipole and quadrupole matrix elements (Sect. 2.6).

2 Oscillator Strengths and Transition Probabilities

The radiative characteristics of atoms and ions such as oscillator strengths, transition probabilities, lifetimes of excited states and others required for fundamental research and physical applications are considered here. The last Section comprises the data on the angular coefficients used for the calculation of oscillator strengths and excitation cross sections by electron impact.

2.1 Basic Relations

There are two main types of radiative transitions: *electric multipole* E$\varkappa$ transitions when a photon has an angular momentum $\hbar\varkappa$ and a parity $(-1)^{\varkappa}$ and *magnetic multipole* transitions when a photon has an angular momentum $\hbar\varkappa$ and a parity $(-1)^{\varkappa-1}$. The corresponding probabilities $W_{\varkappa}^{\mathrm{E}}$ and $W_{\varkappa}^{\mathrm{M}}$ and the dimensionless oscillator strengths $f_{\varkappa}^{\mathrm{E,M}}$ for transition 0–1 are related by

$$g_1 W_{\varkappa}^{\mathrm{E}}(1\text{–}0) = g_0 A_0 (\Delta E/\mathrm{Ry})^2 f_{\varkappa}^{\mathrm{E}}(0\text{–}1)\,, \tag{2.1.1}$$

$$g_1 W_{\varkappa}^{\mathrm{M}}(1\text{–}0) = \alpha^2 g_0 A_0 (\Delta E/\mathrm{Ry})^2 f_{\varkappa}^{\mathrm{M}}(0\text{–}1)\,, \tag{2.1.2}$$

where $\Delta E = |E_0 - E_1|$ is the transition energy and the constant A_0 is given by

$$A_0 = \alpha^3 \mathrm{Ry}/\hbar = 8.033 \times 10^9\ \mathrm{s}^{-1}\,.$$

Here, g_0 is the statistical weight and α is the fine-structure constant.

Usually, $f_{\varkappa}$ is considered being positive $f_{\varkappa} > 0$ for absorbing transitions, i.e., for transitions from lower to upper states, and $f_{\varkappa} < 0$ in the opposite case. The order of magnitude of $f_{\varkappa}$ is defined by the factor

$$(\omega/c)^{2\varkappa-2} \propto (\alpha\Delta E/\mathrm{Ry})^{2\varkappa-2}$$

both for E$\varkappa$ and M$\varkappa$ transitions, where ω is the frequency of the transition.

It is convenient to introduce a reduced oscillator strength $\bar{f}_{\varkappa}$ of the form:

$$f_{\varkappa}^{\mathrm{E,M}}(0\text{–}1) = \frac{\varkappa+1}{2\varkappa[(2\varkappa-1)!!]^2}\left(\frac{\alpha\Delta E}{2\mathrm{Ry}}\right)^{2\varkappa-2} \bar{f}_{\varkappa}^{\mathrm{E,M}}(0\text{–}1)\,. \tag{2.1.3}$$

Then, for E$\varkappa$ and M$\varkappa$ transitions, one has

$$\bar{f}^{\mathrm{E}}(0\text{–}1) = \frac{Q_{\varkappa}(0\text{–}1)}{(2l_0+1)(2\varkappa+1)}(\Delta E/\mathrm{Ry}) R_{\varkappa}^2\,, \tag{2.1.4}$$

$$\bar{f}^{M}(0\text{–}1) = \frac{Q_\varkappa(0\text{–}1)}{(2l_0+1)(2\varkappa+1)}(\Delta E/\mathrm{Ry})R^2_{\varkappa-1}\,, \tag{2.1.5}$$

$$R_\varkappa = (2l_0+1)^{1/2}(2l_1+1)^{1/2}\begin{pmatrix} l_0 & l_1 & \varkappa \\ 0 & 0 & 0 \end{pmatrix}\int_0^\infty P_0(r)P_1(r)r^\varkappa\,dr\,. \tag{2.1.6}$$

Here, $Q_\varkappa$ is the angular coefficient which depends on the angular quantum numbers and $P(r)$ is the radial wave function of the optical electron.

For direct and inverse transitions, the $Q_\varkappa$ factors are related by

$$\frac{g_0}{(2l_0+1)}Q_\varkappa(0\text{–}1) = \frac{g_1}{(2l_1+1)}Q_\varkappa(1\text{–}0)\,. \tag{2.1.7}$$

For transitions $l_0^q L_0 S_0 J_0 - l_0^{q-1} l_1 L_1 S_1 J_1$ one has

$$Q_\varkappa(0\text{–}1) = (2L_0+1)(2J_1+1)\begin{Bmatrix} J_0 & L_0 & S_0 \\ l_1 & S_1 & \varkappa \end{Bmatrix}^2 \delta(S_0,S_1)Q_\varkappa(L_0,L_1)q|G^{L_0S_0}_{L_cS_c}|^2\,,$$

$$Q_\varkappa(L_0,L_1) = (2l_0+1)(2L_1+1)\begin{Bmatrix} L_0 & l_0 & L_c \\ l_1 & L_1 & \varkappa \end{Bmatrix}^2\,,$$

where L_c and S_c are the quantum numbers of the atomic core. More detailed information on Q-factors is given in Sect. 2.7 and monographs [2.1–3].

The reduced electric oscillator strength $\bar{f}_\varkappa$ (2.1.3) is related to the parameter $c_\varkappa$ of the non-diagonal multipole interaction averaged over the magnetic quantum numbers M_0 and M_1:

$$V_{01}(r) = \left(\frac{1}{g_0}\sum_{M_0M_1}|V(\gamma_0J_0M_0,\gamma_1J_1M_1)|^2\right)^{1/2} = \sum_\varkappa V_\varkappa(r)\,, \tag{2.1.8}$$

$$V_\varkappa(r) \approx c_\varkappa/r^{\varkappa+1}\,,\quad c_\varkappa = \left(\bar{f}_\varkappa\frac{\mathrm{Ry}}{\Delta E}\right)^{1/2}\,,\quad r\to\infty\,.$$

For oscillator strength f_1 and transition probability W_1 of dipole E1 transitions one has from (2.1.1–3):

$$f_1(0\text{–}1) = \bar{f}(0\text{–}1) = \frac{Q_1(0\text{–}1)}{3(2l_0+1)}(\Delta E/\mathrm{Ry})R_1^2\,, \tag{2.1.9}$$

$$R_\varkappa^2 = \max\{l_0,l_1\}\left[\int_0^\infty P_0(r)P_1(r)r\,dr\right]^2\,, \tag{2.1.10}$$

$$W_1(1\text{–}0) = A_0\frac{g_0}{g_1}(\Delta E/\mathrm{Ry})^2 f_1(0\text{–}1)\,,\quad A_0 = 8.033\times10^9\ \mathrm{s}^{-1}\,. \tag{2.1.11}$$

2.2 Selection Rules

Selection rules characterize the change of the quantum numbers of an atom or ion making a transition under the interaction with electromagnetic radiation.

Exact and approximate selection rules are distinguished. The first ones follow from the conservation laws and the properties of the angular parts of the operators for electric and magnetic transitions. The exact selection rules are independent of the coupling scheme of the angular momenta and are written for "exact" quantum numbers – parity, the total angular momentum J and its projection M – in the form

$$\Delta J = 0, \pm 1, \ldots, \pm \varkappa\,, \quad J + J' \geqslant \varkappa\,, \tag{2.2.1}$$

$$\Delta M = 0, \pm 1, \ldots, \pm \varkappa\,, \tag{2.2.2}$$

and the change of parity

$$\Delta P = \begin{cases} (-1)^{\varkappa} & \text{for E}\varkappa \text{ transitions}\,, \\ -(-1)^{\varkappa} & \text{for M}\varkappa \text{ transitions}\,, \end{cases} \tag{2.2.3}$$

where $\varkappa$ is the multiplicity of the transition.

The approximate selection rules are formulated for quantum numbers of the transition described in a certain coupling scheme. In the LS (Rassel-Saunders) coupling scheme (Sect. 1.2), one has the additional selection rules, i.e., for electric E$\varkappa$ transitions:

$$\Delta L = 0, \pm 1, \ldots, \pm \varkappa\,, \quad L + L' \geqslant \varkappa\,, \quad \Delta S = 0\,; \tag{2.2.4}$$

for magnetic M$\varkappa$ transitions:

$$\begin{aligned} &\Delta L = 0, \pm 1, \ldots, \pm(\varkappa - 1)\,, \quad L + L' \geqslant \varkappa - 1\,, \\ &\Delta S = 0, \pm 1, \ldots, \pm(\varkappa - 1)\,, \quad S + S' \geqslant \varkappa - 1\,. \end{aligned} \tag{2.2.5}$$

According to (2.2.1–3) the exact selection rules for electric dipole radiative transitions $LSJM$–$L'SJ'M'$ in multielectron atoms are:

$$\Delta J = J' - J = 0, \pm 1\,; \quad J + J' \geqslant 1\,, \tag{2.2.6}$$

$$\Delta M = M' - M = 0, \pm 1 \tag{2.2.7}$$

$$\text{odd term} \leftrightarrow \text{even term}\,. \tag{2.2.8}$$

For hydrogen-like atoms the selection rules for transitions nlm–$n'l'm'$ are

$$\Delta l = l' - l = \pm 1\,, \quad \Delta m = m' - m = 0, \pm 1\,; \tag{2.2.9}$$

and for transitions nlj–$n'l'j'$ between fine-structure components, respectively,

$$\Delta j = j' - j = 0, \pm 1\,, \quad j = l \pm 1/2\,, \tag{2.2.10}$$

where j is the total angular momentum.

There are no limitations on the principal quantum numbers n and n'. If the conditions (2.2.6) are not fulfilled, the dipole transition is forbidden. However, radiation may be possible but with a probability of about 10^5 times smaller than the dipole one. Such transitions are called *forbidden transitions*.

According to (2.2.8), transitions are possible only between terms of different parity. In the case of LS coupling, one has the additional selection rules:

$$\Delta S = S' - S = 0 \tag{2.2.11}$$

$$\Delta L = L' - L = 0, \pm 1 ; \quad L + L' \geqslant 1 , \tag{2.2.12}$$

and S–S transitions are ruled out.

According to (2.2.1), dipole transitions ($\varkappa = 1$) are possible only between terms of one multiplicity, i.e., transitions with change of spin $\Delta S = 1$ (so-called *intercombination transitions*) are forbidden.

In some cases, the selection rule (2.2.11) is violated, e.g., for highly charged ions when the spin–orbit interaction is comparable or larger than the electrostatic one.

For quadrupole transitions ($\varkappa = 2$), one has

$$\Delta J = 0, \pm 1, \pm 2 ; \quad J + J' \geqslant 2 , \tag{2.2.13}$$

$$\Delta M = 0, \pm 1, \pm 2 . \tag{2.2.14}$$

The quadrupole transitions are possible only between states of the same parity. Additionally, in the LS-coupling approximation

$$\Delta S = 0 ; \quad \Delta L = 0, \pm 1, \pm 2 ; \quad L + L' \geqslant 2 , \tag{2.2.15}$$

i.e., transitions between S terms ($L = L' = 0$) and between S and P terms ($L = 0$, $L' = 1$) are ruled out.

The selection rule for spins S and S' is the same for all E$\varkappa$ transitions, i.e., it allows transitions between terms with equal multiplicity $\varkappa$.

Electric dipole transitions between two different hyperfine structure levels γJ and $\gamma J'$ obey the following selection rules:

$$\Delta F = 0, \pm 1 , \quad F + F' \geqslant 1 . \tag{2.2.16}$$

E1 transitions between two F components of the same level γJ are forbidden by the parity selection rules, but M1 and E2 transitions are allowed. For these two cases one has:

$$\text{M1 transitions: } \Delta F = 0, \pm 1 ; \quad F + F' \geqslant 1 , \tag{2.2.17}$$

$$\text{E2 transitions: } \Delta F = 0, \pm 1, \pm 2 ; \quad F + F' \geqslant 2 . \tag{2.2.18}$$

A violation of selection rules is due to the influence of magnetic interactions, mostly the spin-orbit ones [2.4]. The probabilities of forbidden transitions in multicharged ions rapidly increase with increasing ion charge. Forbidden lines can arise from any of the higher multipoles or from multiphoton transitions.

Selection rules also exist for transitions between states described by other types of coupling schemes (LK, jK, jj coupling and others [2.1]).

For E$\varkappa$ and M$\varkappa$ transitions without change of the principal quantum number $\Delta n \neq 0$, the other of magnitude for the transition probability is given by

$$A(\mathrm{E}\varkappa) \approx \alpha(\alpha Z)^{2\varkappa+2}(mc^2/\hbar) , \tag{2.2.19}$$

$$A(\mathrm{M}\varkappa) \approx (\alpha Z)^{2\varkappa+4}(mc^2/\hbar) , \tag{2.2.20}$$

i.e., the probability of a E$\varkappa$ transition is $(\alpha Z)^2$ times higher than that of a M$\varkappa$ transition, where α is the fine-structure constant.

2.3 Oscillator Strengths and Transition Probabilities

The general relations between dipole oscillator strengths f and radiative transition probabilities W were given in Sect. 2.1. Here, we present information on sum rules for oscillator strengths and tables of numerical data for f and W.

2.3.1 Three Representations of the Oscillator Strength. Sum Rules

There are three gauges of presenting the oscillator strength. Equation (2.1.5) corresponds to the so-called "length" gauge, i.e., to the matrix element of distance r. Using the operator relations

$$\mathbf{r} = \hbar\mathbf{p}/(\mathrm{i}m\Delta E)\,, \quad \mathbf{p} = \hbar\dot{\mathbf{p}}/(\mathrm{i}\Delta E)\,, \tag{2.3.1}$$

one can obtain the "velocity" gauge for the dipole radial integral R_1:

$$R_1 = \frac{\hbar^2}{m\Delta E}\int_0^\infty P_0(r)\left(\frac{\mathrm{d}}{\mathrm{d}r} \pm \frac{2l+1 \pm 1}{2r}\right)P_1(r)\,\mathrm{d}r\,, \tag{2.3.2}$$

or the form in terms of the operator p:

$$R_1 = \frac{Ze^2\hbar^2}{m(\Delta E)^2}\int_0^\infty P_0(r)r^{-2}P_1(r)\,\mathrm{d}r\,, \tag{2.3.3}$$

where Z is the nuclear charge, the signs $\pm$ correspond to transitions $l \to l \pm 1$, respectively.

All three representations

$$f^L \sim |\langle 0|\mathbf{r}|1\rangle|^2\,, \quad f^V \sim |\langle 0|\mathbf{p}|1\rangle|^2\,, \quad f^{\dot{p}} \sim |\langle 0|\dot{\mathbf{p}}|1\rangle|^2$$

are equivalent if the wave functions are exact solutions of the Schroedinger equation. The use of approximate wave functions in (2.1.5, 2.3.2, 3), in particular obtained in the central-field approximation, then leads to the different results, and the accuracy of each formula depends on the method used for the calculation of the wave functions. The semiempirical method described in [2.3] provides better accuracy for $P(r)$ at large r and, therefore, it is better to use the expressions (2.1.5) in the form of f^L. The self-consistent field and other variational methods describe the wave functions $P(r)$ quite well at medium distances r; in this case, the form of the velocity gauge (2.3.2) gives better results. The third form requires the determination of $P(r)$ at small r with very high accuracy, which is quite difficult to provide in calculations of multielectron atoms.

Oscillator strengths satisfy the exact and approximate sum rules. The exact sum rule for a certain state a is as follows [2.5]:

$$\sum_{a'} f_{aa'} + \frac{c}{2\pi^2} \int_I^\infty \sigma_a^{\rm abs}(\omega)\,{\rm d}\omega = N\,. \tag{2.3.4}$$

Here, c in the velocity of light, a is the atomic state, I is the first ionization potential, $\sigma_a^{\rm abs}$ is the photoabsorption cross section at frequency ω (Sect. 3.1) and N is the total number of atomic electrons; all values are in atomic units. The sum in (2.3.4) is over all transitions of all atomic electrons including inner-shell electrons.

It is also possible to give approximate sum rules for single-electron transitions in multielectron atoms [2.1]. For transitions from the group of equivalent electrons nl^q outside the filled shells

$$\gamma_0(nl^q)\gamma SLJ \to \gamma_0(nl^{q-1})\gamma_1 L_1 S n'l' SL'J'\,,$$

one has

$$\sum_{n'l'L'J'} f[\gamma_0(nl^q)\gamma SLJ - \gamma_0(nl^{q-1})\gamma_1 L_1 S n'l' SL'J')] = q\,. \tag{2.3.5}$$

For one electron outside filled shells

$$\sum_{n'l'j'} f(nlj - n'l'j') = 1\,. \tag{2.3.6}$$

The approximate sum rule for a specified nl shell is [2.5]

$$\sum_{k\neq 0} \frac{f_{0k}}{\omega_{0k}} + \frac{1}{4\pi\alpha} \int_0^\infty \sigma_{nl}^{\rm ph}(\omega)\frac{{\rm d}\omega}{\omega} = \frac{1}{3}\langle r^2\rangle_{nl}\,, \tag{2.3.7}$$

where $\sigma_{nl}^{ph}(\omega)$ is the photoionization cross section of the nl state, index 0 refers to the nl shell and $\langle r^2\rangle$ is the mean-square radius of the nl shell (1.8.10).

Near the ionization limit, the oscillator strength f for transition $0-nl$ is connected to the continuum oscillator strength ${\rm d}f/{\rm d}\varepsilon$ by the relation

$$\left.\frac{{\rm d}f}{{\rm d}\varepsilon}\right|_{\varepsilon\to 0} = \left.\frac{n_*^3 f_{0-nl}}{2z^2}\right|_{n\to\infty} = \frac{137}{4\pi}\sigma^{\rm ph}[\pi a_0^2]\,, \tag{2.3.8}$$

$$n_* = n - \Delta = (E_{nl}/z^2{\rm Ry})^{-1/2}\,, \tag{2.3.9}$$

where ε is the energy in continuum in Ry units, n_* and E_{nl} are the effective quantum number (1.1.3) and the energy of the nl level, respectively, and Δ is the quantum defect.

2.3.2 Recommended Data for Wavelengths, Energy Levels and Transition Probabilities

In this section, we give the list of refences for recommended data on wavelengths, energy levels and transition probabilities for neutral atoms taken from [2.6]. The recommended bibliography on atomic energy levels is given in [2.7, 8].

Table 2.1. References for the most recent critical data tables on wavelengths, energy levels and transition probabilities [2.6]

Nuclear charge Z	Element	References on: Wavelengths	Energy levels	Trans. Prob.
1	H; D, T	2.9	2.9	2.10
2	He	2.11–13	2.14, 15	2.10
3	Li	2.11–13	2.16	2.10
4	Be	2.11, 12	2.16	2.10
5	B	2.11, 12, 17	2.17	2.10
6	C	2.11–13, 18	2.18	2.10
7	N	2.11–13, 19	2.19	2.10
8	O	2.11–13, 20	2.10	2.10
9	F	2.11–13	2.16	2.10
10	Ne	2.11–13	2.16	2.10
11	Na	2.11–13	2.21	2.22
12	Mg	2.11–13	2.23	2.22
13	Al	2.11–13	2.24	2.22
14	Si	2.11–13	2.25	2.22
15	P	2.11, 12	2.26	2.22
16	S	2.11, 12	2.27	2.23
17	Cl	2.11–13	2.16	2.22
18	Ar	2.11–13	2.16	2.22
19	K	2.11–13	2.28	2.22
20	Ca	2.11–13	2.28	2.22
21	Sc	2.29	2.28	2.30
22	Ti	2.11–13	2.28	2.30
23	V	2.11, 12	2.28	2.30
24	Cr	2.11–13	2.28	2.30
25	Mn	2.11–13	2.28	2.30
26	Fe	2.11–13	2.28	2.31
27	Co	2.11, 12	2.28	2.31
28	Ni	2.11, 12	2.28	2.31
29	Cu	2.11–13	2.32	2.33
30	Zn	2.11, 12	2.16	2.33
31	Ga	2.11, 12	2.16	2.33
32	Ge	2.11, 12	2.16	2.33
33	As	2.11, 12	2.16	2.33
34	Se	2.11, 12	2.16	2.33
35	Br	2.11, 12	2.16	2.33
36	Kr	2.11–13	2.16	2.33

For elements heavier than krypton ($Z = 36$), the following references are recommended:

wavelength data: [2.12, 36, 37],
energy level data: [2.16, 34, 36],
transition probability data: [2.33].

Table 2.2. Oscillator strengths f and transition probabilities W in hydrogen [2.10]

Transition	f	$W\,[\mathrm{s}^{-1}]$
1s–2p	0.416	6.26×10^{8}
1s–3p	7.91×10^{-2}	1.67×10^{8}
1s–4p	2.90×10^{-2}	6.82×10^{7}
1s–4p	1.39×10^{-2}	3.44×10^{7}
1s–6p	7.80×10^{-3}	1.97×10^{7}
2p–3s	1.36×10^{-2}	6.31×10^{6}
2p–4s	3.04×10^{-3}	2.58×10^{6}
2p–5s	1.21×10^{-3}	1.29×10^{6}
2p–6s	6.18×10^{-4}	7.35×10^{5}
2s–3p	0.435	2.24×10^{7}
2s–4p	0.103	9.67×10^{6}
2s–5p	4.19×10^{-2}	4.95×10^{6}
2s–6p	2.16×10^{-2}	2.86×10^{6}
2p–3d	0.606	6.46×10^{7}
2p–4d	0.122	2.06×10^{7}
2p–5d	4.44×10^{-2}	9.42×10^{6}
2p–6d	2.16×10^{-2}	5.14×10^{6}
3s–4p	0.485	3.06×10^{6}
3s–5p	0.121	1.64×10^{6}
3s–6p	5.14×10^{-2}	9.55×10^{5}
3p–4s	3.22×10^{-2}	1.84×10^{6}
3p–5s	7.43×10^{-3}	9.05×10^{5}
3p–6s	3.03×10^{-3}	5.07×10^{5}
3p–4d	0.618	7.04×10^{6}
3p–5d	0.139	3.39×10^{6}
3p–6d	5.61×10^{-2}	1.88×10^{6}
3d–4p	1.10×10^{-2}	3.48×10^{5}
3d–5p	2.21×10^{-3}	1.50×10^{5}
3d–6p	8.42×10^{-4}	7.82×10^{4}
3d–4f	1.02	1.38×10^{7}
3d–5f	0.157	4.54×10^{6}
3d–6f	5.39×10^{-2}	2.15×10^{6}
4s–5p	0.544	7.37×10^{5}
4s–6p	0.138	6.46×10^{5}
4p–5s	5.29×10^{-2}	6.45×10^{5}
4p–6s	1.23×10^{-2}	3.58×10^{5}
4p–5d	0.609	1.49×10^{6}
4p–6d	0.148	8.62×10^{5}
4d–5p	2.78×10^{-2}	1.88×10^{5}
4d–6p	5.84×10^{-3}	9.42×10^{4}
4d–5f	0.890	2.58×10^{6}
4d–6f	0.186	1.29×10^{4}
4f–5d	8.87×10^{-3}	5.05×10^{4}
4f–6d	1.58×10^{-3}	2.14×10^{4}
4f–5g	1.35	4.25×10^{6}
4f–6g	0.182	1.37×10^{6}
5s–6p	0.608	2.43×10^{5}
5p–6s	7.45×10^{-2}	2.68×10^{5}
5p–6d	0.625	4.49×10^{5}
5d–6p	4.80×10^{-2}	9.59×10^{4}
5d–6f	0.844	7.23×10^{5}
5f–6d	2.33×10^{-2}	3.91×10^{4}
5f–6g	1.18	1.11×10^{6}
5g–6f	7.38×10^{-3}	1.14×10^{4}
5g–6h	1.68	1.64×10^{6}

2.3.3 Tables for f and W in H, He and Light Atoms

In Tables 2.2–4, the oscillator strengths and transition probabilities for H and He atoms are given. Table 2.3 presents the l-averaged oscillator strengths

$$f(n_0 - n_1) = \frac{1}{n_0^2} \sum_{l_0 l_1} (2l_0 + 1) f(n_0 l_0 - n_1 l_1) \tag{2.3.10}$$

as well as wavelengths and transition probabilities (2.1.11) in hydrogen. The f and W values for light atoms and ions are given in Table 2.5.

Table 2.3. Wavelengths λ, oscillator strengths f and transition probabilities W for the l-averaged transitions (2.1.11, 2.3.10) in hydrogen [2.10]

Transition $n_1 - n_0$	λ [Å]	f	W [s^{-1}]
1–2	1215.6	0.416	4.70×10^8
–3	1025.7	7.91×10^{-2}	5.57×10^7
–4	972.5	2.90×10^{-2}	1.28×10^7
–5	949.7	1.39×10^{-2}	4.12×10^6
–6	937.8	7.80×10^{-3}	1.64×10^6
–7	930.7	4.81×10^{-3}	7.57×10^5
–8	926.2	3.18×10^{-3}	3.87×10^5
–9	923.1	2.21×10^{-3}	2.14×10^5
–10	920.9	1.60×10^{-3}	1.26×10^5
2–3	6562.8	0.640	4.41×10^7
–4	4861.3	0.119	8.42×10^6
–5	4340.4	4.46×10^{-2}	2.53×10^6
–6	4101.7	2.21×10^{-2}	9.73×10^5
–7	3970.0	1.27×10^{-2}	4.39×10^5
–8	3889.0	8.03×10^{-3}	2.21×10^5
–9	3835.3	5.43×10^{-3}	1.22×10^5
–10	3797.9	3.85×10^{-3}	7.12×10^4
3–4	18751.0	0.842	8.99×10^6
–5	12818.1	0.151	2.20×10^6
–6	10938.1	5.58×10^{-2}	7.78×10^5
–7	10049.4	2.77×10^{-2}	3.36×10^5
–8	9545.9	1.60×10^{-2}	1.65×10^5
–9	9229.0	1.02×10^{-2}	8.90×10^4
–10	9014.9	6.98×10^{-3}	5.15×10^4
4–5	40512	1.038	2.70×10^6
–6	26252	0.179	7.71×10^5
–7	21655	6.55×10^{-2}	3.04×10^5
–8	19445	3.23×10^{-2}	1.42×10^5
–9	18174	1.87×10^{-2}	7.46×10^4
–10	17362	1.20×10^{-2}	4.23×10^4
5–6	74578	1.231	1.03×10^6
–7	46525	0.207	3.25×10^5
–8	37395	7.45×10^{-2}	1.39×10^5
–9	32961	3.64×10^{-2}	6.91×10^4
–10	30384	2.10×10^{-2}	3.80×10^4

Table 2.4. Oscillator strengths f and transition probabilities W for transitions between terms in He. If the quantum numbers J_0 and J_1 are omitted, the values of f and W correspond to the transition between terms

Element	Transition	Terms	J_0–J_1	f	$W\,[\mathrm{s}^{-1}]$
He I	$1s^2$–$1s2p$	^1S–$^1P^0$	0–1	0.276	1.80×10^9
	–$1s3p$	^1S–$^1P^0$	0–1	0.0734	5.66×10^8
	–$1s4p$	^1S–$^1P^0$	0–1	0.0302	2.46×10^8
	–$1s5p$	^1S–$^1P^0$	0–1	0.0153	1.28×10^8
	$1s2s$–$1s2p$	^1S–$^1P^0$	0–1	0.376	1.98×10^6
	–$1s3p$	^1S–$^1P^0$	0–1	0.151	1.34×10^7
	–$1s4p$	^1S–$^1P^0$	0–1	0.0507	0.72×10^7
	–$1s5p$	^1S–$^1P^0$	0–1	0.0221	3.76×10^6
	$1s2s$–$1s2p$	^3S–$^3P^0$	1–2	0.300	1.02×10^7
		^3S–$^3P^0$	1–1	0.180	1.02×10^7
		^3S–$^3P^0$	1–0	0.060	1.02×10^7
	$1s2s$–$1s3p$	^3S–$^3P^0$		0.0645	0.95×10^7
	–$1s4p$	^3S–$^3P^0$		0.0231	5.05×10^6
	–$1s5p$	^3S–$^3P^0$		0.0114	2.93×10^6
	$1s2p$–$1s3s$	$^1P^0$–^1S	1–0	0.0480	1.81×10^7
	–$1s4s$	$^1P^0$–^1S	1–0	0.834×10^{-2}	0.65×10^7
	–$1s5s$	$^1P^0$–^1S	1–0	0.308×10^{-2}	3.13×10^6
	$1s2p$–$1s3d$	$^1P^0$–^1S	1–2	0.711	0.64×10^8
	–$1s4d$	$^1P^0$–^1D	1–2	0.122	2.02×10^7
	–$1s5d$	$^1P^0$–^1D	1–2	0.0436	0.91×10^7
	$1s2p$–$1s3s$	$^3P^0$–^3S	2–1	0.0693	1.54×10^7
	–$1s3s$	$^3P^0$–^3S	1–1	0.0692	0.93×10^7
	–$1s3s$	$^3P^0$–^3S	0–1	0.0692	3.08×10^6
	–$1s4s$	$^3P^0$–^3S		0.0118	1.06×10^7
	–$1s5s$	$^3P^0$–^3S		0.365×10^{-2}	4.30×10^6
	$1s2p$–$1s3d$	$^3P^0$–^3D		0.609	0.70×10^8
	–$1s4d$	$^3P^0$–^3D		0.125	2.51×10^7
	–$1s5d$	$^3P^0$–^3D		0.0474	1.17×10^7
	$1s3s$–$1s3p$	^1S–$^1P^0$	0–1	0.629	2.53×10^5
	–$1s4p$	^1S–$^1P^0$	0–1	0.140	1.37×10^6
	–$1s5p$	^1S–$^1P^0$	0–1	0.0521	0.96×10^6
	$1s3s$–$1s3p$	^3S–$^3P^0$		0.896	1.08×10^6
	–$1s4p$	^3S–$^3P^0$		0.0429	0.61×10^6
	$1s3s$–$1s5p$	^3S–$^3P^0$		0.0245	0.61×10^6
	$1s3p$–$1s4s$	$^1P^0$–^1S	1–0	0.103	4.59×10^6
	–$1s5s$	$^1P^0$–^1S	1–0	0.182	2.02×10^6
	$1s3p$–$1s4d$	$^1P^0$–^1D	1–2	0.647	7.11×10^6
	–$1s5d$	$^1P^0$–^1D	1–2	0.139	3.31×10^6
	$1s3p$–$1s4s$	$^3P^0$–^3S		0.145	6.52×10^6
	–$1s5s$	$^3P^0$–^3S		0.0222	2.69×10^6
	$1s3p$–$1s3d$	$^3P^0$–^3D		0.111	1.28×10^4
	–$1s4d$	$^3P^0$–^3D		0.482	6.68×10^6
	–$1s5d$	$^3P^0$–^3D		0.123	3.43×10^6
	$1s3d$–$1s3p$	^1D–$^1P^0$	2–1	0.0139	1.68×10^2

Table 2.5. The values of f and W in light atoms and ions (see explanation to Table 2.4)

Element	Transition	Terms	J_0–J_1	f	$W\,[\mathrm{s}^{-1}]$
Li I	$2s$–$2p$	^2S–$^2P^0$		0.753	3.72×10^7
	–$3p$	^2S–$^2P^0$		0.55×10^{-2}	1.17×10^6
	–$4p$	^2S–$^2P^0$		0.48×10^{-2}	1.42×10^6
	–$5p$	^2S–$^2P^0$		0.32×10^{-2}	1.07×10^6
	$2p$–$3s$	$^2P^0$–^2S		0.115	3.49×10^7
	–$4s$	$^2P^0$–^2S		0.125	0.1×10^7
	–$5s$	$^2P^0$–^2S		0.42×10^{-2}	4.60×10^6
	$2p$–$3d$	$^2P^0$–^2D		0.67	7.2×10^7
	–$4d$	$^2P^0$–^2D		0.12	2.3×10^7
	–$5d$	$^2P^0$–^2D		0.045	1.06×10^7
	$3s$–$3p$	^2S–$^2P^0$		1.23	3.77×10^6
	$3p$–$4s$	$^2P^0$–^2S		0.223	7.46×10^6
	–$5s$	$^2P^0$–^2S		0.0254	1.44×10^6
	$3p$–$3d$	$^2P^0$–^2D		0.0743	3.81×10^3
	–$4d$	$^2P^0$–^2D		0.527	6.85×10^6
	–$5d$	$^2P^0$–^2D		0.128	3.41×10^6
Li II	$1s^2$–$1s2p$	^1S–$^1P^0$	0–1	0.457	2.56×10^{10}
	–$1s3p$	^1S–$^1P^0$	0–1	0.111	0.78×10^{10}
	$1s^2$–$1s2p$	^1S–$^1P^0$	0–1	0.213	5.18×10^6
	–$1s3p$	^1S–$^1P^0$	0–1	0.256	2.82×10^6
	$1s2s$–$1s2p$	^3S–$^3P^0$		0.308	2.26×10^7
	–$1s3p$	^3S–$^3P^0$		0.186	2.88×10^8
	$1s2p$–$1s3s$	$^1P^0$–^1S	1–0	0.031	2.04×10^8
	–$1s3d$	$^1P^0$–^1D	1–2	0.714	1.01×10^9
	$1s2p$–$1s3p$	$^3P^0$–^3S		0.039	2.85×10^8
	–$1s3d$	$^3P^0$–^3D		0.625	1.12×10^9
Be III	$1s^2$–$1s2p$	^1S–$^1P^0$	0–1	0.552	1.22×10^{11}
	–$1s3p$	^1S–$^1P^0$	0–1	0.127	3.62×10^{10}
	$1s2s$–$1s2p$	^1S–$^1P^0$	0–1	0.149	8.77×10^6
	–$1s3p$	^1S–$^1P^0$	0–1	0.305	4.28×10^9
	$1s2s$–$1s2p$	^3S–$^3P^0$		0.213	3.42×10^7
	–$1s3p$	^3S–$^3P^0$		0.252	1.65×10^9
	$1s2p$–$1s3d$	$^1P^0$–^1D	1–2	0.711	5.10×10^9
	$1s2p$–$1s3d$	$^3P^0$–^3D		0.640	5.61×10^9
B IV	$1s^2$–$1s2p$	^1S–$^1P^0$	0–1	0.609	3.72×10^{11}
	–$1s3p$	^1S–$^1P^0$	0–1	0.135	1.08×10^{11}
	$1s2s$–$1s2p$	^1S–$^1P^0$	0–1	0.114	1.25×10^7
	–$1s3p$	^1S–$^1P^0$	0–1	0.333	0.51×10^{10}
	$1s2s$–$1s2p$	^3S–$^3P^0$		0.163	0.45×10^8
	–$1s3p$	^3S–$^3P^0$		0.291	0.55×10^{10}
	$1s2p$–$1s3d$	$^1P^0$–^1D	1–2	0.709	1.62×10^{10}
	$1s2p$–$1s3d$	$^3P^0$–^3D		0.650	1.75×10^{10}
C V	$1s^2$–$1s2p$	^1S–$^1P^0$	0–1	0.647	8.87×10^{11}
	–$1s3p$	^1S–$^1P^0$	0–1	0.141	2.55×10^{11}
	$1s2s$–$1s2p$	^1S–$^1P^0$	0–1	0.093	1.65×10^7
	–$1s3p$	^1S–$^1P^0$	0–1	0.351	1.28×10^{10}
	$1s2s$–$1s2p$	^3S–$^3P^0$		0.132	0.56×10^8
	–$1s3p$	^3S–$^3P^0$		0.316	1.36×10^{10}
	$1s2p$–$1s3d$	$^1P^0$–^1D	1–2	0.707	3.96×10^{10}
	$1s2p$–$1s3d$	$^3P^0$–^3D		0.657	4.25×10^{10}
N VI	$1s^2$–$1s2p$	^1S–$^1P^0$	0–1	0.674	1.81×10^{12}
	–$1s3p$	^1S–$^1P^0$	0–1	0.144	5.16×10^{11}
	$1s2s$–$1s2p$	^1S–$^1P^0$	0–1	0.078	2.06×10^7
	–$1s3p$	^1S–$^1P^0$	0–1	0.364	2.69×10^{10}
	$1s2s$–$1s2p$	^3S–$^1P^0$		0.110	0.68×10^8
	–$1s3p$	^3S–$^3P^0$		0.334	2.85×10^{10}

2.3.4 Oscillator Strengths in Alkali Atoms

Dipole transitions in the alkali atoms (one optical electron outside the closed shells) are of interest for many reasons. On one hand, alkali spectra play an important role in magneto-optical (Faraday) spectroscopy [2.38, 39]. On the other hand, the transitions from the ground state in alkalis are characterized by an anomalous distribution of the oscillator strengths. According to the sum rule (2.3.6) for atoms with one valence electron, we have $\sum_n f_{0n} \approx 1$. For the alkali atoms the main contribution to this sum is due to the resonance transition with $f_{res} \approx 1$, so that the f values for high members in the sharp s–p series are very small: $f_{0n} < 0.01$. Therefore, the transitions to the high Rydberg states are quite sensitive to effects due to the core polarization, the spin–orbit interaction and blackbody radiation [2.40, 41].

In Tables 2.6–13, we give experimental and theoretical oscillator strengths for Li, Na, K, Rb and Cs atoms. As a rule, the Faraday rotation method gives the best accuracy for the f values.

Table 2.6. Oscillator strengths for the principal series of Li I*

Transition	Experiment				Theory			
	[2.42]	[2.43]	[2.39]	[2.44]	[2.40]	[2.45]	[2.46]	[2.47]
$2s$–$2p$	0.75 – 0	0.75 – 0			0.74 – 0	0.74 – 0	0.74 – 0	0.78 – 0
$3p$	0.55 – 2	0.55 – 2			0.47 – 2	0.51 – 2	0.42 – 2	0.51 – 2
$4p$	0.48 – 2	0.45 – 2			0.43 – 2	0.44 – 2	0.39 – 2	0.45 – 2
$5p$	0.31 – 2	0.27 – 2			0.26 – 2	0.27 – 2	0.24 – 2	0.27 – 2
$6p$	0.19 – 2				0.16 – 2		0.15 – 2	0.16 – 2
$7p$							0.95 – 3	0.11 – 2
$8p$							0.67 – 3	0.73 – 3
$9p$			0.615 – 3	0.615 – 3			0.46 – 3	0.50 – 3
$10p$			0.444 – 3	0.453 – 3			0.34 – 3	
$11p$			0.348 – 3	0.344 – 3			0.35 – 3	
$12p$	0.33 – 3		0.285 – 3	0.266 – 3	0.21 – 3		0.20 – 3	
$13p$			0.242 – 3	0.211 – 3				
$14p$			0.220 – 3	0.169 – 3				
$15p$			0.184 – 3	0.138 – 3				
$16p$			0.159 – 3	0.114 – 3	0.84 – 4			
$17p$			0.157 – 4	0.950 – 4				
$18p$			0.133 – 3	0.802 – 4				
$19p$			0.115 – 3	0.682 – 4				
$20p$			0.102 – 3	0.587 – 4				
$21p$			0.940 – 4	0.507 – 4				
$22p$			0.860 – 4	0.441 – 4	0.34 – 4			
$23p$			0.830 – 4					

*0.55 – 2 means 0.55×10^{-2}

Table 2.7. Oscillator strengths for the principal series of Na I

Transition	Experiment			Theory				
	[2.48]	[2.39]	[2.44]	[2.40]	[2.46]	[2.49]	[2.50]	[2.47]
$3s$–$3p$	0.969 – 0			0.96 – 0	0.96 – 0	0.96 – 0	0.97 – 0	0.99 – 0
$4p$	0.140 – 0			0.13 – 1	0.15 – 1	0.13 – 1	0.14 – 1	0.14 – 1
$5p$	0.205 – 2			0.18 – 2	0.26 – 2	0.19 – 2	0.21 – 2	0.23 – 2
$6p$	0.631 – 3			0.54 – 3	0.84 – 3	0.57 – 3	0.65 – 3	0.72 – 3
$7p$	0.256 – 3			0.22 – 3	0.38 – 3	0.24 – 3	0.28 – 3	0.32 – 3
$8p$					0.21 – 3	0.12 – 3	0.15 – 3	0.17 – 3
$9p$		0.811 – 4	0.811 – 4		0.13 – 3	0.73 – 4	0.87 – 3	0.10 – 3
$10p$		0.532 – 4	0.523 – 4		0.84 – 4	0.47 – 4	0.56 – 4	0.66 – 4
$11p$		0.361 – 4	0.364 – 4		0.54 – 4	0.31 – 4	0.38 – 4	
$12p$		0.255 – 4	0.262 – 4		0.40 – 4	0.22 – 4	0.27 – 4	
$13p$	0.217 – 4	0.187 – 4	0.198 – 4	0.15 – 4		0.17 – 4	0.20 – 4	
$14p$		0.142 – 4	0.151 – 4			0.12 – 4	0.16 – 4	
$15p$		0.110 – 4	0.119 – 4			0.94 – 5	0.12 – 4	
$16p$		0.984 – 5	0.968 – 5			0.74 – 5		
$17p$	0.920 – 5	0.733 – 5	0.784 – 5	0.57 – 5		0.66 – 4		
$18p$		0.604 – 5	0.647 – 5					
$19p$		0.509 – 5	0.542 – 5					
$20p$		0.429 – 5	0.456 – 5					
$21p$		0.372 – 5	0.389 – 5					
$22p$		0.324 – 5	0.335 – 5					
$23p$		0.272 – 5	0.291 – 5	0.21 – 5				
$24p$		0.224 – 5						
$25p$		0.190 – 5						
$26p$		0.209 – 5						
$27p$		0.154 – 5						
$28p$		0.132 – 5						
$29p$		0.108 – 5						
$30p$		0.99 – 6						
$31p$		0.77 – 6						
$32p$		0.84 – 6						

2.4 Lifetimes of Excited States

The radiative lifetime τ_k of an excited k state is the time when a number of particles in this state due to spontaneous transitions decreases e times (e = 2.7183).

The lifetime τ_k is defined as

$$\tau_k = \left(\sum_{i<k} W_{ik} \right)^{-1} , \tag{2.4.1}$$

where W_{ik} is a radiative transition probability (Sect. 2.3). For the resonance transition (the nearest to the ground state, optically allowed) the lifetimes τ_{res} is

Table 2.8. Oscillator strengths for the principal series of K I; f values with * means j-averaged oscillator strengths

	Experiment			Theory			
Transition	[2.60]	[2.44]	[2.39]	[2.40]	[2.47]	[2.49]	[2.50]
$4s_{1/2}$–$4p_{1/2}$	0.102*			0.33 – 0	0.107 + 1*	0.33 – 0	0.97 + 0*
$4p_{3/2}$				0.66 – 0		0.66 – 0	
$5p_{1/2}$	0.91 – 2*			0.28 – 2	0.92 – 2*	0.26 – 2	0.84 – 2*
$5p_{3/2}$				0.62 – 2		0.57 – 2	
$6p_{1/2}$	0.90 – 3*			0.31 – 3	0.93 – 3*	0.26 – 3	0.82 – 3*
$6p_{3/2}$				0.71 – 3		0.61 – 3	
$7p_{1/2}$	0.214 – 3*			0.70 – 4	0.20 – 3*	0.54 – 4	0.18 – 3*
$7p_{3/2}$				0.17 – 3		0.14 – 3	
$8p_{1/2}$		0.58 – 4*	0.58 – 4*	0.24 – 4	0.62 – 4*	0.17 – 3	0.57 – 4*
$8p_{3/2}$	–			0.61 – 4		0.47 – 4	
$9p_{1/2}$		0.23 – 4*	0.214 – 4*		0.25 – 4*	0.69 – 5	0.24 – 4*
$9p_{3/2}$						0.21 – 4	
$10p_{1/2}$	0.11 – 4*	0.664 – 5*			0.12 – 4*	0.34 – 5	0.12 – 4*
$10p_{3/2}$						0.10 – 4	
$11p_{1/2}$	0.63 – 5*	0.374 – 5*			0.64 – 5*	0.18 – 5	0.65 – 5*
$11p_{3/2}$						0.61 – 5	
$12p_{1/2}$	0.39 – 5*	0.213 – 5*				0.12 – 5	0.39 – 5*
$12p_{3/2}$						0.39 – 5	
$13p_{1/2}$	0.25 – 5*	0.139 – 5*				0.78 – 6	0.25 – 5*
$13p_{3/2}$						0.28 – 5	
$14p_{1/2}$	0.18 – 5*	0.766 – 6*	0.82 – 6			0.56 – 6	0.17 – 5*
$14p_{3/2}$			0.25 – 5			0.20 – 5	
$15p_{1/2}$	0.13 – 5*	0.595 – 6*				0.44 – 6	0.12 – 5*
$15p_{3/2}$						0.16 – 5	
$16p_{1/2}$	0.96 – 6*	0.545 – 6*				0.36 – 6	0.91 – 6*
$16p_{3/2}$						0.13 – 5	
$17p_{1/2}$	0.73 – 6*	0.483 – 6*				0.28 – 6	
$17p_{3/2}$	–	–				0.11 – 5	
$18p_{1/2}$	0.57 – 6*	0.424 – 6*	0.27 – 6				
$18p_{3/2}$			0.87 – 6				
$19p_{1/2}$	0.46 – 6*	0.345 – 6*					
$19p_{3/2}$	–	–					
$20p_{1/2}$	0.37 – 6*	0.294 – 6*					
$20p_{3/2}$	–	–					
$21p_{1/2}$	0.31 – 6*	0.247 – 6*					
$21p_{3/2}$	–	–					
$22p_{1/2}$	0.26 – 6*	0.203 – 6*					
$22p_{3/2}$	–	–					
$23p_{1/2}$	0.22 – 6*	0.165 – 6*					
$23p_{3/2}$	–	–					
$24p_{1/2}$		0.132 – 6*	0.86 – 7				
$24p_{3/2}$			0.29 – 6				
$25p_{1/2}$		0.105 – 6*					
$25p_{3/2}$							
$26p_{1/2}$		0.852 – 7*					
$26p_{3/2}$		–					
$27p_{1/2}$		0.689 – 7*					
$27p_{3/2}$		–					
$28p_{1/2}$		0.570 – 7*					
$28p_{3/2}$		–					
$29p_{1/2}$		0.475 – 7*					
$29p_{3/2}$		–					
$30p_{1/2}$		0.403 – 7*					
$30p_{3/2}$		–					
$31p_{1/2}$		0.341 – 7*					
$31p_{3/2}$		–					
$32p_{1/2}$		0.307 – 7*					
$32p_{3/2}$		–					

Table 2.9. Oscillator strengths for the principal series of Rb I (see explanation to Table 2.8)

	Experiment		Theory				
Transition	[2.51]	[2.52]	[2.40]	[2.49]	[2.53]	[2.44]	[2.50]
$5s_{1/2}$–$5p_{1/2}$	0.332 – 0		0.33 – 0	0.336 – 0	0.344 – 0		0.35 – 0
$5p_{3/2}$	0.668 – 0		0.67 – 0	0.682 – 0	0.697 – 0		0.71 – 0
$6p_{1/2}$	0.373 – 2	0.40 – 2	0.39 – 2	0.347 – 2	0.371 – 2		0.35 – 2
$6p_{3/2}$	0.954 – 2	0.937 – 2	0.10 – 1	0.947 – 2	0.981 – 2		0.96 – 2
$7p_{1/2}$	0.487 – 3	0.56 – 3	0.54 – 3	0.428 – 3	0.449 – 3		0.43 – 3
$7p_{3/2}$	0.148 – 2	0.153 – 2	0.17 – 2	0.144 – 2	0.144 – 2		0.14 – 2
$8p_{1/2}$	0.138 – 3	0.15 – 3	0.15 – 3	0.112 – 3	0.118 – 3		0.11 – 3
$8p_{3/2}$	0.468 – 3	0.460 – 3	0.52 – 3	0.438 – 3	0.436 – 3		0.44 – 3
$9p_{1/2}$	0.522 – 4	0.60 – 4	0.59 – 4	0.424 – 4	0.449 – 4		0.42 – 4
$9p_{3/2}$	0.197 – 3	0.200 – 3	0.23 – 3	0.186 – 3	0.186 – 3		0.19 – 3
$10p_{1/2}$	0.261 – 4	0.30 – 4		0.201 – 4			0.20 – 4
$10p_{3/2}$	0.108 – 3	0.101 – 3		0.965 – 4			0.98 – 4
$11p_{1/2}$	0.146 – 4	0.185 – 4		0.562 – 4			0.11 – 4
$11p_{3/2}$	0.638 – 4	0.730 – 4		0.111 – 4			0.58 – 4
$12p_{1/2}$	0.900 – 5	0.995 – 5		0.685 – 5			0.68 – 5
$12p_{3/2}$	0.409 – 4	0.425 – 4		0.360 – 4			0.37 – 4
$13p_{1/2}$	0.582 – 5	0.634 – 5		0.443 – 5			0.44 – 5
$13p_{3/2}$	0.286 – 4	0.295 – 4		0.242 – 4			0.25 – 4
$14p_{1/2}$	0.397 – 5	0.517 – 5		0.302 – 5			0.30 – 5
$14p_{3/2}$	0.200 – 4	0.228 – 4		0.173 – 4			0.18 – 4
$15p_{1/2}$	0.274 – 5	0.343 – 5		0.218 – 5			0.22 – 5
$15p_{3/2}$	0.144 – 4	0.136 – 4		0.125 – 4			0.13 – 4
$16p_{1/2}$	0.133 – 4*	0.242 – 5		0.149 – 5		0.135*	0.16 – 5
$16p_{3/2}$	–	0.113 – 4		0.955 – 5			0.10 – 4
$17p_{1/2}$	0.104 – 4*	0.189 – 5		0.109 – 5		0.104 – 4*	0.13 – 5
$17p_{3/2}$	–	0.889 – 5		0.728 – 5		–	0.80 – 5
$18p_{1/2}$	0.786 – 5*	0.143 – 5		0.853 – 6		0.825 – 5*	
$18p_{3/2}$	–	0.716 – 5		0.596 – 5		–	
$19p_{1/2}$	0.681 – 5*	0.118 – 5	0.13 – 5	0.658 – 6		0.669 – 5*	
$19p_{3/2}$	–	0.556 – 5	0.65 – 5	0.465 – 5		–	
$20p_{1/2}$	0.554 – 5*	0.771 – 6		0.493 – 6		0.548 – 5*	
$20p_{3/2}$	–	0.384 – 5		0.367 – 5		–	
$21p_{1/2}$	0.460 – 5*	–	–	0.364 – 6	–	0.457 – 5*	–
$21p_{3/2}$				0.288 – 5	–	–	
$22p_{1/2}$	0.380 – 5*	–		–		0.385 – 5*	–
$22p_{3/2}$							
$23p_{1/2}$	0.340 – 5*					0.328 – 5*	
$23p_{3/2}$							
$24p_{1/2}$	0.297 – 5*					0.280 – 5*	
$24p_{3/2}$							
$25p_{1/2}$	0.230 – 5*		0.45 – 6			0.240 – 5*	
$25p_{3/2}$			0.24 – 5				
$26p_{1/2}$	0.210 – 5*						

Table 2.10. The ratio of oscillator strengths $f(5s_{1/2}-np_{3/2})/f(5s_{1/2}-np_{1/2})$ in Rb I

	Experiment			Theory			
n	[2.51]	[2.52]	[2.54]	[2.40]	[2.55]	[2.50]	[2.49]
5	2.011	–	2.016	2.03	2.01	2.03	2.03
6	2.56.	2.34	2.58	2.56	2.96	2.76	2.68
7	3.04	2.73	2.98	3.15	3.37	3.37	3.27
8	3.39	3.07	2.16	3.47	3.61	3.94	3.79
9	3.77	3.33	2.30	3.90	3.80	4.43	4.27
10	4.15	3.37	2.43		4.01	4.85	4.70
11	4.37	3.96				5.20	5.04
12	4.54	4.27				5.46	5.32
13	4.91	4.65				5.67	5.58
14	5.04	4.41				5.94	5.88
15	5.26	3.96				6.11	6.04
16		4.67				6.26	6.42
17		6.22				6.34	6.70
18		5.00				–	6.99
19		4.71		5.00			7.07
20		4.98					7.45
21							–
22							
23							
24							
25							
26				5.33			

related to the resonance oscillator strength f_{res} by

$$\tau_{res}[s] = 1.499 \times 10^{-16} \frac{g_1}{g_0 f_{res}} \lambda^2 , \tag{2.4.2}$$

where g is the statistical weight and λ is the transition wavelength in Å.

The lifetime of the nl state in the hydrogen atom increases with increasing n for constant l [2.5, 61]:

$$\tau_{nl} \propto n^3 ,$$

and the l-averaged lifetime is given by

$$\tau_n = \left(\frac{1}{n^2} \sum_l \frac{(2l+1)}{\tau_{nl}} \right)^{-4} \approx n^5 \tag{2.4.3}$$

[cf. (2.6.8)].

The following law is usually satisfied for heavy atoms:

$$\tau_k = \tau_0 n_*^a \quad \text{for } a \approx 3 , \tag{2.4.4}$$

where τ_0 is a constant for a given series and n_* is the effective principal quantum number (1.1.3). However, deviations from the law (2.4.4) are also possible.

Table 2.11. Oscillator strengths for principal series of Cs I

	Experiment			Theory			
Transition	[2.56]	[2.57]	[2.58]	[2.50]	[2.45]	[2.40]	[2.43]
$6s_{1/2}$–$6p_{1/2}$	0.347 – 0	–	0.31 – 0	0.354 – 0	0.34 – 0	0.34 – 0	0.33 – 0
–$6p_{3/2}$	0.721 – 0	–	0.74 – 0	0.740 – 0	0.707 – 0	0.70 – 0	0.69 – 0
–$7p_{1/2}$	0.267 – 2	0.254 – 0	0.277 – 0	0.204 – 2	0.241 – 2	0.30 – 2	0.24 – 2
–$7p_{3/2}$	0.113 – 1	0.117 – 1	0.12 – 1	0.105 – 1	0.104 – 1	0.12 – 1	0.11 – 1
–$8p_{1/2}$	0.254 – 3	0.22 – 3	0.188 – 3	0.155 – 3	0.216 – 3	0.32 – 3	0.20 – 3
–$8p_{3/2}$	0.185 – 2	0.170 – 2	0.178 – 2	0.164 – 2	0.164 – 2	0.21 – 2	0.16 – 2
–$9p_{1/2}$	0.492 – 4	0.50 – 4	0.374 – 4	0.243 – 4	0.404 – 4	0.78 – 4	0.39 – 4
–$9p_{3/2}$	0.570 – 3	0.60 – 3	0.516 – 3	0.512 – 3	0.515 – 3	0.72 – 3	0.53 – 3
$10p_{1/2}$	0.116 – 4	0.12 – 4	0.111 – 4	0.532 – 5	0.110 – 4	0.26 – 4	0.11 – 4
$10p_{3/2}$	0.229 – 3	0.24 – 3	0.215 – 3	0.223 – 3	0.224 – 3	0.33 – 3	0.23 – 3
$11p_{1/2}$	0.374 – 5	0.42 – 5	0.398 – 5	0.139 – 5	0.380 – 5		0.36 – 5
$11p_{3/2}$	0.131 – 3	0.14 – 3	0.11 – 3	0.117 – 3	0.118 – 3		0.12 – 3
$12p_{1/2}$	0.175 – 5	0.19 – 5	0.165 – 5	0.399 – 6	0.154 – 5		0.16 – 5
$12p_{3/2}$	0.784 – 4	0.75 – 4	0.69 – 4	0.689 – 4	0.699 – 4		0.75 – 4
$13p_{1/2}$	–	0.85 – 6	0.78 – 6	0.110 – 6	0.712 – 6		0.72 – 6
$13p_{3/2}$	0.513 – 4	0.53 – 4	0.425 – 4	0.448 – 4	0.454 – 4		0.49 – 4
$14p_{1/2}$	–	0.44 – 6	0.379 – 6	0.262 – 7	0.354 – 6		0.35 – 6
$14p_{3/2}$	0.329 – 4	0.34 – 4	0.240 – 4	0.307 – 4	0.309 – 4		0.34 – 4
$15p_{1/2}$	–	0.21 – 6	0.22 – 6	0.515 – 6	0.191 – 6		0.24 – 6
$15p_{3/2}$	0.241 – 4	0.24 – 4	0.202 – 4	0.220 – 4	0.222 – 4		0.24 – 4
$16p_{1/2}$	–	0.12 – 6	–	0.162 – 9	0.111 – 6	0.88 – 6	0.12 – 6
$16p_{3/2}$	0.178 – 4	0.17 – 4	0.145 – 4	0.160 – 4	0.222 – 4	0.26 – 4	0.18 – 4
$17p_{1/2}$	–	0.75 – 6	–	0.408 – 9	0.696 – 7	–	0.80 – 7
$17p_{3/2}$	0.135 – 4	0.14 – 4	0.106 – 4	0.126 – 4	0.165 – 4	–	0.14 – 4
$18p_{1/2}$	–	0.42 – 7	–	0.215 – 8	–	–	0.50 – 7
$18p_{3/2}$	0.984 – 5	0.10 – 4	–	0.100 – 4	0.127 – 4	–	0.11 – 4
$19p_{1/2}$							0.35 – 7
$19p_{3/2}$	0.796 – 5						0.88 – 5
$20p_{1/2}$						0.28 – 6	0.24 – 7
$20p_{3/2}$	0.648 – 5					0.11 – 4	0.71 – 5
$21p_{1/2}$							0.18 – 7
$21p_{3/2}$	0.510 – 5						0.59 – 5
$22p_{1/2}$							
$22p_{3/2}$	0.380 – 5						
$23p_{1/2}$							
$23p_{3/2}$	0.308 – 5						
$24p_{1/2}$							
$24p_{3/2}$	0.255 – 5						
$25p_{1/2}$							
$25p_{3/2}$	0.218 – 5						
$26p_{1/2}$						0.90 – 7	
$26p_{3/2}$	0.177 – 5					0.39 – 5	
$27p_{1/2}$							
$27p_{3/2}$							

Table 2.12. The ratio of oscillator strengths $f(6s_{1/2}-np_{3/2})/f(6s_{1/2}-np_{1/2})$ in Cs I

	Experiment				Theory			
n	[2.56]	[2.57]	[2.58]	[2.59]	[2.50]	[2.45]	[2.40]	[2.49]
6	2.077	–	2.4	–	2.093	2.077	2.06	2.08
7	4.243	4.6	4.3	4.3	5.1	4.3	4.0	4.40
8	7.30	7.7	9.5	7.1	10.5	7.6	6.6	7.97
9	11.58	12.	14	12.5	21.1	12.7	9.2	13.64
10	19.7	20.	20	19.5	41.9	20.3	12.7	22.16
11	35.0	30.	28	26.9	84.0	31.0		34.53
12	44.8	40.	41	38.3	172.6	45.3		46.65
13	–	62.	54		407.7	63.7		67.64
14		77.	63		1168	87.5		95.72
15		114	91		4288	116		100.2
16		142			99137	149	29.5	154.8
17		186			30937	188		174.2
18		238			4645			218.5
19								253.2
20							39.3	302.6
21								337.0
22								
23								
24								
25								
26							43.3	

Table 2.13. Oscillator strengths for transitions $5D_{5/2}$-nF in Cs I

n	Experiment [2.38]	Theory ATOM code [2.3]
14	3.3 ± 0.5	3.6
15	3.1 ± 0.5	2.9
16	2.1 ± 0.4	2.3
17	1.6 ± 0.3	2.0
18	1.5 ± 0.3	1.7

The lifetimes of atoms in the excited states vary from 10^{-6} to 10^{-9} s, with the exception of metastable states for which $\tau \leqslant 1$ s, and autonionizing states with $\tau \ll 10^{-9}$ s.

Radiative lifetimes for neutral atoms are given in [2.10, 41, 61–65]; lifetimes of autoionizing states in HeI in [2.66]. The most complete data and bibliography on τ for alkali atoms are given in [2.41, 61].

In this Section, the lifetimes from [2.64] for hydrogen, alkalis (Li, Na, K, Rb and Cs atoms) and Cu, Ag and Au atoms are given (Tables 2.14–19).

Table 2.14. Radiative lifetimes τ of the excited states in H I

State	τ[ns]	Method	State	τ[ns]	Method
$3S$	160	T	$5d$	69.5	T
$3S$	135(14)	BF	$6d$	126	T
$3S$	160	BG	$6d$	119	T
$3S$	148(11)	BF	$7d$	187	T
$3S$	159	T	$8d$	277	T
$4S$	230	T	$9d$	394	T
$4S$	186(27)	BF	$10d$	537	T
$4S$	230(17)	BF	$11d$	712	T
$4S$	227	T	$12d$	923	T
$5S$	360	T	$4f$	73	T
$5S$	378(38)	BF	$4f$	72.5	T
$5S$	352	T	$5f$	140	T
$6S$	570	T	$5f$	140	T
$6S$	534	T	$6f$	243	T
$7S$	782	T	$6f$	240	T
$8S$	1103	T	$7f$	378	T
$9S$	1511	T	$8f$	559	T
$10S$	2009	T	$9f$	791	T
$11S$	2610	T	$10f$	1079	T
$12S$	3334	T	$11f$	1426	T
$2p$	1.6	T	$12f$	1849	T
$2p$	1.600(4)	BG	$5g$	235	T
$2p$	1.60(1)	BF	$5g$	235	T
$2p$	1.592(25)	BF	$6g$	405	T
$2p$	1.6	T	$6g$	403	T
$3p$	5.4	T	$7g$	635	T
$3p$	5.4	BG	$8g$	938	T
$3p$	5.58(13)	BG	$9g$	1328	T
$3p$	5.5(2)	BF	$10g$	1810	T
$3p$	5.41(18)	BF	$11g$	2393	T
$3p$	5.8(3)	BF	$12g$	3099	T
$3p$	5.2	T	$6h$	610	T
$4p$	12.4	T	$6h$	608	T
$4p$	14.6(25)	BF	$7h$	960	T
$4p$	11.25(78)	BF	$8h$	1425	T
$4p$	12.4(6)	BF	$9h$	2017	T
$4p$	12.7	T	$10h$	2750	T
$5p$	24	T	$11h$	3634	T
$5p$	21.9(30)	BF	$12h$	4699	T
$5p$	23.8(30)	BF	$7i$	1350	T
$5p$	23.8	T	$8i$	2006	T
$6p$	41	T	$9i$	2836	T
$6p$	40.7	T	$10i$	3872	T
$7p$	64.4	T	$11i$	5118	T
$8p$	95.6	T	$12i$	6612	T
$9p$	136	T	$8k$	2683	T
$10p$	186	T	$9k$	3813	T
$11p$	247	T	$10k$	5211	T
$12p$	321	T	$11k$	6890	T
$3d$	15.6	T	$12k$	8894	T
$3d$	16.1(6)	BF	$9l$	4909	T
$3d$	15.6	BG	$10l$	6708	T
$3d$	15.8(6)	BF	$11l$	8914	T
$3d$	15.5	T	$12l$	11526	T
$4d$	36.5	T	$10m$	8419	T
$4d$	37.7(55)	BF	$11m$	11179	T
$4d$	36.9(15)	BF	$12m$	14432	T
$4d$	36	T	$11n$	13684	T
$5d$	70	T	$12n$	17772	T
$5d$	69.8(70)	BF	$12o$	21310	T

Table 2.15. Radiative lifetimes τ of the ^2P states in alkali atoms

State	τ[ns]	Method	State	τ[ns]	Method
			Li I		
$2P$	31.9(19)	BG	$4P$	479(24)	BG
$2P$	27.2(4)	LC	$4P$	403	T
$2P$	25(1)	BF	$4P$	446.2	T
$2P$	26.2(10)	BF	$5P$	863.5(432)	BG
$2P$	27(3)	DL	$5P$	628	T
$2P$	27.3	T	$5P$	706.2	T
$2P$	26.17	T	$6P$	940	T
$2P$	27.9(10)	DCL	$6P$	1058	T
$3P$	235(12)	BG	$7P$	1359	T
$3P$	182(6)	LC	$7P$	1528	T
$3P$	216	T	$8P$	2176	T
$3P$	223.3	T			
			Na I		
$3P$	15.9(1)	PS	$4P_{3/2}$	90	LC
$3P$	15.9(4)	LC	$4P$	95(4)	DCL
$3P$	15.9	PS	$4P$	12(12)	T
$3P$	15.9(4)	LC	$4P$	102	T
$3P$	16.3(5)	O	$4P$	100	T
$3P_{3/2}$	14.0(2)	PS	$5P$	345(43)	DCL
$3P$	16.1(3)	LC	$5P$	342	T
$3P$	16.1(7)	PS	$5P$	402.9	T
$3P$	16.2	DC	$5P$	330	T
$3P$	16.3(4)	DC	$6P$	890(90)	DCL
$3P$	16.1(10)	DC	$6P$	837	T
$3P$	16.6(2)	LC	$6P$	991.7	T
$3P$	16.0(3)	LC	$6P$	770	T
$3P$	15.9(4)	PS	$7P$	1450(100)	DCL
$3P$	16.6(4)	LC	$7P$	1676	T
$3P_{3/2}$	16.11(5)	LC	$7P$	2073	T
$3P$	16.5(5)	LC	$7P$	1500	T
$3P$	15.6	BF	$8P$	3041	T
$3P$	17(2)	BF	$8P$	2600	T
$3P$	17.3(10)	DC	$9P$	4100	T
$3P$	16.0	DC	$10P$	6000	T
$3P_{3/2}$	16.4(6)	BF	$11P$	8600	T
$3P_{1/2}$	16.1(2)	BF	$12P$	12000	T
$3P_{1/2}$	16.30(16)	H	$13P$	15000	T
$3P$	16.0(2)	T	$17P$	11400^{+5000}_{-1400}	DCL
$3P$	16.5	T			
$3P$	18.17	T	$18P$	13900^{+8000}_{-2900}	DCL
$4P$	17	LC			

Table 2.15 (*Cont.*)

State	τ[ns]	Method	State	τ[ns]	Method
			K I		
$4P_{3/2}$	27.8(8)	LC	$5P_{3/2}$	140.8(10)	LC
$4P_{3/2}$	28(2)	LC	$5P_{3/2}$	120(4)	LC
$4P_{3/2}$	27.8(5)	PS	$5P_{3/2}$	133(3)	LC
$4P_{3/2}$	27.6(8)	DC	$5P$	121	T
$4P_{3/2}$	27.3(10)	H	$6P_{3/2}$	310(15)	LC
$4P_{1/2}$	27.8(8)	DC	$6P$	299	T
$4P_{1/2}$	25.0(3)	H	$7P$	572	T
$4P_{1/2}$	27.3(3)	H	$8P$	957	T
$4P_{3/2}$	28(2)	LC			
$4P_{3/2}$	27.8(5)	PS			
$4P_{3/2}$	27.6(8)	DC			
$4P_{3/2}$	26.3(10)	H			
$4P$	26.6	T			
			Rb I		
$5P_{3/2}$	28.2(9)	LC	$13P$	3230	T
$5P_{3/2}$	27.0(5)	PC	$14P_{3/2}$	2600(400)	DCL
$5P_{3/2}$	27.1(14)	LC	$14P$	4230	T
$5P_{3/2}$	25.5(5)	LC	$15P$	5410	T
$5P_{3/2}$	25.8(8)	LC	$16P$	6800	T
$5P_{3/2}$	27.0(5)	LC	$17P_{3/2}$	6400(1300)	DCL
$5P_{3/2}$	26.0(18)	H	$17P$	8410	T
$5P_{1/2}$	30(3)	LC	$18P$	10200	T
$5P_{1/2}$	28.5(11)	LC	$5P_{3/2}$	28.2(9)	LC
$5P_{1/2}$	29.4(7)	LC	$5P_{3/2}$	27.0(5)	PS
$5P$	26.5	T	$5P_{3/2}$	27.1(14)	LC
$6P_{3/2}$	100	LC	$5P_{3/2}$	25.5(5)	LC
$6P_{3/2}$	114(6)	LC	$5P_{3/2}$	25.8(8)	LC
$6P_{3/2}$	118(4)	LC	$5P_{3/2}$	27.0(5)	LC
$6P_{3/2}$	109(7)	LC	$5P_{3/2}$	26.0(18)	H
$6P_{3/2}$	111(3)	LC	$19P$	12300	T
$6P_{3/2}$	112(8)	LC	$20P$	14700	T
$6P_{3/2}$	97(3)	LC	$21P$	1800	T
$6P_{1/2}$	131(5)	LC	$22P_{3/2}$	14000(5000)	DCL
$6P_{1/2}$	114(13)	LC	$22P$	21100	T
$6P$	99.3	T	$23P$	24500	T
$7P_{3/2}$	240(20)	LC	$24P$	28300	T
$7P_{3/2}$	233(10)	LC	$25P$	32500	T
$7P$	221	T	$26P$	37000	T
$8P_{3/2}$	400(80)	LC	$27P$	41900	T
$8P$	404	T	$28P$	47200	T
$9P$	659	T			
$10P$	1190	T			
$11P$	1710	T			
$12P_{3/2}$	1550(200)	DCL			
$12P$	2400	T			

Table 2.15 (*Cont.*)

State	τ[ns]	Method	State	τ[ns]	Method
			Cs I		
$6P_{3/2}$	30.5(7)	PS	$7P_{1/2}$	155(5)	LC
$6P_{3/2}$	28(2)	LC	$7P_{1/2}$	158(5)	DCL
$6P_{3/2}$	29.4(26)	LC	$7P_{1/2}$	158(3)	DCL
$6P_{3/2}$	29.7(2)	LC	$7P_{1/2}$	165(6)	DCL
$6P_{3/2}$	30.8(15)	PS	$7P$	113	T
$6P_{3/2}$	32.7(15)	LC	$8P_{3/2}$	240(20)	LC
$6P_{3/2}$	31(1)	LC	$8P_{3/2}$	310(15)	LC
$6P_{1/2}$	34(1)	LC	$8P_{3/2}$	300(15)	LC
$6P_{1/2}$	35.0(15)	PS	$8P_{3/2}$	274(12)	DCL
$6P$	30.7	T	$8P_{3/2}$	330(30)	LC
$7P_{3/2}$	160	LC	$8P_{1/2}$	330(30)	LC
$7P_{3/2}$	180	LC	$8P_{1/2}$	307(14)	DCL
$7P_{3/2}$	122(2)	LC	$8P$	265	T
$7P_{3/2}$	111(6)	DC	$9P_{3/2}$	390(30)	LC
$7P_{3/2}$	135(1)	LC	$9P_{3/2}$	502(22)	DCL
$7P_{3/2}$	134.5(28)	LC	$9P_{3/2}$	575(35)	DCL
$7P_{3/2}$	135(1)	LC	$9P$	498	T
$7P_{3/2}$	136(4)	DCL	$10P$	826	T
$7P_{3/2}$	134(3)	DCL	$10P_{3/2}$	900(40)	DCL
$7P_{3/2}$	135(3)	DCL	$10P_{1/2}$	920(50)	DCL
$7P_{3/2}$	136(4)	DCL			

The following notations are used in Tables 2.14–17 for experimental methods and theory:

DC: Delayed Coincidence technique ,
BF: Beam-Foil method ,
BG: Beam-Gas method ,
H: Hanle effect ,
LC: Level Crossing ,
PS: Phase Shift ,
EB: Excited Beams ,
EOC: Electron-Optical Chronography ,
ODR: Optical Double Resonance ,
DO: Direct Oscilloscopic method ,
LE: Laser Excitation ,
T: Theory .

2.5 Autoionizing States

The information about autoionizing states, i.e., discrete states lying above the first ionization potential, for neutral atoms is quite limited although investiga-

Table 2.16. Radiative lifetimes τ of the S, D, F, G, H and I states in alkalis

State	τ[ns]	Method	State	τ[ns]	Method
			Li I		
3S	25.5(13)	BG	4D	330(1)	BF
3S	30.3	T	4D	42.3(15)	O
3S	30.6	T	4D	33.4	T
4S	55.8(28)	BG	4D	34.14	T
4S	48(2)	BF	4D	31(1)	DCL
4S	56.6	T	5D	72.0(36)	BG
4S	56.47	T	5D	56(2)	BF
5S	113(6)	BG	5D	63.7	T
5S	103	T	5D	65.23	T
5S	102.9	T	6D	106(5)	BG
6S	190(10)	BG	6D	108	T
6S	173	T	6D	111.1	T
6S	173.3	T	7D	139.5(70)	BG
7S	220(12)	BG	7D	170	T
7S	273	T	7D	169.7	T
7S	272.4	T	8D	251.3	T
8S	405.7	T	4F	72.4	T
3D	14.0(7)	BG	4F	72.46	T
3D	15.0(1)	BF	5F	140	T
3D	16.7(10)	O	5F	140.2	T
3D	14.60(13)	BGL	6F	239	T
3D	14.7	T	6F	240.4	T
3D	14.86	T	7F	376	T
3D	14.5(7)	DCL	5G	235	T
4D	39.2(20)	BG	6G	403	T
			7G	636	T
			Na I		
4S	39.5	T	9S	618	DCL
4S	39.87	T	9S	713(76)	DCL
4S	37	T	9S	635	T
5S	71(5)	DCL	9S	690	T
5S	80.3	T	10S	1024(49)	DCL
5S	80.57	T	10S	913	T
5S	800	T	10S	1000	T
6S	152	T	10S	900	T
6S	152.4	T	11S	1270(130)	DCL
6S	160	T	11S	1263	T
7S	269(10)	DCL	11S	1400	T
7S	276(14)	DCL	11S	1250	T
7S	263	T	12S	1690	T
7S	262.42	T	12S	1900	T
7S	280	T	13S	2270(170)	DCL
8S	393(20)	DCL	13S	2300	T
8S	465(40)	DCL	13S	2180	T
8S	422	T	3D	19.9	T
8S	418.5	T	3D	20.45	T
8S	450	T	3D	21	T

Table 2.16 (*Cont.*)

State	τ[ns]	Method	State	τ[ns]	Method
		Na I			
4*D*	53.5(30)	DCL	4*F*	71.6	T
4*D*	51.3	O	4*F*	72.22	T
4*D*	57	DCL	4*F*	71	T
4*D*	52.4	T	5*F*	137	T
4*D*	54.81	T	5*F*	132.9	T
4*D*	55	T	5*F*	140	T
5*D*	120(14)	DCL	6*F*	234	T
5*D*	108	T	6*F*	238.6	T
5*D*	113.5	T	6*F*	230	T
5*D*	110	T	7*F*	367	T
6*D*	176(10)	DCL	7*F*	376.1	T
6*D*	206(24)	DCL	7*F*	370	T
6*D*	191	T	8*F*	558.2	T
6*D*	202.2	T	8*F*	540	T
6*D*	200	T	9*F*	760	T
7*D*	279(15)	DCL	10*F*	1030	T
7*D*	324(32)	DCL	13*F*	2270(400)	DCL
7*D*	308	T	13*F*	2260	T
7*D*	326.6	T	14*F*	2640(450)	DCL
7*D*	320	T	14*F*	2810	T
8*D*	449(50)	DCL	15*F*	3540(500)	DCL
8*D*	520(39)	DCL	15*F*	3540	T
8*D*	463	T	5*G*	235	T
8*D*	492.7	T	5*G*	235	T
8*D*	490	T	6*G*	402	T
9*D*	643(47)	DCL	6*G*	402	T
9*D*	720(67)	DCL	7*G*	634	T
9*D*	665	T	7*G*	633	T
9*D*	710	T	8*G*	940	T
10*D*	971(35)	DCL	9*G*	1320	T
10*D*	915	T	6*H*	608	T
10*D*	980	T	7*H*	960	T
10*D*	920	T	8*H*	1420	T
11*D*	1218	T	9*H*	2020	T
11*D*	1300	T	10*H*	2750	T
12*D*	1650(150)	DCL	7*I*	1350	T
12*D*	1572	T	8*I*	2000	T
12*D*	1590	T	9*I*	2840	T
13*D*	2120(400)	DCL	10*I*	3880	T
13*D*	2020	T	11*I*	5130	T
		K I			
5*S*	46.4	T	3*D*	41.6	T
6*S*	68(9)	LC	$3D_{5/2}$	42(3)	DCL
6*S*	87.8	T	$3D_{3/2}$	42(3)	DCL
7*S*	160	T	4*D*	284	T
8*S*	269	T	5*D*	720	T
9*S*	423	T	6*D*	106(6)	T

Table 2.16 (*Cont.*)

State	τ[ns]	Method	State	τ[ns]	Method
			K I		
$7D$	1416	T	$7F$	287	T
$4F$	62.9	T	$5G$	233	T
$5F$	114	T	$6G$	397	T
$6F$	187	T	$7G$	624	T
			Rb I		
$6S$	51.5	T	$5D$	266	T
$7S$	95(9)	DC	$6D_{3/2}$	294(12)	DC
$7S$	91(11)	LC	$6D_{3/2}$	285(16)	DCL
$7S$	97.7	T	$6D$	295	T
$8S$	169(10)	DC	$7D_{3/2}$	370(28)	DC
$8S$	153(8)	DCL	$7D_{3/2}$	388(25)	DCL
$8S$	178	T	$7D$	386	T
$9S$	288(15)	DC	$8D_{3/2}$	515(30)	DCL
$9S$	258(13)	DCL	$8D$	532	T
$9S$	300	T	$9D$	565(120)	DCL
$9S$	245(50)	DCL	$10D$	1070	T
$10S$	471	T	$10D$	720(120)	DCL
$10S$	427	T	$11D$	1410	T
$11S$	628	T	$11D$	975(200)	DCL
$12S$	887	T	$12D$	1830	T
$12S$	770(150)	DCL	$12D$	1250(300)	DCL
$13S$	1210	T	$13D$	2330	T
$14S$	1600	T	$13D$	1400(300)	DCL
$14S$	1260(250)	DCL	$14D$	2910	T
$15S$	2070	T	$15D$	3610	T
$16S$	2620	T	$15D$	3740(700)	DCL
$16S$	2190(500)	DCL	$16D$	4400	T
$17S$	3260	T	$17D$	5300	T
$17S$	2600(600)	DCL	$18D$	6320	T
$18S$	4000	T	$18D$	5300(1100)	DCL
$18S$	3300(700)	DCL	$19D$	7470	T
$19S$	4850	T	$20D$	8750	T
$20S$	5810	T	$21D$	10200	T
$21S$	6880	T	$22D$	11700	T
$22S$	8080	T	$23D$	13500	T
$23S$	9410	T	$24D$	15400	T
$24S$	10900	T	$25D$	17400	T
$25S$	12500	T	$26D$	19700	T
$26S$	14200	T	$27D$	22100	T
$27S$	16200	T	$28D$	24800	T
$28S$	18300	T	$4F$	57.2	T
$4D$	85	T	$5F$	101	T
$4D_{5/2}$	94(6)	DCL	$6F$	165	T
$4D_{3/2}$	86(6)	DCL	$7F$	253	T
$5D_{3/2,5/2}$	240	LC	$9F$	550(80)	DCL
$5D_{3/2}$	242(23)	DC	$10F$	680(100)	DCL
$5D_{3/2}$	205(40)	LC	$11F$	900(140)	DCL

Table 2.16 (*Cont.*)

State	τ[ns]	Method	State	τ[ns]	Method
		Rb I			
$11F$	904	T	$20F$	5260	T
$12F$	1170	T	$21F$	6700(1700)	DCL
$13F$	1620(240)	DCL	$21F$	6080	T
$13F$	1470	T	$22F$	6980	T
$14F$	1830	T	$23F$	7970	T
$15F$	1960(290)	DCL	$24F$	9040	T
$15F$	2240	T	$25F$	10200	T
$16F$	2710	T	$26F$	11500	T
$17F$	2960(440)	DCL	$27F$	12800	T
$17F$	3250	T	$28F$	14300	T
$18F$	3850	T	$5G$	232	T
$19F$	4000(800)	DCL	$6G$	395	T
$19F$	4520	T	$7G$	617	T
		Cs I			
$7S$	57.3	T	$8D_{3/2}$	152(3)	DCL
$8S$	104	T	$8D_{3/2}$	154(5)	DCL
$8S$	101(4)	DC	$8D$	161	T
$8S$	96(14)	LC	$9D_{5/2}$	95(10)	LC
$8S$	87(9)	DCL	$9D_{3/2}$	218(4)	DCL
$8S$	73(5)	O	$9D$	245	T
$8S$	106	T	$9D_{3/2}$	311(6)	DCL
$9S$	196(13)	DC	$11D_{3/2}$	428(12)	DCL
$9S$	231(35)	LC	$12D_{3/2}$	561(18)	DCL
$9S$	167(3)	DCL	$13D_{3/2}$	741(22)	DCL
$9S$	160(8)	DCL	$14D_{3/2}$	980(30)	DCL
$9S$	192	T	$4F$	44.9	T
$10S$	260(12)	DCL	$5F$	76.8	T
$10S$	270(5)	DCL	$5F_{7/2}$	97(6)	DCL
$10S$	319	T	$6F$	124	T
$11S$	343(22)	DCL	$6F_{7/2}$	149(8)	DCL
$11S$	411(8)	DCL	$7F_{7/2}$	238(10)	DCL
$11S$	498	T	$7F_{7/2}$	229(15)	DCL
$12S$	545(30)	DCL	$7F$	189	T
$12S$	571(15)	DCL	$8F_{7/2}$	336(22)	DCL
$13S$	754(35)	DCL	$9F_{7/2}$	473(30)	DCL
$14S$	959(50)	DCL	$10F_{7/2}$	646(35)	DCL
$6D_{5/2}$	64(2)	DC	$11F_{7/2}$	830(15)	DCL
$6D_{5/2}$	60.7(25)	DCL	$11F_{7/2}$	846(40)	DCL
$6D$	69.9	T	$12F_{7/2}$	1060(25)	DCL
$6D_{3/2}$	57(15)	LC	$13F_{7/2}$	1320(35)	DCL
$6D_{3/2}$	60.0(25)	DCL	$14F_{7/2}$	16254(35)	DCL
$7D_{5/2}$	92.5(15)	DC	$15F_{7/2}$	1959(40)	DCL
$7D_{5/2}$	88(9)	DCL	$16F_{7/2}$	2340(50)	DCL
$7D_{3/2}$	98(3)	DC	$5G$	230	T
$7D_{3/2}$	89(4)	DC	$6G$	389	T
$7D$	103	T			

Table 2.17. Radiative lifetimes τ of the excited states in Cu I, Ag I and Au I

State	τ [ns]	Method	State	τ [ns]	Method
			Cu I		
$3d^{10}\ 5s\ ^2S_{1/2}$	21.0(42)	EOC	$4d\ ^2D_{3/2}$	66.2(34)	EOC
$5s\ ^2S_{1/2}$	23.5	T	$4d\ ^2D_{3/2}$	12.2	T
$6s\ ^2S_{1/2}$	86.0(71)	EOC	$4d\ ^2D_{3/2}$	14.5(6)	DC
$6s\ ^2S_{1/2}$	52.8	T	$4d\ ^2D_{5/2}$	45.0(42)	EOC
$6s\ ^2S_{1/2}$	49.3(25)	DC	$4d\ ^2D_{5/2}$	12.3	T
$7s\ ^2S_{1/2}$	101	T	$4d\ ^2D_{5/2}$	14.2(13)	DC
$7s\ ^2S_{1/2}$	86.1(83)	DC	$5d\ ^2D_{3/2}$	54.8(51)	EOC
$8s\ ^2S_{1/2}$	181	T	$5d\ ^2D_{3/2}$	26.8	T
$9s\ ^2S_{1/2}$	306	T	$5d\ ^2D_{3/2}$	30.9(17	DC
$4p\ ^2P^0_{1/2}$	7.1(6)	EOC	$5d\ ^2D_{5/2}$	42.7(43)	EOC
$4p\ ^2P^0_{1/2}$	6.11	T	$5d\ ^2D_{5/2}$	27.1	T
$4p\ ^2P^0_{1/2}$	7.0(9)	DC	$5d\ ^2D_{5/2}$	29.8(27)	DC
$4p\ ^2P^0_{3/2}$	7.0(2)	LC	$6d\ ^2D_{3/2}$	51.1	T
$4p\ ^2P^0_{3/2}$	7.2(3)	PS	$6d\ ^2D_{5/2}$	71.5(91)	EOC
$4p\ ^2P^0_{3/2}$	7.4(7)	BF	$6d\ ^2D_{5/2}$	52.7	T
$4p\ ^2P^0_{3/2}$	7.24(15)	LC	$6d\ ^2D_{5/2}$	55.6(35)	DC
$4p\ ^2P^0_{3/2}$	7.6(7)	BF	$7d\ ^2D_{5/2}$	88.3	T
$4p\ ^2P^0_{3/2}$	7.2(10)	EOC	$7d\ ^2D_{5/2}$	87.9	T
$4p\ ^2P^0_{3/2}$	5.97	T	$8d\ ^2D_{3/2}$	137	T
$4p\ ^2P^0_{3/2}$	7.0(9)	DC	$8d\ ^2D_{5/2}$	137	T
$5p\ ^2P^0_{1/2}$	38.8	T	$6f\ ^2F^0_{5/2}$	70.2	T
$5p\ ^2P_{1/2,3/2}$	12(2)	BF	$4f\ ^2F^0_{7/2}$	70.4	T
$5p\ ^2P^0_{3/2}$	34.5	EOC	$5f\ ^2F^0_{5/2}$	136	T
$5p\ ^2P^0_{3/2}$	38.8	T	$5f\ ^2F^0_{7/2}$	136	T
$5p\ ^2P^0_{3/2}$	27.9(42)	DC	$3d^9\ 4p'\ ^2P^0_{3/2}$	10.7(4)	LC
$6p\ ^2P^0_{1/2}$	185	T	$4p'\ ^2P^0_{3/2}$	13.2(6)	DC
$6p\ ^2P^0_{1/2}$	23(5)	D	$4p'\ ^2D^0_{3/2}$	19.2	LC
$6p\ ^2P^0_{3/2}$	19.1(30)	EOC	$4p'\ ^2P^0_{3/2}$	20.2(23)	DC
$6p\ ^2P^0_{3/2}$	353	T	$4p'\ ^4P^0_{3/2}$	318(16)	ODR
$6p\ ^2P^0_{3/2}$	6.2(6)	DC	$4p'\ ^4P^0_{3/1}$	320(10)	LC
$7p\ ^2P^0_{1/2}$	338	T	$4p'\ ^4D_{3/2}$	361(54)	LC
$7p\ ^2P^0_{1/2}$	5.8(6)	DC	$4p'\ ^2D_{3/2}$	376(28)	DC
$7p\ ^2P^0_{3/2}$	118	T	$4p'\ ^2F^0_{7/2}$	424(10)	DC
$7p\ ^2P^0_{3/2}$	17.0(45)	DC	$4d'\ ^4F_{9/2}$	377(28)	DC
$8p\ ^2P^0_{1/2}$	165	T	$4d'\ ^2G_{9/2}$	61(9)	DC
$8p\ ^2P^0_{3/2}$	220	T	$4d'\ ^4G_{5/2}$	4.8(6)	BF
$9p\ ^2P^0_{1/2}$	343	T	$4d'\ ^4G_{7/2}$	97.9(57)	DC
$9p\ ^2P^0_{3/2}$	407	T	$4d'\ ^4G_{9/2}$	98.9(38)	DC
$4d\ ^2D_{5/2,3/2}$	11(2)	BF			

Table 2.17 (*Cont.*)

State	τ [ns]	Method	State	τ [ns]	Method
Ag I					
$5p\,^2P_{3/2}$	6.3(6)	LC	$5d\,^2D_{5/2}$	13.4(12)	DC
$5p\,^2P_{3/2}$	6.5(6)	DC	$6d\,^2D_{3/2}$	30.2(33)	DC
$5p\,^2P_{3/2}$	7.3(4)	O	$7d\,^2D_{3/2}$	61.6(56)	DC
$5p\,^2P_{1/2}$	7.5(7)	DC	$7d\,^2D_{5/2}$	60.6(43)	DC
$5p\,^2P_{3/2}$	7.5(7)	DC	$8d\,^2D_{5/2}$	106(7)	DC
$7s\,^2S_{1/2}$	41.5(32)	DC	$5p''\,^2P^0_{3/2}$	3.0(2)	DC
$8s\,^2S_{1/2}$	139(13)	DC	$6s'\,^4D_{3/2}$	5.1(9)	DC
$5d\,^2D_{3/2}$	12.9(8)	DC	$5p'\,^4D^0_{7/2}$	3.9(10)	DC
Au I					
$5d^{10}\,6p\,^2P_{3/2}$	4.6	LC	$5d^{10}\,6p\,^2P_{1/2}$	7.4(7)	DC
$5d^{10}\,6p\,^2P_{3/2}$	4.6(3)	LC	$5d^{10}\,6p\,^2P_{3/2}$	5.6(8)	DC

Table 2.18. Constants a and τ_0 (in ns) (2.4.4) for the spectral series of hydrogen

Series	p	d	f	g	h
τ_0	0.209(1)	0.627(2)	1.235(1)	2.068(1)	3.113(1)
a	2.948(8)	2.932(6)	2.941(2)	2.942(1)	2.945(1)
Series	i	k	l	m	
τ_0	4.372(5)	5.766(18)	7.212(14)	9.313(82)	
a	2.946(2)	2.954(6)	2.968(6)	2.956(11)	

Table 2.19. Constants a and τ_0 (2.4.4) for S, P, D, F and G series of alkali atoms

Atom	Constants	S	P	D	F	G
Li I	a			2.96(11)	2.94(3)	2.95(9)
	τ_0			0.74(16)	1.22(5)	2.00(8)
Nd I	a	2.89(3)		3.06(8)	2.93(7)	2.94(4)
	τ_0	1.80(7)		0.94(4)	1.20(4)	2.05(8)
K I	a		3.06(4)		2.70(5)	2.92(7)
	τ_0		3.47(11)		1.48(5)	2.09(4)
Rb I	a	2.95	3.03(9)		2.86(7)	2.90(6)
	τ_0	1.44	2.57(7)		0.97(6)	2.15(8)
Cs I	a	2.60(11)	3.13(6)	–	2.76(4)	2.85(4)
	τ_0	3.11(13)	2.47(5)	–	1.01(6)	2.33(6)

tions of such states is important for many problems in atomic physics: for inner-shell processes, in multiple-electron excitation and ionization processes, in laser excitation reactions, etc.

The general problems of autoionization processes are discussed in [2.67–73], lifetimes of autoionizing states for neutral atoms are given in [2.64, 66, 74–77].

In this Section, we consider the selection rules for transitions from autoionizing states are give tables for the lifetimes of neutral atoms.

If the energy of a discrete atomic state is larger than the first ionization potential, an atom (or ion) can undergo a radiationless transition to a continuum state of the same energy; so, in the final state one has a free electron and an ion in the ground or excited state:

$$X_z^{**}(\gamma_1 nl) \rightarrow X_{z+1}(\gamma_0) + e(\varepsilon l') . \quad (2.5.1)$$

This process is called *Auger* or *autoionization decay*. Here, ε and l' are the kinetic energy and angular momentum of a free electron, respectively, and

$$\varepsilon = E(\gamma_1 nl) - E(\gamma_0) > 0 , \quad (2.5.2)$$

where E are the corresponding atomic energies.

The initial state can be of different type, in particular, a hole in the inner shell or a doubly (or more) excited state. For example, excitation of the inner $1s$ electron of LiI ($1s^2 2s$) to the $2p$ state forms a metastable autoionizing state $(1s2s2p)^4P_{5/2}$. The number of excited electrons can be more than two. In Be-like ions (4 electrons) autoionizing states $1s2s^2 2p$, $1s2s2p^2$, $1s2p^3$, or generally, $1snln'l'n''l''$ are also possible.

In the first-order perturbation theory, the autoionization probability A is given by

$$A = \frac{2\pi}{\hbar} |\langle \gamma_1 nl \| V | \gamma_0 \varepsilon l' \rangle|^2 [\mathrm{s}^{-1}] , \quad (2.5.3)$$

where V is the interaction coupling of the discrete and continuum states and may be either electrostatic ($1/r_{12}$) or magnetic in nature. Typical autoionization rates via the electrostatic (Coulomb) interaction vary in the range 10^{12}–10^{15} s^{-1} [2.69].

The metastable excited state can decay also by a radiative transition to a lower-lying state with radiative probability W (Sect. 2.3). The lifetime of the autoionizing state is given by

$$\tau = (W + A)^{-1} \, [\mathrm{s}] . \quad (2.6.4)$$

The selection rules for non-radiative transitions via different interactions are given in Table 2.20 [2.69]. Here, $\Delta L = L' - L$, $\Delta S = S' - S$, etc., where the angular continuum states, respectively. The parity of the state is defined by the sum $\sum_i l_i$ taken over all electrons.

In Table 2.21, the lifetimes for autoionizing states in alkali atoms are given.

Table 2.20. Selection rules for autoionization transitions [2.69]*

Interaction	ΔL	ΔS	ΔJ	Parity change	Relative order of magnitude
Coulomb	0	0	0	No	1
Spin–orbit, Spin–other–orbit	0, ± 1	0, ± 1	0	No	a^4
Spin–spin	0, ± 1, ± 2	0, ± 1, ± 2	0	No	a^4
Hyperfine	0, ± 1, ± 2	0, ± 1	0, ± 1	No	$a^4(m/m_p)^2$

* α is the fine-structure constant, m_p the proton mass.

Table 2.21. Radiative lifetimes τ of autoionizing states in alkali atoms [2.64]. For the notation of the experimental method see Table 2.14.

State	τ [ns]	Method	State	τ [ns]	Method
			Li I		
$1s\,2s\,3s\,^4S$	9.7(7)	BF	$1s\,2p^2\,^4P$	5.76(5)	T
$1s\,2s\,3s\,^4S$	6.72(2)	T	$1s\,2p\,3p\,^4P$	2.9(3)	BF
$1s\,2s\,3s\,^4S$	7.7(10)	BF	$1s\,2p\,3p\,^4P$	0.8(3)	BF
$1s\,2s\,4s\,^4S$	10.4(20)	BF	$1s\,2p\,3p\,^4P$	30(3)	T
$1s\,2s\,4s\,^4S$	15.4(5)	BF	$1s\,2p\,4p\,^4P$	7.1(30)	BF
$1s\,2s\,2p\,^4P_{5/2}$	5800(1200)	EB	$1s\,2p\,4p\,^4P$	3.5(10)	BF
$1s\,2s\,2p\,^4P_{3/2}$	460(100)	BF	$1s\,2p\,5p\,^4P$	42(4)	T
$1s\,2s\,2p\,^4P_{1/2}$	140(70)	BF	$1s\,2p\,3d\,^4D$	2.3(4)	BF
$1s\,2s\,3p\,^4P$	270(30)	T	$1s\,2p\,3d\,^4D$	3.0(8)	BF
$1s\,2s\,4p\,^4P$	300(30)	T	$1s\,2p\,3d\,^4D$	4.47(6)	T
$1s\,2s\,5p\,^4P$	500	T	$1s\,2p\,3d\,^4D$	1.5(4)	BF
$1s\,2s\,3d\,^4D$	4.5(4)	BF	$1s\,2p\,4d\,^4D$	2.3(5)	BF
$1s\,2s\,3d\,^4D$	5.3(12)	BF	$1s\,2p\,4d\,^4D$	1.11(15)	BF
$1s\,2s\,3d\,^4D$	4.15(4)	T	$1s\,2p\,5d\,^4D$	2.5(10)	BF
$1s\,2s\,3d\,^4D$	4.3(1)	BF	$1s\,2p\,4f\,^4F$	23.0(7)	T
$1s\,2s\,4d\,^4D$	5.9(10)	BF	$1s\,2p(^3P)\,3d\,^4D$	1.45(10)	BF
$1s\,2s\,4d\,^4D$	9.6(6)	BF	$1s\,2p(^3P)\,4d\,^2D$	0.24(10)	BF
$1s\,2s\,4f\,^4F^0$	60.6(23)	T	$1s\,2p('P)\,3d\,^2D$	0.063(50)	BF
$1s\,2p\,3s\,^4P^0$	11.8(2)	BF	$1s\,2p^2\,^2D$	0.177	T
$1s\,2p\,3s\,^4P^0$	12.4(5)	BF	$1s\,2p^2\,^2P$	0.015(10)	BF
$1s\,2p^2\,^4P$	6.4(3)	BF	$1s\,2p^2\,^2P$	0.054	
$1s\,2p^2\,^4P$	5.8(7)	BF	$1s\,2p^2\,^2P$	0.041	
$1s\,2p^2\,^4P$	7(2)	BF	$1s(2s\,2p'\,P)^2P$	0.00075(10)	BF
$1s\,2p^2\,^4P$	6.5(3)	BF			
			K I		
$3p^5\,4s\,3d\,^4F$	90000(20000)	EB			
			Rb I		
$4p^5\,5s\,4d\,^4F$	75000(20000)	EB			

2.6 Asymptotic Formulas

In this Section, the asymptotic formulas for oscillator strengths and transition probabilities are given. Tables for dipole and quadrupole matrix elements calculated in the Bates-Damgaard approximation are also presented.

2.6.1 Quasiclassical Formulas

The oscillator strength f for transition n_0l_0–n_1l_1 can be presented in the form (2.1.9, 10):

$$f = \frac{\Delta E}{3\mathrm{Ry}} \frac{l_m Q(0\text{–}1)}{(2l_0+1)} R^2 , \quad l_m = \max\{l_0, l_1\} ,$$
$$R = \int_0^\infty P_0(r) P_1(r) r \, dr , \tag{2.6.1}$$

where Q is the angular coefficient (Sect. 2.7), $P(r)$ the radial wave function and ΔE the transition energy.

For transitions from the levels with the principal quantum numbers n_0, $n_1 > n_c$ (n_c is the largest principal number of the core electron) and $n_1 < 10$, one can use the Bates-Damgaard tables (Sect. 2.6.2) for dipole matrix elements.

For transitions with $n_0 > 10$ the analytical formulas obtained in the quasiclassical approximation [2.78] can be applied:

$$R = \frac{n_c^2}{2\Delta}\left[J_{\Delta-1}(x)\left(1 - \frac{l_m \Delta l}{n_c}\right) - J_{\Delta+1}(-x)\left(1 + \frac{l_m \Delta l}{n_c}\right) + \frac{2\sin \pi\Delta}{\pi}\left(1 - \frac{x}{\Delta}\right)\right],$$
$$\Delta l = l_1 - l_c , \quad \Delta = n_1^* - n_c^* , \quad x = (1 - (l_m/n_c)^2)^{1/2} \Delta , \tag{2.6.2}$$
$$n_c = \frac{n_0^* n_1^*}{n_0^* - n_1^*} , \quad l_m = \max(l_0, l_1) , \quad n^* = (E_{nl}/z^2 \mathrm{Ry})^{-1/2} ,$$

where $\mathbf{J}_p(x)$ are the Anger functions and n^* is the effective principal quantum number.

For transitions in H-like systems ($n^* = n$), the quantum Δ is an integer and the Anger functions become the Bessel functions $\mathbf{J}_p(x) \to J_p(x)$. If $n_0 = n_1 = n$, one has from (2.6.1) the well-known result [2.5]:

$$R = \frac{3}{2} n \sqrt{n^2 - l_m^2} . \tag{2.6.3}$$

The oscillator strength for transition $n_0 - n_1$, i.e., averaged over l_0 and summed over l_1,

$$f(n_0 - n_1) = n_0^{-2} \sum_{l_0 l_1} (2l_0 + 1) f(n_0 l_1 - n_1 l_1)$$

may also be obtained from (2.6.2) if $\Delta = n_1^* - n_0^*$ is an integer [2.79]:

$$f(n_0 - n_1) = \frac{32\Delta n}{3n_0^2}\left(\frac{n_0 n_1}{\Delta n(n_0 + n_1)}\right)^3 J_{\Delta n}(\Delta n) J'_{\Delta n}(\Delta n)\,;$$
$$\Delta n = n_1 - n_0\,, \quad n_0, n_1 \gg 1\,. \tag{2.6.4}$$

If the additional condition $n_0, n_1 \gg \Delta n \gg 1$ is satisfied, one has from (2.6.4) the Kramers formula [2.80]:

$$f^{\mathrm{Kr}}(n_0 - n_1) = \frac{2C}{n_0^2}\left(\frac{n_0 n_1}{\Delta n(n_0 + n_1)}\right)^3 = \frac{2C(E_0 E_1)^{3/2}}{n_0^2 \Delta E^3}\,, \tag{2.6.5}$$

where $C = 16/3\sqrt{3}\pi \approx 1$.

The corresponding expression for the transition probability has the form

$$W^{\mathrm{Kr}}(n_1 - n_0) = CA_1 \frac{2}{n_1^3 n_0 (n_1^2 - n_0^2)}\,, \tag{2.6.6}$$

$$A_1 = \frac{me^4}{2\hbar^3}\alpha^3 z^4 \approx 0.80 \times 10^{10} z^4\,[\mathrm{s}^{-1}]\,. \tag{2.6.7}$$

The total probability of the radiative decay of the level n is expressed as

$$W^{\mathrm{Kr}}(n) = \frac{3A_1}{n^5}\ln\left(\frac{n - 1/n}{2}\right), \quad n \gg 1\,. \tag{2.6.8}$$

The accurate expressions for bound–bound and bound–continuum oscillator strengths for transitions n_0–n_1 were obtained in [2.61, 81] by carrying the expansion of $f(n_0 - n_1)$ in terms of the inverse powers of n up to $1/n^3$ terms:

$$f(n_0 - n_1) = f^{\mathrm{Kr}}(n_0 - n_1) F(n_0, n_1)\,, \tag{2.6.9}$$

$$F(n_0, n_1) = 1 - 0.17286 \frac{ab^{-2/3}}{n^{2/3}} - 0.0165319 \frac{b^{-4/3} d}{n^{4/3}}$$
$$+ \frac{1}{175}\frac{ab^2 d}{n^2} + O(1/n^{8/3})\,, \tag{2.6.10}$$

with $a = 1 + \gamma^2$, $b = 1 - \gamma^2$, $d = 3 - 4\gamma^2 + 3\gamma^4$, $\gamma = n_0/n_1 \ll 1$, where $f^{\mathrm{Kr}}(n_0 - n_1)$ is given by (2.6.5).

The accuracy of (2.6.9, 10) is better than 0.5% including transition $1 \to 2$, which is seen from Table 2.22 where the asymptotic expansion oscillator strengths are compared with accurate values taken from [2.82].

Similar expressions for the bound-continuum oscillator strengths (the Gaunt factor) are given in Sect. 3.2.

2.6.2 Bates-Damgaard Tables for Dipole and Quadrupole Matrix Elements

According to (2.1.1–5), the probability of E$\varkappa$ transitions $0 \to 1$ can be written in the form

Table 2.22. Oscillator strengths $f(n_0 - n_1)$ in H-like atoms

Transition n_0 n_1	(2.6.9, 10)	[2.82]
1–2	4.16 − 1	4.14 − 1
2–3	6.41 − 1	6.38 − 1
5–6	1.23	1.23
1–40	2.45 − 5	2.44 − 5
2–40	4.41 − 5	5.40 − 5
5–40	1.50 − 4	1.50 − 4
10–11	2.19	2.18
20–21	4.10	4.08
20–40	1.40 − 3	1.40 − 3
30–40	1.04 − 2	1.04 − 2
40–50	1.28 − 2	1.28 − 2
50–51	9.82	9.78
50–98	2.52 − 4	2.52 − 4
100–101	1.94 + 1	1.93 + 1
200–201	3.84 + 1	3.83 + 1
900–901	1.72 + 2	1.71 + 2

2.55 − 2 means 2.55×10^{-2}

$$W_\varkappa = A_\varkappa \left(\frac{\Delta E}{z^2 \mathrm{Ry}}\right)^2 \frac{g_0 Q_\varkappa(0,1)}{g_1(2l_0+1)} F_\varkappa(l_0, l_1)\,, \tag{2.6.11}$$

$$A_\varkappa = \frac{me^4}{2\hbar^3}\left(\frac{e^2}{\hbar c}\right)^{2\varkappa+1} \frac{\varkappa+1}{2^{2\varkappa-1}\varkappa[(2\varkappa-1)!!]^2} z^{2\varkappa+1}\,, \tag{2.6.12}$$

$$F_\varkappa(l_0, l_1) = \frac{|(l_0\|C^\varkappa\|l_1)|^2}{2\varkappa+1}\left(\frac{\Delta E}{z^2 \mathrm{Ry}}\right)^{2\varkappa-1}\left(\frac{z}{a_0}\right)^{2\varkappa}\left|\int_0^\infty P_0(r) r^\varkappa P_1(r)\right|^2\,, \tag{2.6.13}$$

where z is the spectroscopic symbol, $g_{0,1}$ are the statistical weights, ΔE is the transition energy, a_0 the Bohr radius, $P(r)$ the radial wave functions and $(l_0\|C^\varkappa\|l_1)$ is the reduced matrix element [2.1].

For dipole transitions ($\varkappa = 1$), one has

$$(l_0\|C^1\|l_1) = \max\{l_0, l_1\}\,, \quad A_1 = 0.80 \times 10^{10} z^4\,[\mathrm{s}^{-1}]\,, \tag{2.6.14}$$

$$f_1 = \frac{Q_1(0,1)}{2l_0+1} F_1(l_0, l_1)\,. \tag{2.6.15}$$

For quadrupole transitions ($\varkappa = 2$), correspondingly

$$A_2 = 0.89 \times 10^4 z^6\,[\mathrm{s}^{-1}] \tag{2.6.16}$$

The dipole (F_1) and quadrupole (F_2) matrix elements calculated with the Bates-Damgaard radial wave functions $P(r)$ (Sect. 1.8.3) are given in Tables 2.23, 24 as a function of the initial effective quantum number n_0^* and the quantity $\Delta n^* = n_1^* - n_0^*$, where

Table 2.23. Dipole matrix elements F_1 [(2.6.13), $\varkappa = 1$] with Bates-Damgaard radial wavefunctions $P(r)$ (1.8.18–31) [2.83]

	n_0^*									
Δn^*	0.5	1.0	1.5	2.0	2.5	3.0	3.5	4.0	4.5	5.0
					Transition s–p					
0.1			15 − 0	25 − 0	33 − 0	42 − 0	50 + 0	58 + 0	65 + 0	73 + 0
0.2			34 − 0	53 + 0	69 + 0	85 + 0	10 + 1	11 + 1	12 + 1	14 + 1
0.3			55 + 0	80 + 0	10 + 1	12 + 1	14 + 1	16 + 1	18 + 1	19 + 1
0.4			74 + 0	10 + 1	12 + 1	14 + 1	17 + 1	19 + 1	21 + 1	23 + 1
0.5			87 + 0	11 + 1	13 + 1	15 + 1	18 + 1	20 + 1	22 + 1	24 + 1
0.6		60 + 0	91 + 0	11 + 1	13 + 1	15 + 1	17 + 1	19 + 1	21 + 1	23 + 1
0.7		63 + 0	86 + 0	10 + 1	12 + 1	13 + 1	15 + 1	16 + 1	18 + 1	19 + 1
0.8		60 + 0	74 + 0	88 + 0	98 + 0	10 + 1	12 + 1	13 + 1	14 + 1	15 + 1
0.9		52 + 0	57 + 0	66 + 0	71 + 0	78 + 0	84 + 0	91 + 0	97 + 0	10 + 1
1.0		41 − 0	39 − 0	43 − 0	45 − 0	48 − 0	51 + 0	54 + 0	57 + 0	60 + 0
1.1	20 − 0	27 − 0	24 − 0	23 − 0	23 − 0	24 − 0	24 − 0	25 − 0	26 − 0	27 − 0
1.2	14 − 0	15 − 0	11 − 0	96 − 1	86 − 1	81 − 1	78 − 1	76 − 1	76 − 1	75 − 1
1.3	90 − 1	63 − 1	32 − 1	18 − 1	12 − 1	82 − 2	57 − 2	40 − 2	28 − 2	19 − 2
1.4	48 − 1	14 − 1	10 − 2	39 − 3	30 − 2	66 − 2	10 − 1	15 − 1	19 − 1	24 − 1
1.5	20 − 1	81 − 5	10 − 1	23 − 1	36 − 1	49 − 1	62 − 1	74 − 1	86 − 1	98 − 1
1.6	55 − 2	11 − 1	40 − 1	65 − 1	86 − 1	10 − 0	12 − 0	14 − 0	16 − 0	18 − 0
1.7	23 − 3	35 − 1	72 − 1	10 − 0	12 − 0	15 − 0	17 − 0	19 − 0	21 − 0	23 − 0
1.8	99 − 3	59 − 1	93 − 1	12 − 0	14 − 0	17 − 0	18 − 0	20 − 0	22 − 0	24 − 0
1.9	44 − 2	75 − 1	98 − 1	12 − 0	13 − 0	15 − 0	17 − 0	18 − 0	19 − 0	21 − 0
2.0	81 − 2	79 − 1	85 − 1	10 − 0	10 − 0	12 − 0	12 − 0	13 − 0	14 − 0	15 − 0
2.1	28 − 1	64 − 1	68 − 1	69 − 1	71 − 1	74 − 1	77 − 1	81 − 1	84 − 1	87 − 1
2.2	25 − 1	43 − 1	39 − 1	35 − 1	33 − 1	33 − 1	32 − 1	32 − 1	32 − 1	32 − 1
2.3	19 − 1	22 − 1	15 − 1	10 − 1	83 − 2	69 − 2	57 − 2	49 − 2	42 − 2	37 − 2
2.4	12 − 1	72 − 2	19 − 2	29 − 3	38 − 6	20 − 3	69 − 3	13 − 2	20 − 2	28 − 2
2.5	64 − 2	39 − 3	11 − 2	39 − 2	71 − 2	10 − 1	13 − 1	17 − 1	20 − 1	23 − 1
2.6	23 − 2	16 − 2	92 − 2	17 − 1	23 − 1	29 − 1	35 − 1	42 − 1	47 − 1	53 − 1
2.7	35 − 3	84 − 2	20 − 1	32 − 1	40 − 1	49 − 1	57 − 1	65 − 1	72 − 1	79 − 1
2.8	51 − 4	17 − 1	30 − 1	44 − 1	51 − 1	61 − 1	68 − 1	77 − 1	84 − 1	91 − 1
2.9	81 − 3	25 − 1	35 − 1	47 − 1	52 − 1	61 − 1	67 − 1	74 − 1	80 − 1	86 − 1
3.0	19 − 2	28 − 1	32 − 1	41 − 1	44 − 1	51 − 1	54 − 1	59 − 1	62 − 1	67 − 1
3.1	93 − 2	25 − 1	29 − 1	30 − 1	32 − 1	34 − 1	36 − 1	37 − 1	39 − 1	41 − 1
3.2	88 − 2	18 − 1	18 − 1	16 − 1	16 − 1	16 − 1	16 − 1	17 − 1	17 − 1	17 − 1
3.3	71 − 2	10 − 1	78 − 2	58 − 2	48 − 2	43 − 2	37 − 2	34 − 2	30 − 2	28 − 2
3.4	49 − 2	36 − 2	13 − 2	37 − 3	67 − 4	50 − 7	62 − 4	21 − 3	41 − 3	65 − 3
3.5	27 − 2	35 − 3	28 − 3	12 − 2	24 − 2	38 − 2	52 − 2	66 − 2	80 − 2	94 − 2
3.6	11 − 2	46 − 3	35 − 2	71 − 2	98 − 2	12 − 1	15 − 1	18 − 1	21 − 1	24 − 1
3.7	23 − 3	33 − 2	89 − 2	14 − 1	18 − 1	23 − 1	26 − 1	30 − 1	34 − 1	38 − 1
3.8	25 − 5	75 − 2	13 − 1	21 − 1	24 − 1	30 − 1	33 − 1	38 − 1	42 − 1	46 − 1
3.9	25 − 3	11 − 1	16 − 1	23 − 1	26 − 1	31 − 1	34 − 1	38 − 1	41 − 1	45 − 1
4.0	75 − 3	13 − 1	16 − 1	21 − 1	23 − 1	27 − 1	29 − 1	32 − 1	34 − 1	36 − 1
4.1	41 − 2	12 − 1	15 − 1	16 − 1	17 − 1	19 − 1	20 − 1	21 − 1	22 − 1	23 − 1
4.2	40 − 2	92 − 2	99 − 2	94 − 2	94 − 2	97 − 2	99 − 2	10 − 1	10 − 1	10 − 1
4.3	34 − 2	53 − 2	45 − 2	34 − 2	30 − 2	27 − 2	25 − 2	23 − 2	21 − 2	20 − 2
4.4	24 − 2	21 − 2	87 − 3	29 − 3	90 − 4	14 − 4	26 − 5	41 − 4	11 − 3	20 − 3

Table 2.23 (*Cont.*)

Δn^*	n_0^* 1.2	2.0	2.5	3.0	3.5	4.0	4.5	5.0
				p–s				
0.1	11 − 0	20 − 0	29 − 0	37 − 0	45 − 0	53 + 0	61 + 0	68 + 0
0.2	19 − 0	36 − 0	52 + 0	67 + 0	82 + 0	96 + 0	11 + 1	12 + 1
0.3	24 − 0	45 − 0	66 + 0	86 + 0	10 + 1	12 + 1	14 + 1	16 + 1
0.4	25 − 0	48 − 0	71 + 0	93 + 0	11 + 1	13 + 1	15 + 1	17 + 1
0.5	23 − 0	45 − 0	67 + 0	88 + 0	11 + 1	13 + 1	15 + 1	17 + 1
0.6	18 − 0	38 − 0	56 + 0	75 + 0	94 + 0	11 + 1	13 + 1	15 + 1
0.7	13 − 0	28 − 0	42 − 0	57 + 0	72 + 0	87 + 0	10 + 1	11 + 1
0.8	81 − 1	18 − 0	28 − 0	38 − 0	49 − 0	59 + 0	70 + 0	81 + 0
0.9	38 − 1	10 − 0	15 − 0	22 − 0	28 − 0	34 − 0	41 − 0	47 − 0
1.0	11 − 1	40 − 1	64 − 1	96 − 1	12 − 0	15 − 0	18 − 0	22 − 0
1.1	26 − 2	86 − 2	16 − 1	24 − 1	34 − 1	44 − 1	55 − 1	66 − 1
1.2	27 − 3	18 − 4	41 − 4	35 − 3	90 − 3	17 − 2	26 − 2	37 − 2
1.3	51 − 2	71 − 2	88 − 2	96 − 2	10 − 1	10 − 1	11 − 1	11 − 1
1.4	12 − 1	21 − 1	28 − 1	35 − 1	41 − 1	47 − 1	52 − 1	58 − 1
1.5	18 − 1	33 − 1	48 − 1	61 − 1	74 − 1	87 − 1	99 − 1	11 − 0
1.6	19 − 1	39 − 1	58 − 1	77 − 1	95 − 1	11 − 0	13 − 0	14 − 0
1.7	17 − 1	38 − 1	57 − 1	77 − 1	96 − 1	11 − 0	13 − 0	15 − 0
1.8	12 − 1	30 − 1	46 − 1	64 − 1	81 − 1	99 − 1	11 − 0	13 − 0
1.9	60 − 2	19 − 1	29 − 1	43 − 1	55 − 1	69 − 1	81 − 1	95 − 1
2.0	15 − 2	91 − 2	13 − 1	22 − 1	28 − 1	37 − 1	44 − 1	52 − 1
2.1	65 − 3	22 − 2	42 − 2	68 − 2	95 − 2	12 − 1	15 − 1	18 − 1
2.2	64 − 4	70 − 7	45 − 4	21 − 3	47 − 3	83 − 3	12 − 2	17 − 2
2.3	13 − 2	18 − 2	21 − 2	22 − 2	23 − 2	23 − 2	23 − 2	23 − 2
2.4	36 − 2	60 − 2	82 − 2	10 − 1	11 − 1	13 − 1	14 − 1	15 − 1
2.5	55 − 2	10 − 1	14 − 1	19 − 1	23 − 1	26 − 1	30 − 1	34 − 1
2.6	63 − 2	13 − 1	19 − 1	25 − 1	31 − 1	37 − 1	43 − 1	48 − 1
2.7	57 − 2	13 − 1	20 − 1	27 − 1	34 − 1	41 − 1	47 − 1	54 − 1
2.8	40 − 2	11 − 1	16 − 1	23 − 1	30 − 1	37 − 1	43 − 1	50 − 1
2.9	19 − 2	74 − 2	11 − 1	17 − 1	21 − 1	27 − 1	32 − 1	37 − 1
3.0	42 − 3	36 − 2	53 − 2	91 − 2	11 − 1	15 − 1	18 − 1	21 − 1
3.1	25 − 3	95 − 3	18 − 2	29 − 2	41 − 2	54 − 2	68 − 2	82 − 2
3.2	27 − 4	27 − 6	23 − 4	11 − 3	23 − 3	42 − 3	62 − 3	87 − 3
3.3	60 − 3	77 − 3	93 − 3	96 − 3	99 − 3	97 − 3	95 − 3	92 − 3
3.4	15 − 2	26 − 2	36 − 2	44 − 2	52 − 2	58 − 2	64 − 2	70 − 2
3.5	24 − 2	47 − 2	68 − 2	88 − 2	10 − 1	12 − 1	14 − 1	15 − 1
3.6	28 − 2	61 − 2	91 − 2	12 − 1	15 − 1	17 − 1	20 − 1	23 − 1
3.7	26 − 2	63 − 2	96 − 2	13 − 1	16 − 1	20 − 1	23 − 1	27 − 1
3.8	18 − 2	54 − 2	82 − 2	11 − 1	15 − 1	18 − 1	21 − 1	25 − 1
3.9	88 − 3	37 − 2	55 − 2	86 − 2	10 − 1	14 − 1	16 − 1	19 − 1
4.0	15 − 3	18 − 2	26 − 2	47 − 2	59 − 2	80 − 2	94 − 2	11 − 1
4.1	13 − 3	49 − 3	94 − 3	15 − 2	22 − 2	29 − 2	37 − 2	45 − 2
4.2	14 − 4	21 − 6	13 − 4	63 − 4	13 − 3	24 − 3	36 − 3	50 − 3
4.3	31 − 3	40 − 3	50 − 3	51 − 3	52 − 3	51 − 3	50 − 3	47 − 3
4.4	85 − 3	14 − 2	20 − 2	24 − 2	28 − 2	32 − 2	35 − 2	38 − 2

Table 2.23 (*Cont.*)

	n_0^*							
Δn^*	1.5	2.0	2.5	3.0	3.5	4.0	4.5	5.0
				p–d				
0.1			36 − 0	58 + 0	79 + 0	98 + 0	11 + 1	13 + 1
0.2			87 + 0	13 + 1	17 + 1	20 + 1	23 + 1	27 + 1
0.3			14 + 1	20 + 1	25 + 1	30 + 1	35 + 1	39 + 1
0.4			20 + 1	27 + 1	33 + 1	38 + 1	43 + 1	48 + 1
0.5			25 + 1	32 + 1	38 + 1	43 + 1	48 + 1	52 + 1
0.6		22 + 1	28 + 1	35 + 1	39 + 1	44 + 1	48 + 1	52 + 1
0.7		24 + 1	29 + 1	34 + 1	37 + 1	41 + 1	44 + 1	47 + 1
0.8		25 + 1	27 + 1	30 + 1	32 + 1	35 + 1	36 + 1	39 + 1
0.9		24 + 1	23 + 1	25 + 1	25 + 1	26 + 1	27 + 1	28 + 1
1.0		20 + 1	18 + 1	18 + 1	18 + 1	18 + 1	18 + 1	18 + 1
1.1	14 + 1	15 + 1	13 + 1	11 + 1	10 + 1	10 + 1	10 + 1	10 + 1
1.2	12 + 1	10 + 1	76 + 0	61 + 0	52 + 0	46 − 0	42 − 0	39 − 0
1.3	95 + 0	57 + 0	34 − 0	22 − 0	16 − 0	12 − 0	92 − 1	72 − 1
1.4	65 + 0	24 − 0	91 − 1	32 − 1	94 − 2	11 − 2	26 − 3	33 − 2
1.5	38 − 0	50 − 1	96 − 3	82 − 2	29 − 1	56 − 1	84 − 1	11 − 0
1.6	15 − 0	51 − 3	40 − 1	10 − 0	15 − 0	20 − 0	25 − 0	30 − 0
1.7	32 − 1	56 − 1	14 − 0	24 − 0	30 − 0	37 − 0	42 − 0	47 − 0
1.8	21 − 3	16 − 0	26 − 0	36 − 0	41 − 0	48 − 0	52 + 0	57 + 0
1.9	31 − 1	28 − 0	35 − 0	43 − 0	46 − 0	50 + 0	53 + 0	56 + 0
2.0	93 − 1	36 − 0	37 − 0	41 − 0	42 − 0	44 − 0	45 − 0	47 − 0
2.1	19 − 0	36 − 0	34 − 0	33 − 0	32 − 0	32 − 0	32 − 0	32 − 0
2.2	23 − 0	30 − 0	25 − 0	22 − 0	20 − 0	18 − 0	17 − 0	16 − 0
2.3	22 − 0	21 − 0	15 − 0	11 − 0	88 − 1	73 − 1	61 − 1	52 − 1
2.4	19 − 0	11 − 0	60 − 1	31 − 1	18 − 1	98 − 2	49 − 2	21 − 2
2.5	14 − 0	41 − 1	83 − 2	42 − 3	70 − 3	43 − 2	96 − 2	15 − 1
2.6	75 − 1	36 − 2	26 − 2	15 − 1	29 − 1	45 − 1	59 − 1	74 − 1
2.7	26 − 1	59 − 2	31 − 1	62 − 1	83 − 1	10 − 0	12 − 0	14 − 0
2.8	23 − 2	39 − 1	77 − 1	11 − 0	13 − 0	16 − 0	18 − 0	20 − 0
2.9	30 − 2	87 − 1	12 − 0	15 − 0	17 − 0	19 − 0	20 − 0	22 − 0
3.0	20 − 1	13 − 0	14 − 0	16 − 0	17 − 0	18 − 0	19 − 0	20 − 0
3.1	60 − 1	14 − 0	14 − 0	14 − 0	14 − 0	14 − 0	14 − 0	15 − 0
3.2	82 − 1	13 − 0	12 − 0	10 − 0	10 − 0	95 − 1	90 − 1	87 − 1
3.3	91 − 1	10 − 0	77 − 1	60 − 1	50 − 1	43 − 1	37 − 1	32 − 1
3.4	85 − 1	61 − 1	35 − 1	20 − 1	13 − 1	88 − 2	55 − 2	33 − 2
3.5	69 − 1	26 − 1	71 − 2	13 − 2	35 − 4	45 − 3	18 − 2	39 − 2
3.6	40 − 1	43 − 2	22 − 3	45 − 2	95 − 2	16 − 1	22 − 1	29 − 1
3.7	16 − 1	95 − 3	11 − 1	25 − 1	34 − 1	46 − 1	55 − 1	65 − 1
3.8	28 − 2	14 − 1	33 − 1	53 − 1	63 − 1	77 − 1	87 − 1	98 − 1
3.9	40 − 3	38 − 1	57 − 1	77 − 1	84 − 1	97 − 1	10 − 0	11 − 0
4.0	72 − 2	64 − 1	73 − 1	87 − 1	90 − 1	99 − 1	10 − 0	10 − 0
4.1	26 − 1	74 − 1	80 − 1	80 − 1	81 − 1	83 − 1	84 − 1	85 − 1
4.2	38 − 1	71 − 1	67 − 1	61 − 1	57 − 1	56 − 1	54 − 1	52 − 1
4.3	45 − 1	57 − 1	45 − 1	35 − 1	30 − 1	27 − 1	24 − 1	21 − 1
4.4	45 − 1	37 − 1	22 − 1	13 − 1	95 − 2	67 − 2	46 − 2	31 − 2

Table 2.23 (*Cont.*)

Δn^*	n_0^* 2.5	3.0	3.5	4.0	4.5	5.0
			d–p			
0.1	24 − 0	44 − 0	63 + 0	81 + 0	99 + 0	11 + 1
0.2	39 − 0	74 + 0	10 + 1	14 + 1	17 + 1	20 + 1
0.3	46 − 0	90 + 0	13 + 1	17 + 1	21 + 1	25 + 1
0.4	45 − 0	92 + 0	13 + 1	18 + 1	22 + 1	27 + 1
0.5	38 − 0	83 − 0	12 + 1	16 + 1	21 + 1	25 + 1
0.6	29 − 0	66 + 0	10 + 1	13 + 1	17 + 1	21 + 1
0.7	19 − 0	47 − 0	73 + 0	10 + 1	13 + 1	15 + 1
0.8	10 − 0	29 − 0	46 − 0	65 + 0	84 + 0	10 + 1
0.9	44 − 1	15 − 0	23 − 0	35 − 0	45 − 0	57 + 0
1.0	10 − 1	55 − 1	88 − 1	13 − 0	18 − 0	24 − 0
1.1	11 − 2	84 − 2	15 − 1	27 − 1	39 − 1	54 − 1
1.2	19 − 2	10 − 2	82 − 3	27 − 3	51 − 4	27 − 4
1.3	10 − 1	17 − 1	23 − 1	27 − 1	31 − 1	34 − 1
1.4	19 − 1	40 − 1	59 − 1	76 − 1	92 − 1	10 − 0
1.5	24 − 1	59 − 1	88 − 1	11 − 0	14 − 0	17 − 0
1.6	23 − 1	65 − 1	10 − 0	13 − 0	17 − 0	21 − 0
1.7	18 − 1	59 − 1	91 − 1	13 − 0	16 − 0	20 − 0
1.8	11 − 1	44 − 1	68 − 1	10 − 0	13 − 0	16 − 0
1.9	43 − 2	26 − 1	40 − 1	64 − 1	83 − 1	10 − 0
2.0	59 − 3	11 − 1	16 − 1	29 − 1	38 − 1	51 − 1
2.1	57 − 4	18 − 2	34 − 2	65 − 2	96 − 2	13 − 1
2.2	77 − 3	25 − 3	22 − 3	47 − 4	22 − 5	42 − 4
2.3	31 − 2	45 − 2	64 − 2	73 − 2	82 − 2	87 − 2
2.4	56 − 2	11 − 1	17 − 1	22 − 1	26 − 1	30 − 1
2.5	68 − 2	18 − 1	27 − 1	36 − 1	45 − 1	54 − 1
2.6	66 − 2	21 − 1	32 − 1	45 − 1	57 − 1	69 − 1
2.7	51 − 2	20 − 1	30 − 1	45 − 1	57 − 1	71 − 1
2.8	28 − 2	15 − 1	23 − 1	37 − 1	47 − 1	60 − 1
2.9	91 − 3	96 − 2	14 − 1	24 − 1	31 − 1	40 − 1
3.0	23 − 4	42 − 2	59 − 2	11 − 1	14 − 1	20 − 1
3.1	56 − 6	72 − 3	13 − 2	26 − 2	38 − 2	55 − 2
3.2	44 − 3	11 − 3	11 − 3	25 − 4	30 − 5	15 − 4
3.3	14 − 2	20 − 2	28 − 2	32 − 2	36 − 2	38 − 2
3.4	25 − 2	51 − 2	77 − 2	99 − 2	12 − 1	13 − 1
3.5	29 − 2	81 − 2	12 − 1	17 − 1	21 − 1	25 − 1
3.6	28 − 2	97 − 2	15 − 1	21 − 1	27 − 1	33 − 1
3.7	20 − 2	94 − 2	14 − 1	21 − 1	27 − 1	34 − 1
3.8	10 − 2	75 − 2	11 − 1	18 − 1	23 − 1	29 − 1
3.9	26 − 3	47 − 2	69 − 2	12 − 1	15 − 1	20 − 1
4.0	35 − 5	21 − 2	28 − 2	57 − 2	73 − 2	10 − 1
4.1	24 − 5	36 − 3	64 − 3	13 − 2	19 − 2	28 − 2
4.2	28 − 3	66 − 4	73 − 4	17 − 4	36 − 5	59 − 5
4.3	85 − 3	10 − 2	15 − 2	17 − 2	20 − 2	21 − 2
4.4	13 − 2	28 − 2	42 − 2	55 − 2	67 − 2	77 − 2

Table 2.23 (*Cont.*)

Δn^*	n_0^* 2.5	3.0	3.5	4.0	4.5	5.0
			d–f			
0.1			58 + 0	96 + 0	13 + 1	16 + 1
0.2			14 + 1	22 + 1	28 + 1	34 + 1
0.3			25 + 1	36 + 1	45 + 1	53 + 1
0.4			37 + 1	50 + 1	60 + 1	69 + 1
0.5			48 + 1	62 + 1	70 + 1	80 + 1
0.6		49 + 1	56 + 1	69 + 1	76 + 1	84 + 1
0.7		56 + 1	61 + 1	70 + 1	75 + 1	81 + 1
0.8		59 + 1	60 + 1	66 + 1	68 + 1	72 + 1
0.9		57 + 1	54 + 1	56 + 1	56 + 1	58 + 1
1.0		50 + 1	45 + 1	44 + 1	42 + 1	42 + 1
1.1	50 + 1	40 + 1	34 + 1	30 + 1	28 + 1	26 + 1
1.2	43 + 1	28 + 1	22 + 1	18 + 1	15 + 1	13 + 1
1.3	33 + 1	16 + 1	12 + 1	83 + 0	63 + 0	48 − 0
1.4	23 + 1	81 + 0	47 − 0	22 − 0	12 − 0	60 − 1
1.5	14 + 1	24 − 0	81 − 1	48 − 2	24 − 2	26 − 1
1.6	62 + 0	13 − 1	73 − 2	84 − 1	15 − 0	25 − 0
1.7	14 − 0	52 − 1	16 − 0	34 − 0	45 − 0	58 + 0
1.8	43 − 3	26 − 0	42 − 0	63 + 0	73 + 0	87 + 0
1.9	86 − 1	54 + 0	67 + 0	85 + 0	91 + 0	10 + 1
2.0	29 − 0	78 + 0	81 + 0	93 + 0	93 + 0	98 + 0
2.1	65 + 0	84 + 0	85 + 0	83 + 0	81 + 0	79 + 0
2.2	80 + 0	77 + 0	72 + 0	63 + 0	57 + 0	53 + 0
2.3	82 + 0	59 + 0	50 + 0	38 − 0	32 − 0	26 − 0
2.4	74 + 0	37 − 0	27 + 0	16 − 0	11 − 0	78 − 1
2.5	57 + 0	17 − 0	92 − 1	31 − 1	11 − 1	14 − 2
2.6	32 − 0	41 − 1	67 − 2	23 − 2	13 − 1	34 − 1
2.7	12 − 0	18 − 3	16 − 1	64 − 1	96 − 1	14 − 0
2.8	20 − 1	46 − 1	96 − 1	17 − 0	21 − 0	27 − 0
2.9	38 − 2	15 − 0	20 − 0	29 − 0	31 − 0	37 − 0
3.0	58 − 1	26 − 0	29 − 0	36 − 0	37 − 0	40 − 0
3.1	19 − 0	33 − 0	35 − 0	36 − 0	36 − 0	36 − 0
3.2	28 − 0	33 − 0	33 − 0	30 − 0	28 − 0	26 − 0
3.3	33 − 0	28 − 0	25 − 0	20 − 0	17 − 0	15 − 0
3.4	33 − 0	20 − 0	15 − 0	10 − 0	79 − 1	57 − 1
3.5	28 − 0	10 − 0	64 − 1	28 − 1	15 − 1	54 − 2
3.6	18 − 0	35 − 1	10 − 1	75 − 4	12 − 2	78 − 2
3.7	86 − 1	13 − 2	24 − 2	20 − 1	33 − 1	54 − 1
3.8	22 − 1	12 − 1	35 − 1	73 − 1	91 − 1	12 − 0
3.9	15 − 4	61 − 1	90 − 1	13 − 0	15 − 0	18 − 0
4.0	17 − 1	12 − 0	14 − 0	18 − 0	18 − 0	21 − 0
4.1	81 − 1	16 − 0	18 − 0	19 − 0	19 − 0	20 − 0
4.2	13 − 0	18 − 0	18 − 0	17 − 0	16 − 0	15 − 0
4.3	17 − 0	16 − 0	15 − 0	12 − 0	11 − 0	97 − 1
4.4	18 − 0	12 − 0	97 − 1	66 − 1	53 − 1	40 − 1

Table 2.23 (*Cont.*)

	n_0^*			
Δn^*	3.5	4.0	4.5	5.0
		f–d		
0.1	34 − 0	67 + 0	98 + 0	12 + 1
0.2	52 + 0	11 + 1	16 + 1	21 + 1
0.3	57 + 0	13 + 1	19 + 1	26 + 1
0.4	52 + 0	13 + 1	19 + 1	26 + 1
0.5	42 − 0	11 + 1	17 + 1	24 + 1
0.6	30 − 0	89 + 0	13 + 1	19 + 1
0.7	18 − 0	61 + 0	95 + 0	13 + 1
0.8	98 − 1	37 − 0	57 + 0	85 + 0
0.9	38 − 1	18 − 0	28 − 0	44 − 0
1.0	87 − 2	62 − 1	96 − 1	16 − 0
1.1	23 − 2	76 − 2	14 − 1	26 − 1
1.2	42 − 3	24 − 2	24 − 2	21 − 2
1.3	47 − 2	24 − 1	33 − 1	45 − 1
1.4	93 − 2	53 − 1	77 − 1	11 − 0
1.5	10 − 1	74 − 1	10 − 0	16 − 0
1.6	10 − 1	79 − 1	11 − 0	18 − 0
1.7	69 − 2	69 − 1	10 − 0	16 − 0
1.8	33 − 2	50 − 1	74 − 1	12 − 0
1.9	83 − 3	28 − 1	41 − 1	74 − 1
2.0	18 − 8	11 − 1	15 − 1	31 − 1
2.1	16 − 3	14 − 2	26 − 2	54 − 2
2.2	86 − 4	65 − 3	71 − 3	57 − 3
2.3	62 − 3	63 − 2	86 − 2	12 − 1
2.4	10 − 2	14 − 1	20 − 1	31 − 1
2.5	98 − 3	21 − 1	30 − 1	48 − 1
2.6	79 − 3	23 − 1	34 − 1	57 − 1
2.7	36 − 3	21 − 1	31 − 1	54 − 1
2.8	39 − 4	16 − 1	22 − 1	42 − 1
2.9	65 − 4	97 − 2	12 − 1	26 − 1
3.0	46 − 3	39 − 2	45 − 2	11 − 1
3.1	17 − 5	49 − 3	81 − 3	19 − 2
3.2	66 − 4	29 − 3	36 − 3	29 − 3
3.3	18 − 3	26 − 2	37 − 2	52 − 2
3.4	20 − 3	61 − 2	87 − 2	13 − 1
3.5	94 − 4	92 − 2	12 − 1	21 − 1
3.6	32 − 4	10 − 1	14 − 1	26 − 1
3.7	18 − 5	98 − 2	13 − 1	25 − 1
3.8	10 − 3	75 − 2	10 − 1	20 − 1
3.9	38 − 3	45 − 2	55 − 2	12 − 1
4.0	76 − 3	19 − 2	18 − 2	56 − 2
4.1	13 − 4	23 − 3	34 − 3	96 − 3
4.2	66 − 4	15 − 3	22 − 3	17 − 3
4.3	93 − 4	14 − 2	19 − 2	28 − 2
4.4	56 − 4	32 − 2	46 − 2	75 − 2

Table 2.24. The same as in Table 2.23 for quadrupole matrix elements F_2 ($\varkappa = 2$)

	n_0^*									
Δn^*	0.5	1.0	1.5	2.0	2.5	3.0	3.5	4.0	4.5	5.0
					Transition *s*–*d*					
0.1					15 − 2	18 − 2	19 − 2	18 − 2	18 − 2	16 − 2
0.2					14 − 1	16 − 1	16 − 1	15 − 1	14 − 1	13 − 1
0.3					55 − 1	58 − 1	56 − 1	53 − 1	49 − 1	45 − 1
0.4					13 − 0	14 − 0	13 − 0	12 − 0	11 − 0	10 − 0
0.5					27 − 0	26 − 0	24 − 0	21 − 0	19 − 0	17 − 0
0.6				44 − 0	46 − 0	42 − 0	37 − 0	33 − 0	29 − 0	26 − 0
0.7				70 + 0	68 + 0	59 + 0	51 + 0	45 − 0	39 − 0	35 − 0
0.8				99 + 0	89 + 0	74 + 0	63 + 0	54 + 0	47 − 0	41 − 0
0.9				12 + 1	10 + 1	85 + 0	71 + 0	59 + 0	51 + 0	44 − 0
1.0				14 + 1	11 + 1	89 + 0	72 + 0	59 + 0	50 + 0	43 − 0
1.1			19 + 1	15 + 1	11 + 1	84 + 0	66 + 0	54 + 0	45 − 0	38 − 0
1.2			21 + 1	14 + 1	10 + 1	72 + 0	55 + 0	43 − 0	35 − 0	29 − 0
1.3			21 + 1	13 + 1	81 + 0	55 + 0	40 − 0	30 − 0	24 − 0	20 − 0
1.4			19 + 1	10 + 1	57 + 0	36 − 0	25 − 0	18 − 0	13 − 0	11 − 0
1.5			15 + 1	68 + 0	33 − 0	19 − 0	12 − 0	80 − 1	57 − 1	41 − 1
1.6		25 + 1	10 + 1	37 − 0	14 − 0	67 − 1	33 − 1	18 − 1	99 − 2	56 − 2
1.7		20 + 1	59 + 0	14 − 0	31 − 1	65 − 2	58 − 3	11 − 3	11 − 2	23 − 2
1.8		14 + 1	23 − 0	19 − 1	43 − 3	75 − 2	15 − 1	20 − 1	23 − 1	24 − 1
1.9		86 + 0	41 − 1	73 − 2	39 − 1	54 − 1	61 − 1	62 − 1	61 − 1	59 − 1
2.0		42 − 0	39 − 2	82 − 1	11 − 0	12 − 0	11 − 0	10 − 0	10 − 0	92 − 1
2.1	18 + 1	84 − 1	74 − 1	19 − 0	20 − 0	18 − 0	16 − 0	14 − 0	12 − 0	11 − 0
2.2	10 + 1	11 − 2	21 − 0	29 − 0	25 − 0	21 − 0	17 − 0	15 − 0	13 − 0	11 − 0
2.3	45 − 0	93 − 1	36 − 0	36 − 0	27 − 0	20 − 0	16 − 0	13 − 0	11 − 0	96 − 1
2.4	13 − 0	26 − 0	47 − 0	36 − 0	24 − 0	17 − 0	13 − 0	10 − 0	81 − 1	67 − 1
2.5	13 − 1	42 − 0	51 + 0	30 − 0	17 − 0	11 − 0	82 − 1	60 − 1	45 − 1	35 − 1
2.6	57 − 2	52 + 0	42 − 0	21 − 0	10 − 0	60 − 1	37 − 1	24 − 1	16 − 1	11 − 1
2.7	45 − 1	53 + 0	29 − 0	11 − 0	40 − 1	17 − 1	80 − 2	35 − 2	14 − 2	50 − 3
2.8	88 − 1	46 − 0	16 − 0	37 − 1	51 − 2	32 − 3	24 − 3	13 − 2	25 − 2	35 − 2
2.9	11 − 0	34 − 0	55 − 1	14 − 2	32 − 2	83 − 2	12 − 1	15 − 1	16 − 1	17 − 1
3.0	11 − 0	20 − 0	39 − 2	12 − 1	29 − 1	34 − 1	37 − 1	37 − 1	36 − 1	35 − 1
3.1	39 − 0	66 − 1	70 − 2	54 − 1	68 − 1	66 − 1	63 − 1	59 − 1	54 − 1	50 − 1
3.2	24 − 0	39 − 2	53 − 1	10 − 0	10 − 0	91 − 1	81 − 1	71 − 1	63 − 1	56 − 1
3.3	12 − 0	13 − 1	12 − 0	15 − 0	12 − 0	10 − 0	84 − 1	71 − 1	61 − 1	53 − 1
3.4	47 − 1	69 − 1	18 − 0	17 − 0	12 − 0	92 − 1	73 − 1	59 − 1	48 − 1	41 − 1
3.5	80 − 2	13 − 0	23 − 0	15 − 0	99 − 1	69 − 1	51 − 1	39 − 1	31 − 1	24 − 1
3.6	30 − 3	19 − 0	21 − 0	12 − 0	64 − 1	40 − 1	27 − 1	19 − 1	13 − 1	10 − 1
3.7	10 − 1	21 − 0	15 − 0	73 − 1	29 − 1	15 − 1	84 − 2	47 − 2	25 − 2	14 − 2
3.8	24 − 1	20 − 0	96 − 1	29 − 1	63 − 2	15 − 2	17 − 3	16 − 4	29 − 3	65 − 3
3.9	35 − 1	16 − 0	40 − 1	35 − 2	25 − 3	18 − 2	38 − 2	52 − 2	64 − 2	70 − 2
4.0	39 − 1	11 − 0	63 − 2	28 − 2	11 − 1	13 − 1	16 − 1	17 − 1	17 − 1	17 − 1
4.1	14 − 0	42 − 1	71 − 3	22 − 1	31 − 1	32 − 1	31 − 1	30 − 1	29 − 1	27 − 1
4.2	10 − 0	51 − 2	19 − 1	52 − 1	54 − 1	48 − 1	44 − 1	40 − 1	36 − 1	33 − 1
4.3	55 − 1	29 − 2	55 − 1	80 − 1	69 − 1	57 − 1	49 − 1	42 − 1	37 − 1	33 − 1

Table 2.24 (*Cont.*)

Δn^*	n_0^* 1.5	2.0	2.5	3.0	3.5	4.0	4.5	5.0
				p–p				
0.1	24 − 2	33 − 2	33 − 2	30 − 2	28 − 2	25 − 2	23 − 2	21 − 2
0.2	19 − 1	25 − 1	24 − 1	23 − 1	21 − 1	19 − 1	17 − 1	16 − 1
0.3	62 − 1	78 − 1	76 − 1	70 − 1	64 − 1	58 − 1	53 − 1	49 − 1
0.4	13 − 0	16 − 0	15 − 0	14 − 0	13 − 0	12 − 0	11 − 0	10 − 0
0.5	22 − 0	26 − 0	25 − 0	23 − 0	21 − 0	19 − 0	18 − 0	16 − 0
0.6	32 − 0	37 − 0	35 − 0	32 − 0	30 − 0	27 − 0	25 − 0	23 − 0
0.7	40 − 0	45 − 0	43 − 0	40 − 0	36 − 0	33 − 0	30 − 0	28 − 0
0.8	44 − 0	49 − 0	47 − 0	43 − 0	39 − 0	36 − 0	33 − 0	30 − 0
0.9	44 − 0	47 − 0	45 − 0	41 − 0	38 − 0	34 − 0	32 − 0	29 − 0
1.0	39 − 0	41 − 0	39 − 0	36 − 0	33 − 0	30 − 0	27 − 0	25 − 0
1.1	30 − 0	32 − 0	30 − 0	27 − 0	25 − 0	23 − 0	21 − 0	19 − 0
1.2	20 − 0	21 − 0	19 − 0	18 − 0	16 − 0	14 − 0	13 − 0	12 − 0
1.3	11 − 0	11 − 0	10 − 0	94 − 1	85 − 1	77 − 1	70 − 1	64 − 1
1.4	45 − 1	41 − 1	37 − 1	32 − 1	28 − 1	25 − 1	23 − 1	21 − 1
1.5	72 − 2	50 − 2	38 − 2	29 − 2	23 − 2	19 − 2	16 − 2	14 − 2
1.6	11 − 2	27 − 2	35 − 2	39 − 2	40 − 2	40 − 2	39 − 2	37 − 2
1.7	19 − 1	25 − 1	26 − 1	26 − 1	25 − 1	24 − 1	22 − 1	21 − 1
1.8	48 − 1	57 − 1	59 − 1	57 − 1	54 − 1	50 − 1	47 − 1	44 − 1
1.9	76 − 1	85 − 1	86 − 1	83 − 1	78 − 1	73 − 1	68 − 1	64 − 1
2.0	92 − 1	98 − 1	10 − 0	95 − 1	89 − 1	83 − 1	77 − 1	72 − 1
2.1	84 − 1	94 − 1	94 − 1	90 − 1	84 − 1	78 − 1	73 − 1	68 − 1
2.2	69 − 1	75 − 1	74 − 1	71 − 1	66 − 1	61 − 1	57 − 1	53 − 1
2.3	46 − 1	48 − 1	47 − 1	45 − 1	41 − 1	38 − 1	35 − 1	33 − 1
2.4	22 − 1	23 − 1	22 − 1	20 − 1	18 − 1	17 − 1	16 − 1	14 − 1
2.5	60 − 2	55 − 2	51 − 2	45 − 2	40 − 2	36 − 2	32 − 2	29 − 2
2.6	15 − 4	15 − 4	47 − 4	88 − 4	12 − 3	15 − 3	16 − 3	17 − 3
2.7	42 − 2	57 − 2	62 − 2	63 − 2	63 − 2	61 − 2	59 − 2	56 − 2
2.8	15 − 1	18 − 1	19 − 1	19 − 1	18 − 1	17 − 1	16 − 1	16 − 1
2.9	28 − 1	31 − 1	33 − 1	32 − 1	31 − 1	29 − 1	28 − 1	26 − 1
3.0	37 − 1	40 − 1	42 − 1	41 − 1	39 − 1	37 − 1	35 − 1	33 − 1
3.1	36 − 1	41 − 1	43 − 1	42 − 1	40 − 1	38 − 1	36 − 1	34 − 1
3.2	32 − 1	35 − 1	36 − 1	35 − 1	34 − 1	32 − 1	30 − 1	28 − 1
3.3	22 − 1	24 − 1	25 − 1	24 − 1	23 − 1	21 − 1	20 − 1	19 − 1
3.4	12 − 1	12 − 1	12 − 1	12 − 1	11 − 1	10 − 1	10 − 1	96 − 2
3.5	36 − 2	36 − 2	36 − 2	34 − 2	31 − 2	28 − 2	26 − 2	24 − 2
3.6	86 − 4	28 − 4	20 − 4	95 − 5	23 − 5	49 − 6	94 − 8	10 − 6
3.7	16 − 2	22 − 2	23 − 2	24 − 2	25 − 2	24 − 2	23 − 2	23 − 2
3.8	71 − 2	84 − 2	90 − 2	89 − 2	89 − 2	86 − 2	82 − 2	78 − 2
3.9	14 − 1	15 − 1	16 − 1	16 − 1	16 − 1	15 − 1	15 − 1	14 − 1
4.0	19 − 1	20 − 1	22 − 1	22 − 1	21 − 1	20 − 1	19 − 1	18 − 1
4.1	19 − 1	22 − 1	23 − 1	23 − 1	23 − 1	22 − 1	21 − 1	20 − 1
4.2	17 − 1	19 − 1	20 − 1	20 − 1	20 − 1	19 − 1	18 − 1	17 − 1
4.3	12 − 1	14 − 1	14 − 1	14 − 1	14 − 1	13 − 1	13 − 1	12 − 1

Table 2.24 (*Cont.*)

	n_0^*							
Δn^*	1.5	2.0	2.5	3.0	3.5	4.0	4.5	5.0
				p–f				
0.1					12 − 2	17 − 2	20 − 2	21 − 2
0.2					12 − 1	16 − 1	18 − 1	18 − 1
0.3					51 − 1	62 − 1	66 − 1	66 − 1
0.4					13 − 0	15 − 0	16 − 0	15 − 0
0.5					29 − 0	31 − 0	31 − 0	29 − 0
0.6				42 − 0	51 + 0	53 + 0	51 + 0	48 − 0
0.7				72 + 0	81 + 0	79 + 0	74 + 0	68 + 0
0.8				11 + 1	11 + 1	10 + 1	96 + 0	86 + 0
0.9				15 + 1	14 + 1	12 + 1	11 + 1	10 + 1
1.0				19 + 1	17 + 1	14 + 1	12 + 1	10 + 1
1.1			23 + 1	22 + 1	18 + 1	14 + 1	12 + 1	10 + 1
1.2			28 + 1	23 + 1	18 + 1	14 + 1	11 + 1	91 + 0
1.3			30 + 1	22 + 1	16 + 1	12 + 1	91 + 0	72 + 0
1.4			30 + 1	20 + 1	13 + 1	91 + 0	66 + 0	49 − 0
1.5			28 + 1	15 + 1	94 + 0	60 + 0	40 − 0	28 − 0
1.6		39 + 1	22 + 1	11 + 1	56 + 0	31 − 0	18 − 0	11 − 0
1.7		35 + 1	16 + 1	63 + 0	25 − 0	11 − 0	48 − 1	20 − 1
1.8		29 + 1	96 + 0	26 − 0	63 − 1	10 − 1	14 − 3	22 − 2
1.9		21 + 1	43 − 0	48 − 1	44 − 6	13 − 1	32 − 1	47 − 1
2.0		13 + 1	10 − 0	40 − 2	53 − 1	93 − 1	11 − 0	12 − 0
2.1	27 + 1	60 + 0	27 − 5	10 − 0	18 − 0	21 − 0	21 − 0	21 − 0
2.2	19 + 1	14 − 0	90 − 1	27 − 0	32 − 0	32 − 0	29 − 0	27 − 0
2.3	11 + 1	35 − 4	31 − 0	46 − 0	44 − 0	38 − 0	33 − 0	28 − 0
2.4	56 + 0	11 − 0	57 + 0	59 + 0	48 − 0	38 − 0	31 − 0	25 − 0
2.5	19 − 0	38 − 0	78 + 0	62 + 0	45 − 0	33 − 0	24 − 0	18 − 0
2.6	34 − 2	68 + 0	83 + 0	55 + 0	35 − 0	23 − 0	15 − 0	11 − 0
2.7	67 − 1	90 + 0	76 + 0	41 − 0	22 − 0	12 − 0	74 − 1	45 − 1
2.8	25 − 0	98 + 0	58 + 0	24 − 0	10 − 0	43 − 1	17 − 1	63 − 2
2.9	45 − 0	91 + 0	35 − 0	99 − 1	21 − 1	27 − 2	11 − 3	26 − 2
3.0	59 + 0	73 + 0	15 − 0	13 − 1	65 − 3	97 − 2	21 − 1	29 − 1
3.1	71 + 0	42 − 0	33 − 1	59 − 2	33 − 1	54 − 1	66 − 1	72 − 1
3.2	62 + 0	17 − 0	16 − 2	63 − 1	10 − 0	11 − 0	11 − 0	11 − 0
3.3	46 − 0	28 − 1	63 − 1	15 − 0	17 − 0	16 − 0	15 − 0	14 − 0
3.4	28 − 0	53 − 2	18 − 0	24 − 0	22 − 0	19 − 0	16 − 0	14 − 0
3.5	13 − 0	84 − 1	32 − 0	30 − 0	23 − 0	18 − 0	14 − 0	11 − 0
3.6	18 − 1	21 − 0	39 − 0	30 − 0	20 − 0	14 − 0	10 − 0	80 − 1
3.7	53 − 2	35 − 0	40 − 0	25 − 0	14 − 0	94 − 1	61 − 1	40 − 1
3.8	62 − 1	44 − 0	34 − 0	17 − 0	80 − 1	41 − 1	21 − 1	10 − 1
3.9	14 − 0	47 − 0	24 − 0	84 − 1	26 − 1	80 − 2	14 − 2	56 − 5
4.0	22 − 0	42 − 0	12 − 0	20 − 1	10 − 2	77 − 3	47 − 2	90 − 2
4.1	30 − 0	27 − 0	42 − 1	17 − 4	88 − 2	18 − 1	27 − 1	32 − 1
4.2	29 − 0	13 − 0	11 − 2	21 − 1	42 − 1	52 − 1	58 − 1	59 − 1
4.3	23 − 0	36 − 1	17 − 1	71 − 1	87 − 1	88 − 1	85 − 1	80 − 1

Table 2.24 (*Cont.*)

	n_0^*					
Δn^*	2.5	3.0	3.5	4.0	4.5	5.0
			d–s			
0.1	10 − 2	14 − 2	15 − 2	15 − 2	15 − 2	14 − 2
0.2	66 − 2	94 − 2	10 − 1	10 − 1	10 − 1	10 − 1
0.3	17 − 1	25 − 1	29 − 1	30 − 1	29 − 1	29 − 1
0.4	31 − 1	47 − 1	54 − 1	56 − 1	57 − 1	56 − 1
0.5	44 − 1	68 − 1	79 − 1	84 − 1	86 − 1	85 − 1
0.6	53 − 1	83 − 1	99 − 1	10 − 0	10 − 0	10 − 0
0.7	56 − 1	89 − 1	10 − 0	11 − 0	12 − 0	12 − 0
0.8	51 − 1	83 − 1	10 − 0	11 − 0	11 − 0	12 − 0
0.9	42 − 1	68 − 1	86 − 1	96 − 1	10 − 0	10 − 0
1.0	29 − 1	48 − 1	63 − 1	72 − 1	77 − 1	80 − 1
1.1	15 − 1	28 − 1	38 − 1	44 − 1	49 − 1	52 − 1
1.2	62 − 2	12 − 1	17 − 1	21 − 1	24 − 1	26 − 1
1.3	10 − 2	27 − 2	45 − 2	61 − 2	75 − 2	86 − 2
1.4	10 − 3	13 − 4	18 − 4	13 − 3	33 − 3	56 − 3
1.5	22 − 2	27 − 2	26 − 2	23 − 2	19 − 2	15 − 2
1.6	58 − 2	83 − 2	93 − 2	94 − 2	91 − 2	86 − 2
1.7	91 − 2	13 − 1	16 − 1	17 − 1	17 − 1	17 − 1
1.8	11 − 1	17 − 1	21 − 1	23 − 1	24 − 1	24 − 1
1.9	11 − 1	17 − 1	22 − 1	24 − 1	26 − 1	27 − 1
2.0	93 − 2	14 − 1	19 − 1	21 − 1	23 − 1	24 − 1
2.1	50 − 2	94 − 2	12 − 1	15 − 1	17 − 1	18 − 1
2.2	22 − 2	47 − 2	68 − 2	85 − 2	99 − 2	10 − 1
2.3	50 − 3	12 − 2	21 − 2	29 − 2	37 − 2	43 − 2
2.4	93 − 5	66 − 5	76 − 4	20 − 3	38 − 3	56 − 3
2.5	69 − 3	80 − 3	72 − 3	57 − 3	42 − 3	30 − 3
2.6	20 − 2	29 − 2	32 − 2	32 − 2	31 − 2	29 − 2
2.7	35 − 2	53 − 2	64 − 2	68 − 2	69 − 2	68 − 2
2.8	46 − 2	70 − 2	89 − 2	98 − 2	10 − 1	10 − 1
2.9	49 − 2	74 − 2	98 − 2	11 − 1	11 − 1	12 − 1
3.0	46 − 2	64 − 2	88 − 2	10 − 1	11 − 1	11 − 1
3.1	23 − 2	44 − 2	62 − 2	75 − 2	85 − 2	92 − 2
3.2	10 − 2	23 − 2	34 − 2	43 − 2	51 − 2	57 − 2
3.3	25 − 3	67 − 3	11 − 2	16 − 2	20 − 2	23 − 2
3.4	23 − 5	76 − 5	56 − 4	14 − 3	25 − 3	36 − 3
3.5	32 − 3	37 − 3	32 − 3	24 − 3	17 − 3	11 − 3
3.6	10 − 2	14 − 2	16 − 2	16 − 2	15 − 2	14 − 2
3.7	18 − 2	27 − 2	33 − 2	36 − 2	36 − 2	36 − 2
3.8	25 − 2	37 − 2	48 − 2	52 − 2	56 − 2	57 − 2
3.9	28 − 2	40 − 2	54 − 2	60 − 2	66 − 2	69 − 2
4.0	27 − 2	35 − 2	50 − 2	57 − 2	64 − 2	67 − 2
4.1	12 − 2	25 − 2	35 − 2	43 − 2	49 − 2	54 − 2
4.2	61 − 3	13 − 2	19 − 2	25 − 2	30 − 2	34 − 2
4.3	14 − 3	39 − 3	68 − 3	97 − 3	12 − 2	14 − 2

Table 2.24 (*Cont.*)

Δn^*	n_0^*					
	2.5	3.0	3.5	4.0	4.5	5.0
			d–d			
0.1	11 – 2	18 – 2	21 – 2	21 – 2	21 – 2	20 – 2
0.2	92 – 2	14 – 1	16 – 1	16 – 1	16 – 0	15 – 1
0.3	30 – 1	44 – 1	49 – 1	51 – 1	50 – 1	48 – 1
0.4	67 – 1	94 – 1	10 – 0	10 – 0	10 – 0	10 – 0
0.5	11 – 0	15 – 0	17 – 0	17 – 0	17 – 0	16 – 0
0.6	17 – 0	22 – 0	24 – 0	24 – 0	24 – 0	23 – 0
0.7	21 – 0	28 – 0	30 – 0	30 – 0	29 – 0	28 – 0
0.8	25 – 0	31 – 0	33 – 0	33 – 0	32 – 0	31 – 0
0.9	25 – 0	31 – 0	33 – 0	33 – 0	32 – 0	30 – 0
1.0	23 – 0	28 – 0	29 – 0	29 – 0	28 – 0	26 – 0
1.1	18 – 0	22 – 0	23 – 0	23 – 0	22 – 0	20 – 0
1.2	13 – 0	15 – 0	16 – 0	15 – 0	14 – 0	14 – 0
1.3	80 – 1	92 – 1	92 – 1	87 – 1	81 – 1	75 – 1
1.4	35 – 1	39 – 1	37 – 1	34 – 1	31 – 1	28 – 1
1.5	77 – 2	81 – 2	68 – 2	53 – 2	42 – 2	34 – 2
1.6	31 – 4	14 – 3	55 – 3	10 – 2	15 – 2	18 – 2
1.7	80 – 2	10 – 1	13 – 1	15 – 1	16 – 1	17 – 1
1.8	25 – 1	30 – 1	36 – 1	39 – 1	40 – 1	40 – 1
1.9	43 – 1	51 – 1	59 – 1	61 – 1	62 – 1	61 – 1
2.0	55 – 1	63 – 1	72 – 1	74 – 1	74 – 1	72 – 1
2.1	54 – 1	65 – 1	73 – 1	74 – 1	73 – 1	70 – 1
2.2	47 – 1	56 – 1	61 – 1	61 – 1	60 – 1	57 – 1
2.3	33 – 1	39 – 1	42 – 1	41 – 1	40 – 1	38 – 1
2.4	18 – 1	21 – 1	22 – 1	21 – 1	20 – 1	18 – 1
2.5	56 – 2	71 – 2	71 – 2	63 – 2	56 – 2	50 – 2
2.6	26 – 3	38 – 3	25 – 3	11 – 3	37 – 4	66 – 5
2.7	16 – 2	17 – 2	24 – 2	29 – 2	34 – 2	37 – 2
2.8	83 – 2	90 – 2	11 – 1	12 – 1	13 – 1	13 – 1
2.9	17 – 1	18 – 1	22 – 1	23 – 1	24 – 1	24 – 1
3.0	24 – 1	25 – 1	30 – 1	31 – 1	32 – 1	32 – 1
3.1	25 – 1	29 – 1	33 – 1	34 – 1	35 – 1	34 – 1
3.2	23 – 1	26 – 1	30 – 1	31 – 1	31 – 1	30 – 1
3.3	17 – 1	20 – 1	22 – 1	23 – 1	22 – 1	21 – 1
3.4	99 – 2	11 – 1	13 – 1	13 – 1	12 – 1	11 – 1
3.5	34 – 2	45 – 2	50 – 2	46 – 2	42 – 2	39 – 2
3.6	27 – 3	46 – 3	46 – 3	35 – 3	23 – 3	15 – 3
3.7	66 – 3	55 – 3	71 – 3	90 – 3	11 – 2	12 – 2
3.8	41 – 2	40 – 2	48 – 2	54 – 2	60 – 2	62 – 2
3.9	90 – 2	90 – 2	10 – 1	11 – 1	12 – 1	12 – 1
4.0	13 – 1	13 – 1	16 – 1	17 – 1	17 – 1	17 – 1
4.1	14 – 1	15 – 1	18 – 1	19 – 1	20 – 1	19 – 1
4.2	13 – 1	15 – 1	17 – 1	18 – 1	18 – 1	18 – 1
4.3	10 – 1	11 – 1	13 – 1	14 – 1	14 – 1	13 – 1

$$n_i^* = (E_i/z^2\mathrm{Ry})^{-1/2} . \tag{2.6.17}$$

The energies E_i of the initial and final states are counted from the ionization limit.

The Bates-Damgaard tables give reasonable results for not very high transitions with $n_1 \leqslant 10$. For transitions from the states with $n_0 \geqslant 10$, the quasiclassical formulas given in Sect. 2.6.1 give better results.

If $\Delta n^* < 0.1$, the following expression for F_1 can be used:

$$F_1(l_0, l_1) = \frac{3}{4}\frac{\Delta E}{z^2\mathrm{Ry}} \max\{l_0, l_1\}(n_m^*)^2[(n_m^*)^2 - l_m^2] , \quad \Delta n^* < 0.1 , \tag{2.6.18}$$

where n_m^* is the effective quantum number n^* corresponding to the maximum value of l_0 and l_1.

The quantities F_1 and F_2 given in Tables 2.23, 24 can also be used for the determination of the dipole and quadrupole excitation cross sections using the model potential (Sect. 4.1.4).

2.7 Angular Coefficients

In this Section, the angular coefficients are given for generalized oscillator strengths $f_\varkappa$ (1.4.1, 2) and excitation cross sections by electron impact. The different types of coupling schemes are used. More detailed information about angular coefficients is given in [2.1, 2, 83].

The angular coefficients $Q_\varkappa$ are defined in such a way that for direct (0–1) and inverse (1–0) transitions one has:

$$\frac{g_0}{2l_0 + 1} Q_\varkappa(0\text{–}1) = \frac{g_1}{2l_1 + 1} Q_\varkappa(1\text{–}0) . \tag{2.7.1}$$

2.7.1 *LS*-coupling

In the *LS*-coupling scheme an atomic state is defined by the set of quantum numbers $(L_pS_p)nlLSJ$, where L_p, S_p are the angular and spin momenta of the core, n and l are the quantum numbers of the optical electron and LSJ are the quantum numbers of the total momenta of an atom. In the one-electron approximation used here, L_p and S_p are not changed: $\Delta L_p = \Delta S_p = 0$.

For transitions between fine structure components $L_0S_0J_0$–$L_1S_1J_1$,

$$Q_\varkappa(J_0, J_1) = [L_0J_1]^2 \begin{Bmatrix} L_0 & J_0 & S_0 \\ J_1 & L_1 & \varkappa \end{Bmatrix}^2 A_0 Q_\varkappa(L_0, L_1) , \quad A_0 = \delta(S_0, S_1)$$

$$Q_\varkappa''(J_0, J_1) = B \cdot A_2 Q_\varkappa(L_0, L_1) , \quad A_2 = \frac{[S_1]^2}{2[S_p]^2} , \tag{2.7.2}$$

$$B = \sum_{x,q} [L_0J_1S_0S_pxq]^2 \begin{Bmatrix} L_0 & J_0 & S_0 \\ x & L_1 & \varkappa \end{Bmatrix}^2 \begin{Bmatrix} L_1 & J_1 & S_1 \\ q & S_0 & x \end{Bmatrix}^2 \begin{Bmatrix} S_0 & \frac{1}{2} & S_p \\ \frac{1}{2} & S_1 & q \end{Bmatrix}^2 ,$$

where $[x] = (2x+1)^{1/2}$ and $Q_\varkappa(L_0, L_1)$ is the orbital angular factor for transitions between LS-terms. It may depend on the spin quantum numbers only through the Racah fractional parentage coefficients $G^{LS}_{L_pS_p}$.

According to the sum rule for 6j-symbols, one has

$$\begin{aligned}\sum_{J_1} Q_\varkappa(J_0, J_1) &= \sum_{J_0 J_1} \frac{[J_0]^2}{[L_0 S_0]^2} Q_\varkappa(J_0, J_1) = A_0 Q_\varkappa(L_0, L_1)\,,\\ \sum_{J_1} Q''_\varkappa(J_0, J_1) &= \sum_{J_0 J_1} \frac{[J_0]^2}{[L_0 S_0]^2} Q''_\varkappa(J_0, J_1) = A_2 Q_\varkappa(L_0, L_1)\,.\end{aligned} \tag{2.7.3}$$

For $S_p = 0$ (or $S_0 = 0$ or $L_0 = 0$), the factor B in (2.7.2) is

$$B = [J_1]^2/[L_1 S_1]^2\,.$$

For one electron out of the closed shells of the core $L_p = S_p = 0$, $L = l$, $S = 1/2$ and one has

$$Q_\varkappa(l_0, l_1) = 1\,, \quad Q_\varkappa(l_0 j_0, l_1 j_1) = [l_0 j_1]^2 \begin{Bmatrix} l_0 & j_0 & \frac{1}{2} \\ j_1 & l_1 & \varkappa \end{Bmatrix}^2 . \tag{2.7.4}$$

In particular, for dipole transitions ($\varkappa = 1$), one has

$$\begin{aligned} &Q_1(0\ \ 1/2, 1\ \ 1/2) = 1/3\,, \quad Q_1(1\ \ 3/2, 2\ \ 5/2) = 3/5\,,\\ &Q_1(0\ \ 1/2, 1\ \ 3/2) = 2/3\,, \quad Q_1(1\ \ 3/2, 2\ \ 3/2) = 2/30\,,\\ &Q_1(1\ \ 3/2, 0\ \ 1/2) = 1\,, \qquad Q_1(1\ \ 1/2, 2\ \ 3/2) = 1/3\,. \end{aligned}$$

Below, the formulas for $Q_\varkappa(L_0, L_1)$ for the most important cases are given:

1) Transitions from the shell with one electron:

$$\gamma_0 = (L_p S_p) l_0 L_0 S_0\,, \quad \gamma_1 = (L_p S_p) l_1 L_1 S_0\,,$$

$$Q_\varkappa(L_0, L_1) = [l_0 L_1]^2 \begin{Bmatrix} l_0 & L_0 & L_p \\ L_1 & l_1 & \varkappa \end{Bmatrix}^2 , \tag{2.7.5}$$

$$\sum_{L_1} Q_\varkappa(L_0, L_1) = \sum_{L_0 L_1} \frac{[L_0]^2}{[l_0 L_0]^2} Q_\varkappa(L_0, L_1) = 1\,. \tag{2.7.6}$$

2) Transitions from the shell with equivalent electrons:

$$\gamma_0 = l_0^m L_0 S_0\,, \quad \gamma_1 = l_0^{m-1}(L_p S_p) l_1 L_1 S_0\,,$$

$$Q_\varkappa(L_0, L_1) = m |G^{L_0 S_0}_{L_p S_p}|^2 [l_0 L_1]^2 \begin{Bmatrix} l_0 & L_0 & L_p \\ L_1 & l_1 & \varkappa \end{Bmatrix}^2 , \tag{2.7.7}$$

$$\begin{aligned}\sum_{L_1} Q_\varkappa(L_0, L_1) &= m |G^{L_0 S_0}_{L_p S_p}|^2\,,\\ \sum_{L_1} Q_\varkappa(L_0, L_1) &= \sum_{L_1 L_p} \frac{g(\gamma_0)}{g(l_0^m)} Q_\varkappa(L_0, L_1) = m\,,\end{aligned} \tag{2.7.8}$$

where G^b_a is the Racah fractional parentage coefficient.

3) Transitions between configurations $l_0^N l_1^m - l_0^{N-1} l_1^{m+1}$. The general formulas for $Q_\varkappa(L_0, L_1)$ are quite complicated, so we consider only the case of the closed shell $N = 2(2l_0 + 1)$:

$$\gamma_0 = l_0^N l_1^m L_0 S, \quad \gamma_1 = l_0^{N-1}\left(l_0, \frac{1}{2}\right) l_1^{m+1}(L_p S_p) L_1 S_1,$$
$$Q_\varkappa(L_0, L_1) = (4l_1 + 2 - q)|G^{L_0 S_0}_{L_p S_p}(l_1^{M-m})|^2 [l_0 L_1]^2 \begin{Bmatrix} l_0 & L_1 & L_p \\ L_0 & l_1 & \varkappa \end{Bmatrix}^2, \tag{2.7.9}$$

where $M = 2l_1 + 1$, and $L_p S_p$ correspond to the l_1^{m+1} configuration and the parent term l_1^{M-m}, respectively. Summation over the quantum numbers gives:

$$\sum_{L_1} Q_\varkappa(L_0, L_1) = (M - m)|G^{L_0 S_0}_{L_p S_p}|^2 \frac{2l_0 + 1}{2l_1 + 1},$$
$$\sum_{L_1 L_p S_p} Q_\varkappa(L_0, L_1) = (M - m)\frac{2l_0 + 1}{2l_1 + 1}. \tag{2.7.10}$$

In the more general case one has:

$$Q(l_0^q l_1^m - l_0^{q-1} l_1^{m+1}) = \frac{M - m}{M} q, \quad q \leqslant 4l_0 + 2, \quad m \leqslant 4l_1 + 2. \tag{2.7.10a}$$

4) Transitions between terms of one configuration:

$$\gamma_0 = l_0^m L_0 S_0, \quad \gamma_1 = l_0^m L_1 S_0,$$
$$Q_\varkappa(L_0, L_1) = \frac{2l_0 + 1}{2l_1 + 1} |(l_0^m L_0 S \| U^\varkappa \| l^m L_1 S)|^2, \tag{2.7.11}$$

where $(\gamma_0 \| U^\varkappa \| \gamma_1)$ is the reduced matrix element:

$$(l_0^m L_0 S \| U^\varkappa \| l_0^m L_1 S)$$
$$= m \sum_{L_p S_p} G^{L_0 S_0}_{S_p L_p} G^{L_0 S_0}_{S_p L_p} (-1)^{L_p + \varkappa - l - L_0} [L_0 L_1]^2 \begin{Bmatrix} l & L_0 & L_p \\ L_1 & l & \varkappa \end{Bmatrix}^2. \tag{2.7.12}$$

Here, the index $\varkappa$ has only even values. If $\varkappa = 0$, one has

$$(l_0^m L_0 S \| U^0 \| l_0^m L_1 S) = m \left(\frac{2L_0 + 1}{2l_0 + 1}\right)^2 \delta_{L_0 L_1}$$

and, as follows from (2.7.2),

$$Q_0(\gamma_0 J_0, \gamma_1 J_1) \sim \delta_{J_0 J_1},$$

i.e., none of the angular quantum numbers can be changed. This means that the case $\varkappa = 0$ does not correspond to any inelastic transition. Transitions with $\varkappa = 0$ are possible only if the spin is changed, $\Delta S \neq 0$.

Table 2.25. Reduced matrix elements

$(p^2L_0S_1\|U^2\|p^2L_1S_1)$				$(p^3L_0S_1\|U^2\|p^3L_1S_1)$			
	^1S	^3P	^1D		^4S	^4P	^2D
^1S	0	0	$2/\sqrt{3}$	^4S	0	0	0
^3P	0	-1	0	^2P	0	0	$-\sqrt{3}$
^1D	$2/\sqrt{3}$	0	$\sqrt{7/3}$	^2D	0	$\sqrt{3}$	0

Quadrupole transitions with $\varkappa = 2$ are of main interest. For configurations l_0^{N-m}, $N = 2(2l_0 + 1)$ and l_0^m, the reduced matrix elements are related by

$$(l_0^{N-m}L_0S\|U^2\|l_0^{N-m}L_1S) = -(l_0^mL_0S\|U^2\|l_0^mL_1S)\,, \quad N = 2(2l_0 + 1)\,. \tag{2.7.13}$$

The values of $(\gamma_0\|U^2\|\gamma_1)$ for configurations p^2 and p^3 are given in Table 2.25.

5) Intercombination Transitions:
Here, the angular coefficients for excitation cross sections of intercombination transitions are given for two-electron atoms.

For transitions from the ground ns^2 state

$$\gamma_0 = n_0s^2\,{}^1S_0 \to \gamma_1 = n_0sn_1l_1\,{}^{2S_1+1}L_1\,,$$

one has:

$$Q''(\gamma_0J_0 = 0, \quad \gamma_1J_1) = \frac{1}{2}\frac{2J_1 + 1}{2L_1 + 1}\,.$$

For the L_1-averaged transitions

$$Q''(n_0s^2, l_1S_1) = \frac{2S_1 + 1}{2}\,.$$

For transitions between excited states

$$\gamma_0 = ns[0, 1/2]n_0l_0S_0 \to \gamma_1 = ns[0, 1/2]n_1l_1S_1\,,$$

one has

$$Q''(\gamma_0, \gamma_1) = \frac{2S_1 + 1}{4}\,.$$

For singlet–triplet transitions, one has

$$Q''(l_0S_0 = 0, \quad l_1S_1J_1) = \frac{2J_1 + 1}{4(2l_1 + 1)}\,.$$

2.7.2 *jl*-coupling

For transitions between fine structure components, one has

$$Q_\varkappa(J_0,J_1) = [K_0J_1]^2 \begin{Bmatrix} K_0 & J_0 & \frac{1}{2} \\ J_1 & K_1 & \varkappa \end{Bmatrix}^2 Q_\varkappa(K_0,K_1)\,,$$
$$Q''_\varkappa(J_0,J_1) = 2[l_0J_1]Q_\varkappa(K_0,K_1)\,, \tag{2.7.14}$$

$$\sum_{J_1} Q_\varkappa(J_0,J_1) = \sum_{J_0J_1} \frac{[J_0]^2}{2[K_0]^2} Q_\varkappa(J_0,J_1) = Q_\varkappa(K_0,K_1)\,. \tag{2.7.15}$$

For transitions between terms $\gamma_0 = (L_pS_pj)l_0K_0$ and $\gamma_1 = (L_pS_pj)l_1K_1$:

$$Q_\varkappa(K_0,K_1) = [l_0K_1]^2 \begin{Bmatrix} l_0 & K_0 & j \\ K_1 & l_1 & \varkappa \end{Bmatrix}^2 , \tag{2.7.16}$$

$$\sum_{K_1} Q_\varkappa(K_0,K_1) = Q_\varkappa(l_0K_0,l_1) = Q_\varkappa(l_0,l_1) = 1\,. \tag{2.7.17}$$

If the initial state γ_0J_0 is described by the *LS*-coupling scheme and the final state γ_1J_1 by the *jl*-coupling scheme, i.e., $\gamma_0 = (L_pS_p)l_0L_0S_0$, $\gamma_1 = (L_pS_pj)l_1K_1$, then the angular coefficients are

$$Q_\varkappa(J_0,J_1) = [l_0J_1S_0L_0jK_1]^2$$
$$\times \sum_v [v]^2 \left| \begin{Bmatrix} \varkappa & J_0 & J_1 \\ \frac{1}{2} & K_1 & v \end{Bmatrix} \begin{Bmatrix} L_0 & J_0 & S_0 \\ \frac{1}{2} & S_p & v \end{Bmatrix} \begin{Bmatrix} L_0 & l_0 & L_p \\ J_p & S_p & v \end{Bmatrix} \begin{Bmatrix} \varkappa & l_0 & l_1 \\ J_p & K_1 & v \end{Bmatrix} \right|^2 ,$$
$$Q''_\varkappa(J_0,J_1) = [l_0J_1S_0L_0j\tfrac{1}{2}]^2$$
$$\times \sum_v [v]^2 \left| \begin{Bmatrix} L_0 & J_0 & S_0 \\ \frac{1}{2} & S_p & v \end{Bmatrix} \begin{Bmatrix} L_0 & l_0 & L_p \\ J_p & S_p & v \end{Bmatrix} \begin{Bmatrix} \varkappa & l_0 & l_1 \\ J_p & K_1 & v \end{Bmatrix} \right|^2 . \tag{2.7.18}$$

For transitions between terms K_0 and K_1, one has

$$Q_\varkappa(K_0,K_1) = \frac{[l_0jK_1]^2}{[S_p]^2} \sum_v (2v+1) \begin{Bmatrix} L_p & l_1 & v \\ \varkappa & L_0 & l_0 \end{Bmatrix}^2 \begin{Bmatrix} L_p & l_1 & v \\ K_1 & S_p & j \end{Bmatrix}^2 , \tag{2.7.19}$$

$$\sum_{K_1} Q_\varkappa(K_0,K_1) = Q_\varkappa((L_pS_p)l_0L_0S_0,(L_pS_pj)l_1)$$
$$= \frac{2j+1}{(2L_p+1)(2S_p+1)}\,. \tag{2.7.20}$$

The factor $Q_\varkappa$ in (2.7.20) summed over j is given by

$$Q_\varkappa[(L_pS_p)l_0L_0S_0,(L_pS_p)l_1] = 1\,. \tag{2.7.21}$$

For transitions involving equivalent electrons l_0^m–$l_0^{m-1}l_1$, the right-hand parts of (2.7.16, 18–20) should be multiplied by the factor $m|G^{L_0S_0}_{L_pS_p}|^2$.

2.7.3 Arbitrary Coupling

For the excitation cross section (transitions 0–1) in an arbitrary coupling scheme, the angular factors $Q_\varkappa$ and $Q''_\varkappa$ can be also expressed in terms of angular coefficients $b(a_0, a_1)$ in the LS-coupling scheme:

$$Q_\varkappa = \frac{2l_0+1}{2J_0+1} b_\varkappa^2(a_0, a_1)\,, \quad Q''_\varkappa = \frac{2l_0+1}{2J_0+1} \sum_{qv} b^2_{\varkappa qv}\,, \tag{2.7.22}$$

$$b_\varkappa(a_0, a_1) \sum_{L_0 S_0 L_1 S_1} (a_0 | L_0 S_0 J_0) b_\varkappa(L_0 S_0 J_0, L_1 S_1 J_1)(L_1 S_1 J_1 | a_1)\,,$$

$$b_{\varkappa qv}(a_0, a_1) \sum_{L_0 S_0 L_1 S_1} (a_0 | L_0 S_0 J_0) b_{\varkappa qv}(L_0 S_0 J_0, L_1 S_1 J_1)(L_1 S_1 J_1 | a_1)\,, \tag{2.7.23}$$

where

$$b_\varkappa(L_0 S_0 J_0, L_1 S_1 J_1) = (-1)^{J_1 - S_0} [L_0 J_1] \begin{Bmatrix} L_0 & J_0 & S_0 \\ J_1 & L_1 & \varkappa \end{Bmatrix} A_0^{1/2} b_\varkappa(L_0 L_1)\,,$$

$$b_{\varkappa qv}(L_0 S_0 J_0, L_1 S_1 J_1) = (-1)^{S_p - S_1 + 1/2 + L_1} \frac{1}{\sqrt{2}} [L_0 J_1 S_0 S_1 x q] \tag{2.7.24}$$

$$\times \begin{Bmatrix} L_0 & J_0 & S_0 \\ x & L_1 & \varkappa \end{Bmatrix} \begin{Bmatrix} L_1 & J_1 & S_1 \\ q & S_0 & x \end{Bmatrix} \begin{Bmatrix} S_0 & \frac{1}{2} & S_p \\ \frac{1}{2} & S_1 & q \end{Bmatrix} b_\varkappa(L_0 L_1)\,,$$

$$b_\varkappa(L_0 L_1) = (-1)^{L_p} [L_0 L_1] \begin{Bmatrix} \varkappa & L_0 & L_1 \\ L_p & l_1 & l_0 \end{Bmatrix} G^{L_0 S_0}_{L_p S_p} Vm\,. \tag{2.7.25}$$

In the case of the "pure" coupling scheme (jj, jl etc.), the coefficients $(a|LSJ)$ are expressed in terms of $3nj$-symbols, while for the intermediate coupling scheme these coefficients are obtained numerically.

3 Radiative Characteristics

Elementary processes involving photons and/or free electrons in the initial and final states – photoionization, radiative recombination, bremsstrahlung – as well as the multiple electric polarizabilities of atoms and ions are considered in this Chapter.

3.1 Photoionization and Radiative Recombination

Photoionization is one of the basic radiative processes characterizing the interaction of radiation with atoms, ions, molecules and solid. The properties of photoionization and the inverse process – photorecombination or radiative recombination – have been considered in many reviews and monographs [3.1–11].

Photoionization processes play an important role in applied problems of atomic physics and atomic spectroscopy, solid state physics, astrophysics etc. In particular, many problems of diagnostics of laser-produced plasmas and its radiation in the VUV and x-ray region, energy transport problems and radiative cooling in laser thermonuclear targets, the use of EXAFS (EXtended Absorption Fine Structure) and XANES (X-ray Absorption Near-Edge Structure), spectroscopical methods for the investigation of solid-state structures [3.12, 13], require the knowledge of high accuracy photoionization cross sections and photorecombination rates of atoms and ions.

Experimental data and theoretical calculations of photoionization cross sections for neutral and weakly ionized atoms are given in [3.14–20].

The photoionization process consists of photon absorption and the ejection of a bound electron into the continuum

$$X_z(a_0) + \hbar\omega \rightarrow X_{z+1}(\alpha_1) + e(\varepsilon, \lambda)\,, \tag{3.1.1}$$

where a_0 and α_1 define the quantum numbers, ε and λ are the energy and the angular momentum of the photoelectron.

Cross sections $\sigma_k(\omega)$ of multiple-electron photoionization

$$X_z + \hbar\omega \rightarrow X_{z+k} + ke\,, \quad k \geqslant 1 \tag{3.1.2}$$

are measurable quantities [3.21, 22] as well as the *total* (or *photoabsorption*) cross sections [3.23]

$$\sigma^{\text{abs}}(\omega) = \sum_{k=1}^{N} \sigma_k(\omega) , \tag{3.1.3}$$

where N is the total number of target electrons. The cross section $\sigma^{\text{abs}}(\omega)$ is related to the oscillator strength by the sum rule (Sect. 2.3):

$$\sum_{n'} f_{nn'} + \frac{c}{2\pi^2} \int_I^\infty \sigma^{\text{abs}}(\omega) \, d\omega = N , \tag{3.1.4}$$

where c is the speed of light, I the first ionization potential of the target, and $\beta(\omega)$ is the dynamic polarizability (Sect. 3.3):

$$\sigma^{\text{abs}}(\omega) = \frac{4\pi\omega}{c} \operatorname{Im}\{\beta(\omega)\} . \tag{3.1.5}$$

The dynamic polarizabilitie $\beta(\omega)$ defines the van der Waals constant of two interacting atoms in their ground state [3.24]:

$$C_6 = \frac{3}{\pi} \int_0^\infty \beta_1(\mathrm{i}\omega)\beta_2(\mathrm{i}\omega) \, d\omega . \tag{3.1.6}$$

The total probability of photoionization is given by

$$W = \int_0^\infty c N_\omega \sigma(\omega) \, d\omega [\mathrm{s}^{-1}] ,$$

where N_ω is the density of photons at a given frequency ω.

The inverse process to photoionization is the photorecombination or Radiative Recombination (RR)

$$X_{z+1} + e \to X_z + \hbar\omega . \tag{3.1.7}$$

Photoionization σ_v and RR cross sections σ_r are mutually related by the detailed balance principle (the Milne formula)

$$g_{z+1}\sigma_r = \frac{(\hbar\omega)^2}{2mc^2\varepsilon} g_z \sigma_v , \tag{3.1.8}$$

where g_z and g_{z+1} are the statistical weights and ε is the energy of photoelectron.

The photoionization cross section from the state $a_0 = nl^q LS$ in the non-relativistic dipole-length approximation can be expressed as

$$\sigma_v(a_0) = \frac{Q_v}{2l_0 + 1} \frac{4\varepsilon_\omega}{3 \times 137} \sum_{\lambda = 1 \pm 1} R^2(l\lambda) [\pi a_0^2] ,$$

$$R(1\lambda) = \left(\max\{l, \lambda\} \right)^{1/2} \int_0^\infty P_{nl}(r) P_{\varepsilon\lambda}(r) r \, dr , \tag{3.1.9}$$

$$\varepsilon_0 + \varepsilon = \varepsilon_\omega = \hbar\omega/\mathrm{Ry} , \quad \varepsilon_0 > 0 , \quad \varepsilon > 0 ,$$

where ε_0 is the binding energy of the state a_0, q the number of equivalent electrons in the shell, $P(r)$ are the radial wave functions of the optical electron in the discrete and continuous spectra with normalization

$$\int_0^\infty P_{nl}^2(r)\,dr = 1\,,$$
$$\int_0^\infty P_{\varepsilon\lambda}(r)P_{\varepsilon'\lambda}(r)\,dr = \pi\delta(\varepsilon-\varepsilon')\,. \qquad (3.1.10)$$

The angular coefficients Q_v depend on the type of the transition, for example,

$$Q_v(nl^qLS - nl^{q-1}L_iS_i) = q|G_{L_iS_i}^{LS}|^2\,,$$
$$Q_v(nl^q - nl^{q-1}) = q\,, \qquad (3.1.11)$$

where G_a^b are the Racah fractional parentage coefficients [3.6].

In the case of photoionization of H-like ions from the ground state, the formulas for σ_v and σ_r can be written in closed analytical form [3.6–8]:

$$\sigma_v(1s) = \frac{2^9\pi}{3\times 137z^2}\left(\frac{\varepsilon_0}{\varepsilon_\omega}\right)^4 \frac{\exp(-4x\,\mathrm{arccot}\,x)}{1-\exp(-2\pi x)}[\pi a_0^2]\,, \qquad (3.1.12)$$

$$\sigma_r(1s) = \frac{2^8\pi}{3\times 137^3}\frac{\varepsilon_0^3}{\varepsilon_\omega^2(\varepsilon_\omega-\varepsilon_0)}\frac{\exp(-4x\,\mathrm{arccot}\,x)}{1-\exp(-2\pi x)}[\pi a_0^2]\,, \qquad (3.1.13)$$

$$x^2 = \varepsilon_0/(\varepsilon_\omega-\varepsilon_0)\,, \quad \varepsilon_0 = z^2/\mathrm{Ry}\,. \qquad (3.1.14)$$

The cross section σ_v of H-like ions has its maximum at threshold:

$$\sigma_v^{\max}(1s) = \frac{2\pi^2 a_0^2}{3\times 137z^2}\left(\frac{4}{\mathrm{e}}\right)^4 = \frac{0.0726}{z^2}[\pi a_0^2]\,, \quad \varepsilon_\omega = \varepsilon_0\,, \quad \mathrm{e} = 2.718\,, \qquad (3.1.15)$$

whereas $\sigma_r \to \infty$ at $\varepsilon \to 0$.

If $\varepsilon_\omega - \varepsilon_0 \gg \varepsilon_0$, one has from (3.1.12–14):

$$\sigma_v(1s) \approx \frac{2^8}{3\times 137z^2}\left(\frac{\varepsilon_0}{\varepsilon_\omega}\right)^{7/2}[\pi a_0^2]\,,$$
$$\sigma_r(1s) \approx \frac{2^7}{3\times 137^3}\left(\frac{\varepsilon_0}{\varepsilon_\omega}\right)^{5/2}[\pi a_0^2]\,. \qquad (3.1.16)$$

In the case of photoionization of hydrogen-like ions, photoionization cross sections can be presented in terms of special mathematical functions (see [3.25–28]).

The photoionization cross section σ_v is related to the continuum oscillator strength $df/d\varepsilon$ (Sect. 2.3):

$$\frac{df}{d\varepsilon/\mathrm{Ry}} = 13.6\frac{df}{d\varepsilon/\mathrm{eV}} = 10.9\sigma_v[\pi a_0^2]\,. \qquad (3.1.17)$$

Near the ionization limit, the quantity $df/d\varepsilon$ is also related with the usual oscillator strength f for transition $a_0 - \alpha nl$ by [3.2]

$$\left.\frac{df}{d\varepsilon/\mathrm{Ry}}\right|_{\varepsilon\to 0} = \left.\frac{n_*^3 f_{a_0-nl}}{2z^2}\right|_{n\to\infty} = \frac{137}{4\pi}\sigma_v[\pi a_0^2]\,, \qquad n_* = n - \Delta = z(E_{nl}/\mathrm{Ry})^{-1/2}\,, \tag{3.1.18}$$

where n_* and E_{nl} are the effective quantum number and the energy of the nl level counted from the ionization limit, respectively, and Δ is the quantum defect. The oscillator strength $df/d\varepsilon$ for the sharp series in Cs is shown in Fig. 3.1.

The RR cross section for reaction (3.1.7) can be written as

$$\sigma_r = \frac{2Q_r\varepsilon_\omega^2}{3\times 137^3\varepsilon}\sum_{\lambda=l\pm 1} R^2(l\lambda)[a_0^2]\,, \tag{3.1.19}$$

where the radial integral $R(l\lambda)$ is defined by (3.1.9, 10).

The angular factor Q_r is given by

$$Q_r(l^{q-1}L_iS_i \to l^qLS) = q\frac{(2S+1)(2L+1)}{2(2l+1)(2S_i+1)(2L_i+1)}\,, \qquad Q_r(l^{q-1}\to l^q) = 1 - \frac{m-1}{2(2l+1)}\,, \tag{3.1.20}$$

where g_{z+1} and g_z are the statistical weights of the ions X_{z+1} and X_z, respectively.

The RR rate $\varkappa_r$ averaged over the Maxwellian energy distribution of the incident electrons with temperature T is given by

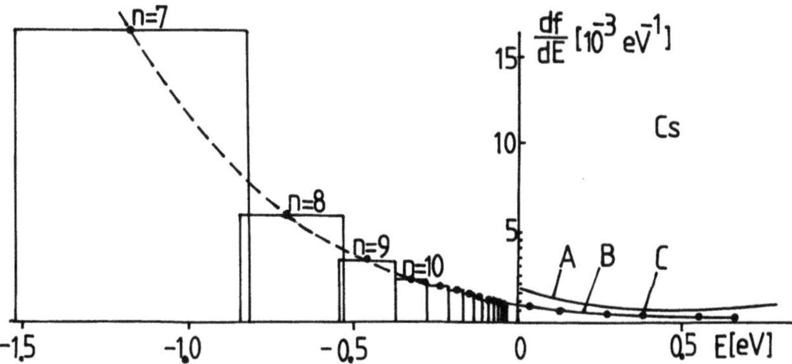

Fig. 3.1. The function df/dE[eV] (3.1.17, 18) in discrete and continuous absorption spectra of Cs (Discrete spectrum $E < 0$: experiment [3.29]; continuous spectrum $E > 0$: A experiment [3.30]; B theory [3.31]; C experiment [3.32])

$$\varkappa_r = K\varepsilon_0^{1/2}\beta^{3/2} \int_0^\infty u \frac{\sigma_r(u)}{\pi a_0^2} e^{-\beta u} du ,$$

$$u = \varepsilon/\varepsilon_0 , \quad \beta = \varepsilon_0/T , \quad K = \frac{2\sqrt{\pi}\hbar a_0}{m} = 2.18 \times 10^{-8} [\text{cm}^3 \text{s}^{-1}] \tag{3.1.21}$$

Photoionization cross sections and RR rates are scaled as

$$\sigma_v(u) , \quad u = \omega/Z , \quad z = 1 ; \tag{3.1.22}$$

$$z^2\sigma_v(u) , \quad u = \omega/z^2 , \quad z > 1 ; \tag{3.1.23}$$

$$z^{-1}\varkappa_r(\beta) , \quad \beta = T/(z^2 \text{Ry}) , \tag{3.1.24}$$

where ω is the photon frequency, z the spectroscopic symbol and Z the nuclear charge.

The contribution of different subshells in neutral atoms to the photoionization cross sections calculated in the Hartree–Fock approximation for the photon energy range up to 1500 eV [3.18] are shown in Figs. 3.2–6. Theoretical

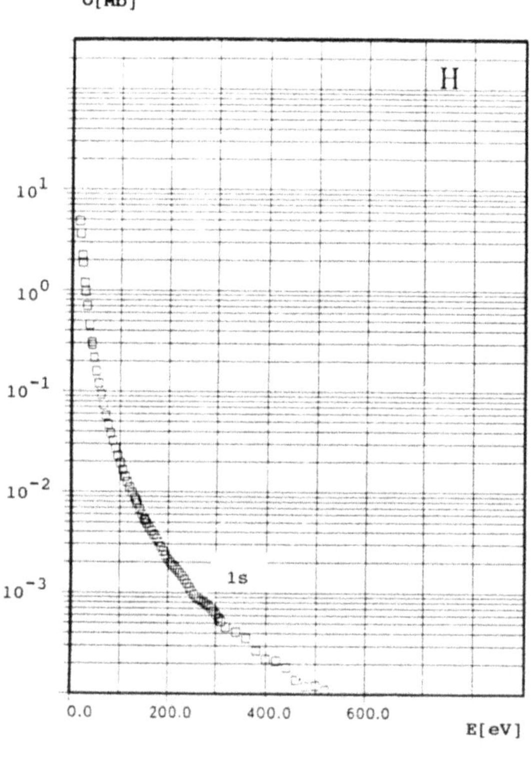

Fig. 3.2

Figs. 3.2–6. Photoionization cross sections (in Mb) of neutral atoms [H (3.2); He (3.3); Li (3.4); Be (3.5); Ar (3.6)] calculated in the Hartree–Fock approximation [3.18]; in the case of complex atoms, the contribution from different subshells is shown (1 b = 10^{-24} cm^2)

σ[Mb]
He
10^1
10^0
10^{-1}
10^{-2}
10^{-3}
1s
0.0
200.0
400.0
600.0
800.0
1000.0
1200.0
E[eV]

Fig. 3.3

σ[Mb]
Li
10^1
10^0
10^{-1}
10^{-2}
10^{-3}
1s
2s
0.0
200.0
400.0
600.0
800.0
1000.0
1200.0
E[eV]

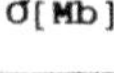

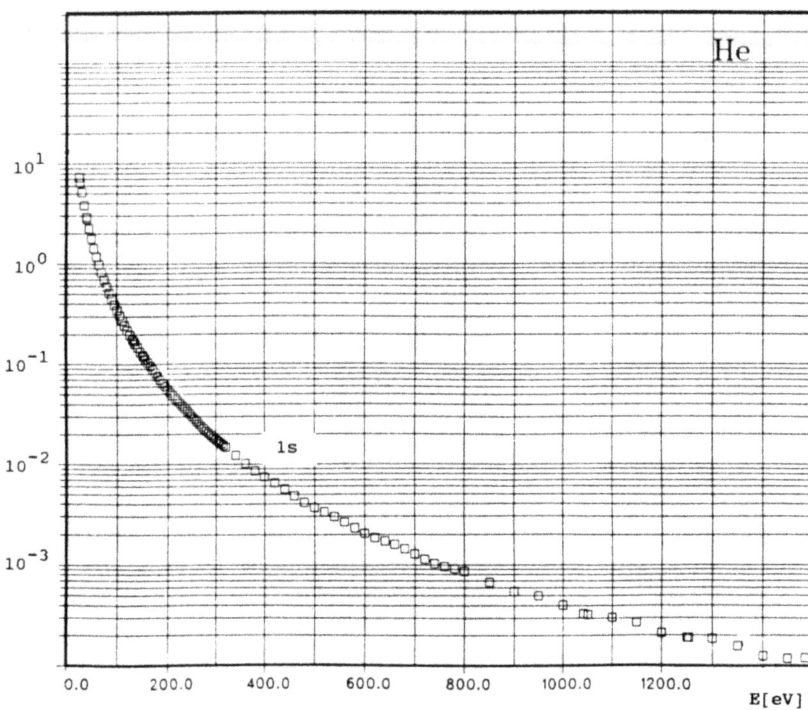

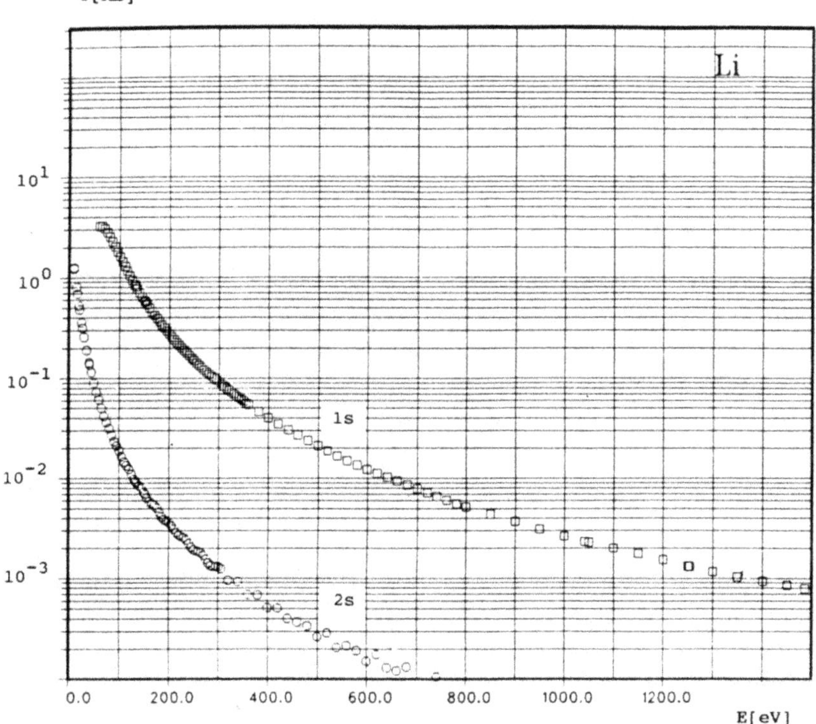

Fig. 3.4

σ[Mb]
Be
10^1
10^0
10^{-1}
10^{-2}
10^{-3}
1s
2s
0.0
200.0
400.0
600.0
800.0
1000.0
1200.0
E[eV]

Fig. 3.5

σ[Mb]
Ar
10^1
10^0
10^{-1}
10^{-2}
10^{-3}
2p
3p
2s
3s
0.0
200.0
400.0
600.0
800.0
1000.0
1200.0
E[eV]

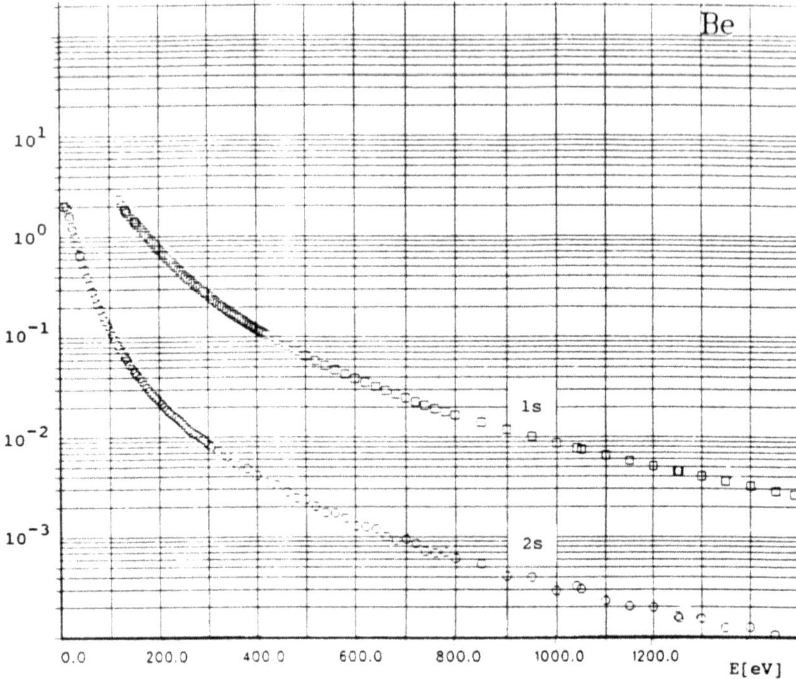

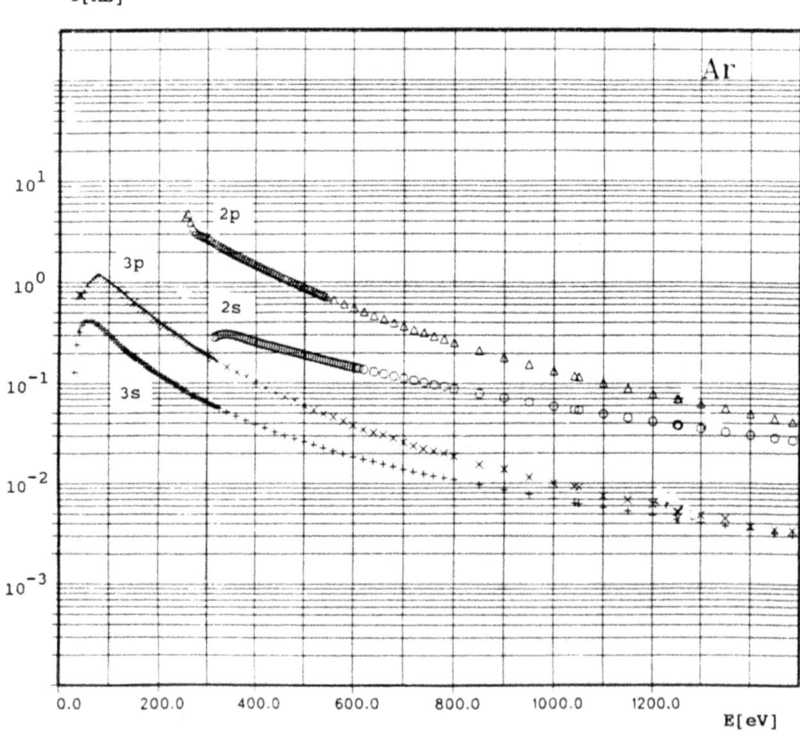

Fig. 3.6

and experimental photoabsorption cross sections taken from [3.23] are given in Figs. 3.7–13 in the photon energy range 0.1–1 keV. Photoabsorption cross section of the high-temperature semiconductor $YBa_2Cu_3O_7$ cauclated in the dipole-length approximation [3.33] is shown in Fig. 3.14.

Calculations of σ_v of complex ions [3.34–36] showed the significant importance of Photo-Excitation of the Core electrons (PEC). PEC processes lead to the appearance of large resonances, followed by autoionization, and can strongly modify the photoionization cross section over a wide range of frequencies (Fig. 3.15). PEC resonances are especially important for photoionization from excited states of ions with a low ion charge.

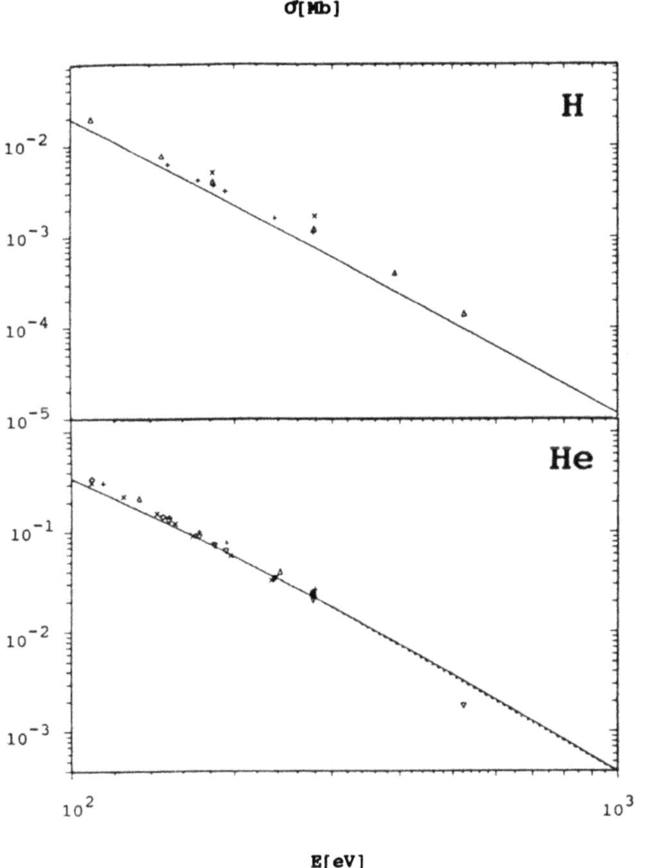

Fig. 3.7

Figs. 3.7–13. X-ray attenuation cross sections (in Mb) of H, He (3.7); Li, Be (3.8); N, O (3.9); Ne, Ar (3.10); Mn, Fe (3.11); Cs, Ba (2.12), Xe, U (3.13) atoms for the energy range 10–100 keV (Symbols: experimental data [3.23]; solid curves: calculations [3.23]; dashed curves: semiempirical values [3.17]) (1 b = 10^{-24} cm^2)

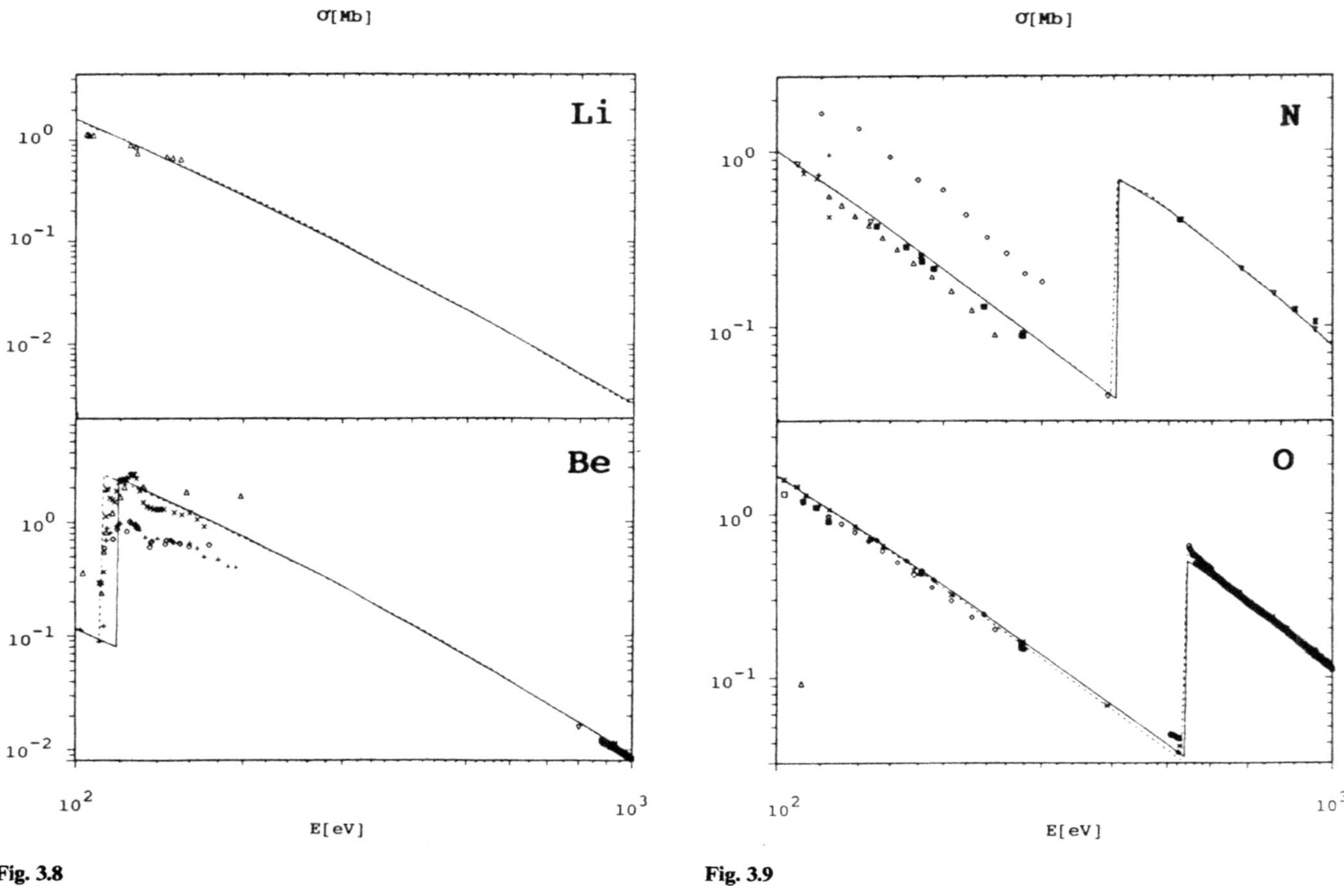

Fig. 3.8

Fig. 3.9

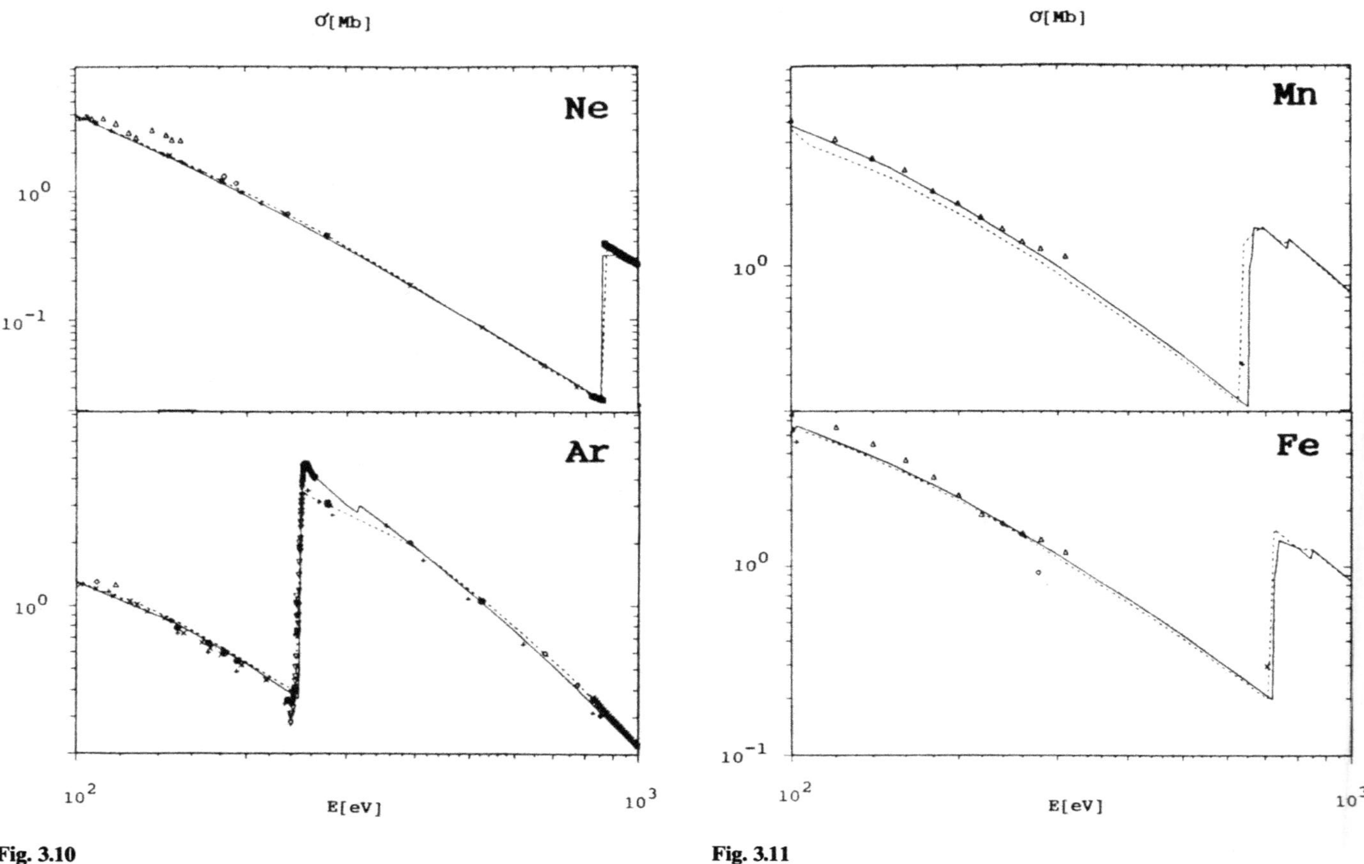

Fig. 3.10

Fig. 3.11

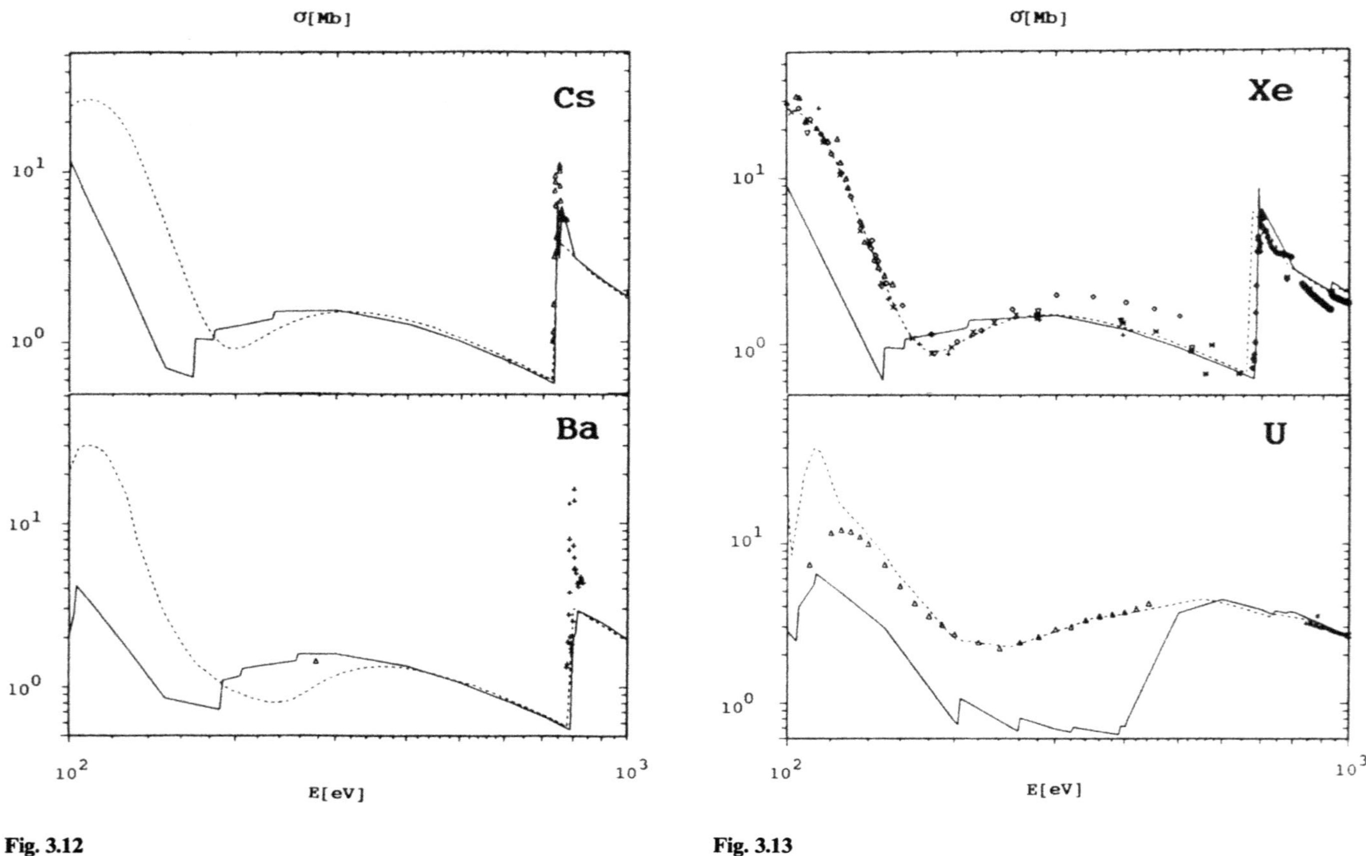

Fig. 3.12

Fig. 3.13

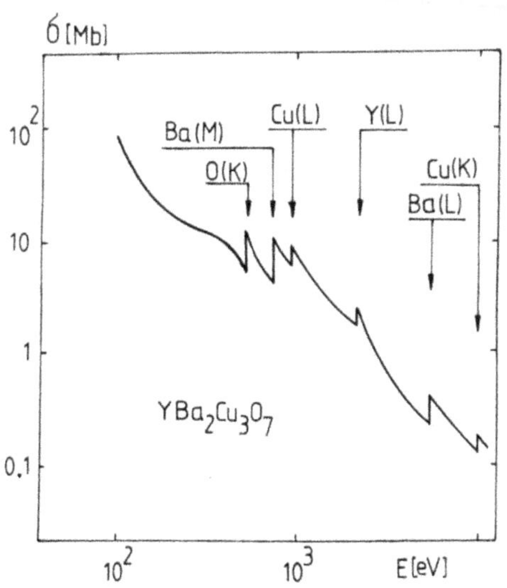

Fig. 3.14. Calculated total photoionization cross section of the superconductor $YBa_2Cu_3O_7$ [3.33]. Atoms and inner-shell edges are indicated with corresponding arrows

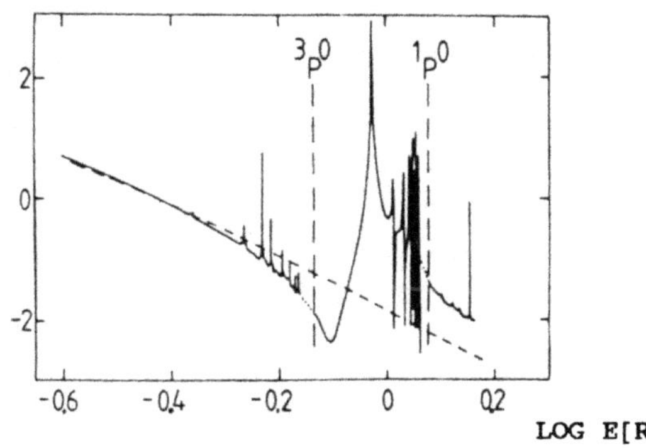

Fig. 3.15. Calculated photoionization cross sections from $4f$ states for He II (dashed curve) and C II (full curve) ions [3.34]; $^3P^0$ and $^1P^0$ are threshold positions in C III ions

3.2 The Kramers Formulas and the Gaunt Factor

The photoionization $\sigma_\nu(nl)$ and RR cross sections $\sigma_r(nl)$ for the nl state are often written in the form

$$\begin{aligned} \sigma_\nu(nl) &= \frac{Q_\nu}{2l+1} n^2 G_{nl}(\varepsilon) \sigma_\nu^{\mathrm{Kr}}(n) , \\ \sigma_r(nl) &= Q_r G_{nl}(\varepsilon) \sigma_r^{\mathrm{Kr}}(n) , \end{aligned} \tag{3.2.1}$$

where ε is the energy of the photoelectron. The dimensionless quantity G_{nl} is terms the *bound–free Gaunt factor*. The angular coefficients Q were defined in Sect 3.1. The quantities $\sigma_{\nu,r}^{\mathrm{Kr}}(n)$ are the Kramers cross sections obtained from classical considerations. They describe the l-averaged cross sections

$$\sigma_\nu(n) = n^{-2}\sum_l (2l+1)\sigma_\nu(nl)$$

and have the form

$$\sigma_\nu^{\mathrm{Kr}}(n) = \frac{64}{3\sqrt{3}\cdot 137 z^2}\left(\frac{\omega_0}{\omega}\right)^3 \frac{\pi a_0^2}{n^5}, \tag{3.2.2}$$

$$\sigma_r^{\mathrm{Kr}}(n) = \frac{32\omega_0^2}{3\sqrt{3}\cdot 137^3\omega(\omega-\omega_0/n^2)}\frac{\pi a_0^2}{n^3}, \tag{3.2.3}$$

$$\omega_0 = z^2\mathrm{Ry}/\hbar .$$

These two Kramers cross sections are related by

$$\sigma_r^{\mathrm{Kr}}(n) = \frac{2mc^2\varepsilon}{\hbar^2\omega^2 2n^2}\sigma_\nu^{\mathrm{Kr}}(n) \tag{3.2.4}$$

in accordance with (3.1.8).

The Kramers formulas for the photoionization and RR (3.2.2–3) cross sections contain the main dependence on atomic parameters: the spectroscopic symbol z, the binding energy ε_0 and the energy of the photoelectron ε, but do not describe the transitions from the different l-states.

The RR rate coefficient corresponding to (3.2.3) has the form

$$\varkappa_r^{\mathrm{Kr}}(n) = 2K_1 z\beta^{3/2}\mathrm{e}^{\beta}|\mathrm{Ei}(-\beta)| ,$$

$$\beta = \frac{z^2\mathrm{Ry}}{n^2T}, \quad K_1 = \frac{32\sqrt{\pi}a_0^2c}{3\sqrt{3}\cdot 137^4} = 2.60\times 10^{-14}\,\mathrm{cm^3\,s^{-1}} , \tag{3.2.5}$$

where T is the electron temperature and $\mathrm{Ei}(x)$ is the exponential integral. According to (3.2.5), $\varkappa_r^{\mathrm{Kr}}(n) \propto n^{-1}$ for $\beta \gg 1$ and $\varkappa_r^{\mathrm{Kr}}(n) \propto n^{-3}$ for $\beta \ll 1$.

The function $\mathrm{Ei}(x)$ is well fitted by

$$\mathrm{e}^x|\mathrm{Ei}(-x)| \approx \ln\left(1+\frac{0.562+1.4x}{x(1+1.4x)}\right), \quad x>0 . \tag{3.2.6}$$

The Kramers formulas are very useful for the eatimation of the contribution from l-averaged H-like states, especially when calculating the total RR rate

$$\varkappa_r^{\mathrm{tot}} = \sum_{n=n_0}^{\bar{n}-1}\sum_{l<n}\varkappa_r(nl) + \sum_{n=\bar{n}}^{\infty}\varkappa_r^{\mathrm{Kr}}(n) , \tag{3.2.7}$$

where $\varkappa_r(nl)$ is obtained numerically with (3.1.21). The second sum in (3.2.7) describes the contribution from all excited states n above a certain state $\bar{n}$. Using (3.2.5) and changing the summation over n by integration, one has

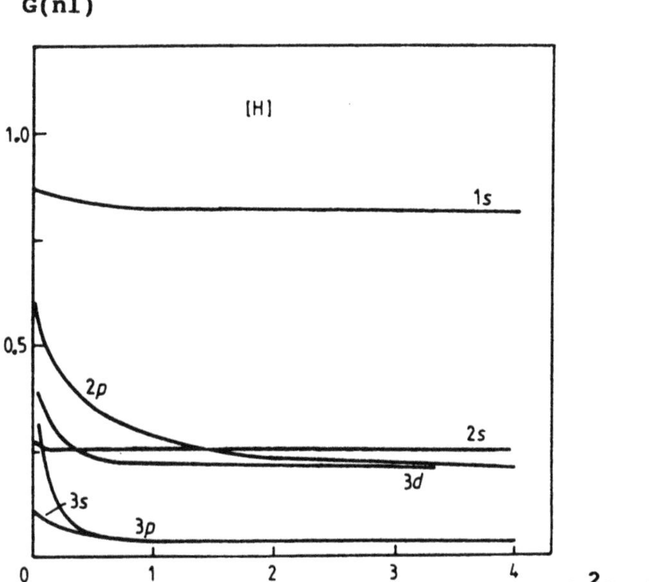

Fig. 3.16. Modified Gaunt factor $\overline{G}(nl_0)$ for the states $n = 1, 2, 3$ in H-like ions with $z = 50$ as a function of the scaled photoelectron energy (3.2.9)

$$\sum_{n=\bar{n}}^{\infty} \varkappa_r^{\mathrm{Kr}}(n) = K_1 \bar{n} \beta_1^{1/2} z[\ln(1.78\beta_1) + \mathrm{e}^{\beta_1}(1 + \beta_1/\bar{n})|\mathrm{Ei}(-\beta_1)|] ,$$
$$\beta_1 = \frac{z^2 \mathrm{Ry}}{\bar{n}^2 T} . \tag{3.2.8}$$

The Gaunt factor for transitions $nl \to \varepsilon l'$ exhibits the following asymptotic behavior

$$G_{nl}(\varepsilon) = \frac{1}{(1 + x^2)^{l+1}}$$
$$\times \exp\left[\frac{4}{\sqrt{\varepsilon}}\left(1 - \frac{\arctan x}{x}\right)\right](1 - \mathrm{e}^{-2\pi/\sqrt{\varepsilon_0 x}})^{-1} \prod_{s=1}^{l'} \frac{1 + (xs/n)^2}{1 + x^2} \overline{G}_{nl}$$
$$x = (\varepsilon/\varepsilon_0)^{1/2} , \quad x \gg 1 , \tag{3.2.9}$$

where $\overline{G}_{nl}$ is the *modified* Gaunt factor. The product $\Pi = 1$ for $l' = 0$ and $\overline{G}_{nl} = G_{nl}$ for $x = 0$. The modified $\overline{G}_{nl}$ factor is practically constant at high energies ε, which is illustrated in Fig. 3.16 for photoionization of H-like systems. In Fig. 3.17, the scaled photoionization cross sections for n states are given in comparison with the Kramers formula.

For the l-averaged Gaunt factor $G(n, \varepsilon)$ [3.37]:

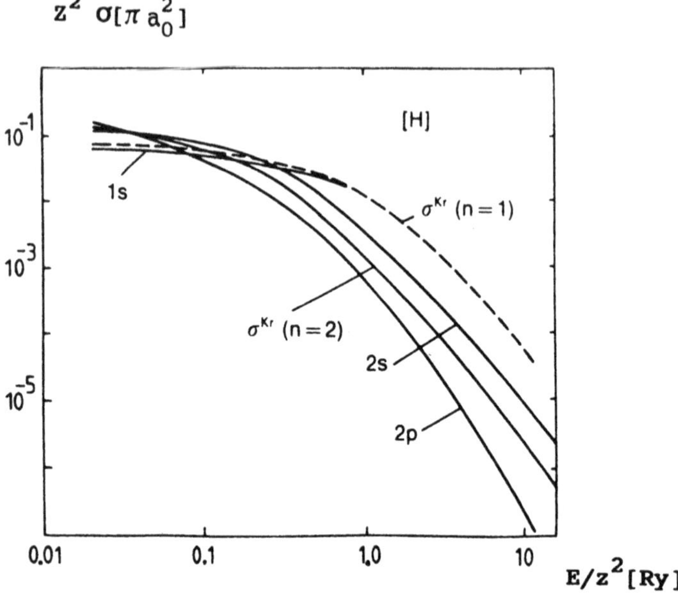

Fig. 3.17. Scaled photoionization cross sections $z^2\sigma_\nu(nl)$ for the states $n = 1, 2, 3$, and 4 in H-like ions with $z = 50$. Solid curves: (3.1.9); dashed curve: Kramers formula (3.2.2). For the states $n = 2$, 3 and 4, the averaged cross sections $\sigma(n)$ coincide with the Kramers ones: $\sigma_\nu(n) \approx \sigma_\nu^{\mathrm{Kr}}(n)$ on the same scale

$$G_n(\varepsilon) \approx [1 - \exp(-2\pi/\sqrt{\varepsilon})]^{-1}[F(n, n')]_{(n')^2 \to 1/\varepsilon'} , \qquad (3.2.10)$$

$\varepsilon \ll 1$

$$G_n(\varepsilon) \approx \frac{4\sqrt{3}}{\varepsilon^{1/2}}\left[1 - \frac{\pi}{\varepsilon^{1/2}} + \frac{1}{3\varepsilon}(\pi^2 + 10 - 5/n^2) - \frac{\pi}{3\varepsilon^{3/2}}(7 - 2/n^2)\right], \qquad (3.2.11)$$

$\varepsilon \gg 1$

where the function $F(n, n')$ is given by (2.6.10).

A comparison of (3.2.10) with the accurate calculations for the Gaunt factor $G_n(\varepsilon)$ is given in Table 3.1. For the intermediate values of ε no satisfactory formulas for $G_n(\varepsilon)$ are available.

3.3 Polarizabilities

In an external electric field with strength $\mathscr{E}$ and frequency ω, the energy E_a of an atom or an ion in the state a is given by [3.1]:

Table 3.1. The Gaunt factor $G_n(\varepsilon)$ for hydrogen [3.39]

n	ε[Ry]	[3.38]	(3.2.10)	(3.2.11)
1	0.01	0.800	0.7971	–
	1.00	0.942	0.934	–
	20	0.821	0.749	0.754
	100	0.513	–	0.506
5	1	1.09	1.09	–
10	0	9.61	9.61	–
	16	8.75	8.46	–
	25	0.790	–	0.800
	2500	0.130	–	0.130
20	0	0.976	0.796	–
	250	0.362	–	0.362
50	0.04	1.05	1.05	–

$$E_a = E_a^0 - \frac{1}{2!}\beta_a(\omega)\frac{1}{2}\mathscr{E}^2 + \frac{1}{4!}\gamma_a(\omega)\frac{3}{8}\mathscr{E}^4 + \mathscr{O}(\mathscr{E}^6)\,, \tag{3.3.1}$$

where $E_a^0 < 0$ is a field-free atomic level, $\beta_a(\omega)$ and $\gamma_a(\omega)$ are the *dynamic* dipole and *hyper polarizabilities* (*susceptibilities*), respectively. For degenerate states, there are also terms with odd powers of $\mathscr{E}$ in expansion (3.3.1). The hyper polarizability $\gamma_a(\omega)$ is expressed in units of a_0^7/e^2 and is given by the fourth-order perturbation theory.

The induced dipole moment of an atomic system is

$$d = \beta(\omega)\mathscr{E} + \tfrac{1}{6}\gamma(\omega)\mathscr{E}^3 + \cdots\,. \tag{3.3.2}$$

At zero frequency ($\omega = 0$), the corresponding values of $\beta_a(0)$ and $\gamma_a(0)$ are called the *static polarizabilities.* In the static case ($\omega \to 0$), expansion (3.3.1) reduces to the Stark shift in a constant electric field:

$$E_a = E_a^0 - \frac{1}{2!}\beta_a(0)\mathscr{E}^2 + \frac{1}{4!}\gamma_a(0)\mathscr{E}^4 + \mathscr{O}(\mathscr{E}^6)\,. \tag{3.3.3}$$

For the H(1s) atom and $\omega = 0$, one has [3.41]:

$$\begin{aligned}\mathrm{E}(1s) = &-1/2 - 9/4\mathscr{E}^2 - 3555/64\mathscr{E}^4 - 4908\mathscr{E}^6 - 794237\mathscr{E}^8\\ &- 1.945 \times 10^8\mathscr{E}^{10} - \cdots\,.\end{aligned} \tag{3.3.4}$$

Higher orders of $\mathscr{E}$ should be accounted for when $\mathscr{E} \geqslant 0.06\mathscr{E}_0$, where $\mathscr{E}_0 = 5.14 \times 10^9$ V/cm is the atomic unit of the field strength.

3.3.1 Dipole Polarizability. Basic Relations

The dipole polarizability $\beta(\omega)$ is necessary for the determination of the induced dipole moment of an atomic system, oscillator strengths of atoms and ions with one valence electron, the energy level shift (Stark effect), Van der Waals constants and characteristics describing the interaction of atomic systems with an external electric monochromatic field [3.40–42].

The dipole polarizability $\beta_a(\omega)$ is given in the second-order perturbation theory by

$$\beta_a(\omega) = \sum_{k \neq a} \int |\langle a|d_z|k\rangle|^2 [(E_k - E_a - \hbar\omega - \mathrm{i}0)^{-1} + (E_k - E_a + \hbar\omega + \mathrm{i}0)^{-1})], \tag{3.3.5}$$

where d_z is the z-component of the dipole moment, E_k are the atomic state energies and ω is the frequency of the external field. The sum-integral in (3.3.5) means the summation over discrete and integration over continuum states.

The real and imaginary parts of the dipole polarizability (3.3.5) can be expressed as

$$\mathrm{Re}\{\beta_a(\omega)\} = 2 \sum_k |\langle a|d_z|k\rangle|^2 \frac{(E_k - E_a)}{(E_k - E_a)^2 - (\hbar\omega)^2}, \tag{3.3.6}$$

$$\mathrm{Im}\{\beta_a(\omega)\} = \pi \sum_k |\langle a|d_z|k\rangle|^2 \delta(E_k - E_a - \hbar\omega). \tag{3.3.7}$$

Equation (3.3.6) for $\mathrm{Re}\{\beta_a(\omega)\}$ can be formulated in terms of the oscillator strengths f_{ak} for discrete states and the continuum as

$$\mathrm{Re}\{\beta_a(\omega)\} = \sum_{k \neq a} \int \frac{f_{ak}}{(E_k - E_a)^2 - (\hbar\omega)^2}. \tag{3.3.8}$$

The theoretical methods of calculation of $\beta(\omega)$ in the ground state are described in [3.43].

The dipole polarizability is related to many physical quantities. For the imaginary part, one has the dipole sum rule

$$\int_0^\infty \omega \, \mathrm{Im}\{\beta(\omega)\} \, \mathrm{d}\omega = \frac{\pi}{2} \frac{Ne^2}{m}, \tag{3.3.9}$$

where N is the total number of atomic electrons.

The photoionization cross section $\sigma^{\mathrm{ph}}(\omega)$ is related to $\mathrm{Im}\{\beta(\omega)\}$ by [3.2]

$$\sigma^{\mathrm{ph}}(\omega) = \frac{4\pi\omega}{c} \mathrm{Im}\{\beta(\omega)\}, \quad \omega \geqslant I, \tag{3.3.10}$$

where I is the binding energy of the atom and c is the speed of light. The elastic photon scattering cross section can be written as

$$\sigma^{\mathrm{el}}(\omega) = \frac{8\pi}{3} (\omega/c)^4 |\beta(\omega)|^2. \tag{3.3.11}$$

The energy Q absorbed by an atomic system per unit of time is given by

$$Q = \frac{\omega}{2} \mathrm{Im}\{\beta(\omega)\} \mathscr{E}^2 . \tag{3.3.12}$$

The dynamic polarizabilities also define the Van der Waals constant of two interacting atoms in their ground states [3.24]:

$$C_6 = \frac{3}{\pi} \int_0^\infty \beta_1(\mathrm{i}\omega)\beta_2(\mathrm{i}\omega)\,\mathrm{d}\omega \tag{3.3.13}$$

The dielectric susceptibility $\varepsilon(\omega)$ of matter is related to $\beta(\omega)$ by

$$\varepsilon(\omega) = 1 + 4\pi N_0 \beta(\omega) , \tag{3.3.14}$$

where N_0 is the particle density.

At very high frequencies ($\omega \to \infty$) one has for the real part

$$\mathrm{Re}\{\beta(\omega)\} = N\beta_\mathrm{e}(\omega) , \quad \beta_\mathrm{e}(\omega) = -\frac{e^2}{m\omega^2} , \tag{3.3.15}$$

where $\beta_\mathrm{e}(\omega)$ is the polarizability of a free electron.

For H-like ions with the spectroscopic symbol z, the dipole polarizability of the nl state scales as

$$\beta_{nl}^z(\omega) = z^{-4}\beta^\mathrm{H}(\omega/z^2) , \tag{3.3.16}$$

where $\beta^\mathrm{H}(\omega)$ is the polarizability of the hydrogen atom. For the ground $1s$-state one has [3.44]:

$$\beta_{1s}^\mathrm{H}(\omega) = f(\omega) + f(-\omega) ,$$

$$f(\pm\omega) = \frac{2^{11}x^3}{(x+1)^{10}} \left\{ (3-1/x)^{-1}(2-1/x)^{-1} {}_2F_1\left[5, 2-1/x; 4-1/x; \left(\frac{x-1}{x+1}\right)^2\right] \right.$$

$$\left. + \frac{5x^2}{(x+1)^2(3-1/x)} {}_2F_1\left[6, 3-1/x; 4-1/x; \left(\frac{x-1}{x+1}\right)^2\right] \right\} ,$$

$$x = (1 \pm \hbar\omega/\mathrm{Ry})^{1/2} , \tag{3.3.17}$$

where ${}_2F_1(x)$ is the Gauss hypergeometric function. In the static limit $\omega = 0$, $x = 1$, ${}_2F_1 = 1$ and

$$\beta_{1s}^\mathrm{H}(0) = \frac{9a_0^3}{2} . \tag{3.3.18}$$

In the frequency region above the first resonance:

$$\beta_{1s}^\mathrm{H}(\omega) = -4.9107 \cot\left(\frac{\pi}{\sqrt{1 - \hbar\omega/\mathrm{Ry}}}\right) - 4.3105 , \quad \hbar\omega > \frac{3\mathrm{Ry}}{4} . \tag{3.3.19}$$

At very high frequencies:

$$\beta_{nl}^z(\omega) \approx -\frac{1}{\omega^2} + \frac{4z^4}{3n^3\omega^4} + \cdots . \tag{3.3.20}$$

At high frequencies, the dynamic dipole polarizability of H-like ions in the nl state behaves as [3.45]

$$\beta_{ns}(\omega) = -\frac{4a_0^3}{z^4}\left(\Omega^{-2} + \frac{16}{3}n^{-3}\Omega^{-4} - \frac{32}{3}n^{-2}\Omega^{-4.5}\right),$$

$$\beta_{np}(\omega) = -\frac{4a_0^3}{z^4}\left(\Omega^{-2} + \frac{32(n^2-1)}{9n^5}\Omega^{-5.5}\right), \tag{3.3.21}$$

$$\Omega = \hbar\omega/(z^2\mathrm{Ry}), \quad \omega \to \infty .$$

The behavior of the dynamic dipole polarizability of atoms in the Rydberg states is described in [3.46].

3.3.2 Static Dipole Polarizabilities of Atoms and Ions

The static dipole polarizabilities of atoms and ions in the ground state are given in Table 3.2. Experimental values, the quantum mechanical and, statistical dipole polarizabilities are taken from [3.47–50] and recent publications.

The non-relativistic polarizability of H-like atoms with nuclear charge Z in the nl state reads [3.49]:

$$\beta_{nl}^z(0) = Z^{-4}[n^6 + 7/4n^4(l^2 + l + 2)][a_0^3], \tag{3.3.22}$$

which, for the 1s state, yields

$$\beta_{1s}^z(0) = \frac{9}{2Z^4}[a_0^3]. \tag{3.3.23}$$

The inclusion of relativistic corrections in (3.3.23) gives [3.52]

$$\beta_{1s}^z(0) = \frac{9}{2Z^4}[1 - 28(\alpha Z)^2/27][a_0^3],$$

where α is the fine-structure constant.

With increasing Z, the contribution of relativistic corrections to $\beta_{1s}^z(0)$ is quite important and reaches about 50% at $Z = 100$.

3.3.3 Multipole Static Polarizabilities. Boundary Radii

It is also possible to introduce the *multipole polarizability* $\beta_\varkappa$ which describes $2^\varkappa$-pole moments of atoms and ions induced by an external electric field, the dipole polarizability $\beta_1 (\varkappa = 1)$, the quadrupole one $\beta_2 (\varkappa = 2)$, etc. The multipole polarizability $\beta_\varkappa$ is expressed in units of $a_0^{2\varkappa+1}$ and is given by the second-order perturbation theory.

Table 3.2. Static dipole polarizabilities $\beta_1(0)$ (in a_0^3) of atoms and ions in their ground state

Nuclear charge	Atom, ion	Experiment	Theory	
			Quantum	Statistical
1	H	–	4.5	
	H^-	206	215.5	
2	He	1.384	1.383	
3	Li	160–169	164	
	Li^+	0.20	0.189–0.22	0.12
	Li^-	–	750	
4	Be	37.8	37.8	
	Be^+	–	23.9	
	Be^{2+}	0.27	0.154	0.17
5	B	20.4	20.4	
	B^+	–	11.38	
	B^{2+}	–	7.67	
	B^{3+}	0.135	0.02	0.06
6	C	7.0; 11.88	11.9	
	C^{2+}	3.9	4.51; 7.9	
	C^{3+}	–	3.39	
	C^{4+}	0.081	0.009	
7	N	7.62	7.43	
	N^{3+}	–	2.237–2.4	
	N^{4+}	–	1.78	
	N^{5+}	–	0.0048	
8	O	5.0–5.41	5.41	
	O^{4+}	–	1.27	
	O^{5+}	–	1.05	
	O^{6+}	–	0.0026	0.0087
	O^{2-}	22		
9	F	3.76	3.76	
	F^{5+}	–	0.79	
	F^{6+}	–	0.668	
	F^{7+}	–	0.0016	0.0054
	F^-	6.6	8.35	
10	Ne	2.7	2.67	8.2–13.3
	Ne^{6+}	–	0.525	0.39
	Ne^{7+}	–	0.450	
	Ne^{8+}	–	0.001	0.0032
11	Na	158–165	159	
	Na^+	0.998; 1.5–1.75	0.934–1.72	3.73
	Na^{8+}	–	0.317	
12	Mg	71.55	71.6	
	Mg^+	34.62	33.9	
	Mg^{2+}	0.81	–	1.84–2.7
	Mg^{7+}	0.29–0.47	0.47	
	Mg^{9+}	–	0.231	
13	Al	45.9	56.3	
	Al^+	24.3–28	26.6	
	Al^{2+}	–	13.8	
	Al^{3+}	0.45	0.26	1.07
	Al^{9+}	–	0.198	
	Al^{10+}	–	0.173	

Table 3.2 (*Cont.*)

Nuclear charge	Atom, ion	Experiment	Theory	
			Quantum	Statistical
14	Si	42	36.3	
	Si^{2+}	11.2–14.0	12.6	
	Si^{3+}	6.86	7.22	
	Si^{4+}	0.15–0.27	0.16	0.63
	Si^{11+}	–	0.132	
15	P	30	24.5	
	P^{3+}	–	7.16	
	P^{4+}	4.4	4.3	
	P^{5+}	0.0948	–	0.39
	P^{12+}	–	0.103	
16	S	23	19.6	
	S^{4+}	–	4.52	
	S^{5+}	–	2.79	
	S^{6+}	0.0631	–	
	S^{13+}	–	0.0813	
	S^{2-}	64	–	
17	Cl	15	14.7	15.4
	Cl^{5+}	–	3.06	
	Cl^{6+}	–	1.92	
	Cl^{7+}	0.0438	–	0.20
	Cl^{14+}	–	0.0655	
	Cl^{-}	24	21.2	
18	Ar	11.08	19.4	14.0
	Ar^{+}	7.2	7.4	7.0
	Ar^{2+}	7	5.3	3.78
	Ar^{3+}	–	3.9	1.63
	Ar^{6+}	–	2.16	
	Ar^{7+}	–	1.38	
	Ar^{8+}	0.0313	–	0.14
	Ar^{14+}	–	0.0639	
	Ar^{15+}	–	0.054	
19	K	293	290	
	K^{+}	5.4–8.1	4.96; 8.1–9.26	6.62
	K^{8+}	0.55	1.02	
	K^{9+}	0.023	–	
	K^{16+}	–	0.043	
20	Ca	169	154	
	Ca^{+}	70.89		
	Ca^{2+}	3.3	3.2	3.5–4.86
	Ca^{9+}	0.47	0.782	
	Ca^{10+}	0.0174	–	0.0246
	Ca^{17+}	–	0.0358	
21	Sc	160–180	120	
	Sc^{3+}	2.16	–	2.21
	Sc^{10+}	0.40	0.605	
	Sc^{11+}	0.0134	–	0.0202
	Sc^{18+}	–	0.0300	

Table 3.2 (*Cont.*)

Nuclear charge	Atom, ion	Experiment	Theory: Quantum	Theory: Statistical
22	Ti	–	98.6	
	Ti^{4+}	1.35	–	1.17
	Ti^{10+}	0.31	0.478	
	Ti^{12+}	0.0104	–	0.0168
	Ti^{18+}	–	0.0325	
	Ti^{19+}	–	0.0249	
23	V	–	83.7	
	V^{5+}	–	–	0.83
	V^{12+}	–	0.384	
	V^{13+}	0.00828	–	0.0141
	V^{20+}	–	0.0210	
24	Cr	–	78.3	
	Cr^{+}	–	9.045	7.7
	Cr^{6+}	–	–	0.567
	Cr^{13+}	–	0.312	
	Cr^{14+}	0.00665	–	0.0119
25	Mn	–	63.4	
	Mn^{2+}	–	3.06	4.84
	Mn^{7+}	0.29	–	0.51
	Mn^{14+}	–	0.257	
	Mn^{15+}	0.00541	–	0.0101
	Mn^{22+}	–	0.015	
26	Fe	–	56.7	
	Fe^{2+}	–	–	4.64
	Fe^{3+}	1.67	0.792	
	Fe^{8+}	0.37	–	0.41
	Fe^{14+}	–	0.352	
	Fe^{15+}	–	0.215	
	Fe^{16+}	0.00444	–	0.00862
	Fe^{22+}	–	0.0188	
	Fe^{23+}	–	0.0133	
27	Co	–	50.6	
	Co^{+}	–	–	5.7
	Co^{3+}	–	–	3.7
	Co^{4+}	–	0.847	
	Co^{9+}	0.26	–	0.30
	Co^{16+}	–	0.181	
	Co^{17+}	0.00240–0.0369	–	0.00741
28	Ni	142	45.9	
	Ni^{+}	–	–	8.97
	Ni^{2+}	–	–	5.2
	Ni^{5+}	–	0.544	
	Ni^{10+}	–	–	0.28
	Ni^{17+}	–	0.153	
	Ni^{18+}	0.0031–0.0040	–	0.006
29	Cu	150	41.2; 49.3	
	Cu^{+}	10.8	5.06–6.55	8.7
	Cu^{2+}	1.35	–	5.06

Table 3.2 (*Cont.*)

Nuclear charge	Atom, ion	Experiment	Theory	
			Quantum	Statistical
	Cu^{18+}	–	0.130	
	Cu^{19+}	0.0026	–	0.0056
30	Zn	47.8	37.8; 47.9	
	Zn^{2+}	5.4	2.3	4.9
	Zn^{19+}	–	0.112	
	Zn^{20+}	0.00221	–	0.0041
31	Ga	66.0	54.8	
	Ga^{+}	18.1–22.5	18	
	Ga^{3+}	–	1.35	3.22
	Ga^{11+}	0.00189	–	0.0042
	Ga^{20+}	–	0.095	
32	Ge	40.97	41.0	
	Ge^{2+}	–	10–12.7	
	Ge^{4+}	–	0.68	2.64
	Ge^{21+}	–	0.082	
	Ge^{22+}	0.00162	–	0.037
33	As	27	29.1	
	As^{3+}	–	8.24	
	As^{5+}	–	–	1.64
	As^{22+}	–	0.072	
	As^{23+}	0.00140	–	0.0033
34	Se	30.0	25.4	26.4
	Se^{4+}	–	5.77	
	Se^{5+}	–	–	1.28
	Se^{23+}	–	0.062	
	Se^{24+}	0.00122	–	0.0029
35	Br	23	20.6	28.8
	Br^{+}	–	–	9.75
	Br^{5+}	–	4.26	
	Br^{7+}	–	–	0.89
	Br^{25+}	0.00106	–	0.026
36	Kr	16.77	16.5–24.8	21.1–27.4
	Kr^{+}	–	–	9.25
	Kr^{6+}	–	3.26	
	Kr^{8+}	–	0.203	0.76
	Kr^{26+}	0.000931	–	0.00234
37	Rb	319	–	
	Rb^{+}	11.5–12.4	12.1–13.6	11.88
	Rb^{27+}	0.000823	–	0.00210
38	Sr	186	230	
	Sr^{2+}	6.6	8.76	7.51
	Sr^{28+}	0.000728	–	0.00188
39	Y	–	153	
	Y^{3+}	3.71	–	3.84
	Y^{29+}	0.000647	–	0.00170
40	Zr	–	121.0	
	Zr^{4+}	2.50	–	2.41
	Zr^{30+}	0.000576	–	0.00154

Table 3.2 (*Cont.*)

Nuclear charge	Atom, ion	Experiment	Theory: Quantum	Theory: Statistical
41	Nb	–	106	
	Nb^{5+}	–	–	1.98
	Nb^{31+}	0.000515	–	0.00139
42	Mo	61	86.4	
	Mo^{+}	–	–	7.1
	Mo^{12+}	–	1.007	
	Mo^{14+}	–	0.100	
	Mo^{30+}	–	0.0518	
	Mo^{32+}	0.000462		0.00127
43	Tc	–	77.0	
44	Ru	–	64.8	
	Ru^{+}	–	67.0	
45	Rh	–	58.0	
46	Pd	46.6	32.4	
47	Ag	–	48.6; 57.8	
	Ag^{+}	–	–	18.1
48	Cd	49.65	48.6	
	Cd^{2+}	–	12.2	6.25
49	In	68.8	61.4	
	In^{3+}	–	–	4.74
50	Sn	–	52.0	
	Sn^{4+}	–	2.29	3.27
51	Sb	35	44.6	
	Sb^{5+}	–	–	2.20
52	Te	–	37.1	
	Te^{6+}	–	–	1.98
	Te^{2-}	100		
53	I	36.1	31.7	29.4
	I^{+}	–	–	16.9
	I^{7+}	–	–	1.55
	I^{-}	50	–	
54	Xe	27.30	27.4–31.2	
	Xe^{+}	–	–	20
	Xe^{8+}	–	–	1.48
55	Cs	402	310	
	Cs^{+}	16.3	17.9–21.2	15.3
56	Ba	268	330	
	Ba^{2+}	11.4	11.5	9.66
57	La		210	
58	Ce		200	
59	Pr		190	
60	Nd		212	
61	Pm		203	
62	Sm		194	
63	Eu	182.2	187	
64	Gd		159	
65	Tb		172	
66	Dy		165	

Table 3.2 (*Cont.*)

Nuclear charge	Atom, ion	Experiment	Theory	
			Quantum	Statistical
67	Ho		159	
68	Er		153	
69	Tm		147	
70	Yb	148.5	142	
	Yb^{2+}	–	–	9.73
71	Lu		148	
72	Hf		109	
73	Ta		88.4	
74	W	–	74.9	
75	Re		65.5	
76	Os		57.4	
77	Ir		51.3	
78	Pt		43.9	
	Pt^{+}	–	–	17.4
79	Au	38.5	39.1; 43.74	
	Au^{+}	–	–	11.7
	Au^{12+}	1.8	–	1.44
	Au^{13+}	0.78–0.9	–	1.1
	Au^{14+}	0.69	–	0.72
	Au^{15+}	0.65	–	0.62
80	Hg	34.4	38.5	
	Hg^{2+}	–	18.7	18.3
81	Tl	51.3	50.6	
	Tl^{3+}	–	–	6.32
82	Pb	–	45.9	
	Pb^{4+}	–	–	4.64
83	Bi	–	50.0	
	Bi^{5+}	–	–	3.91
84	Po		45.9	
85	At		40.5	
86	Rn	42.5	35.8	
87	Fr	452	329	
88	Ra	310.4	258	
89	Ac		217	
90	Th		217	
91	Pa		171	
92	U	137	185	
93	Np		167	
94	Pu		165	
95	Am		157	
96	Cm		155	
97	Bk		153	
98	Cf		138	
99	Es		133	
100	Fm		161	
101	Md		123	
102	No		118	

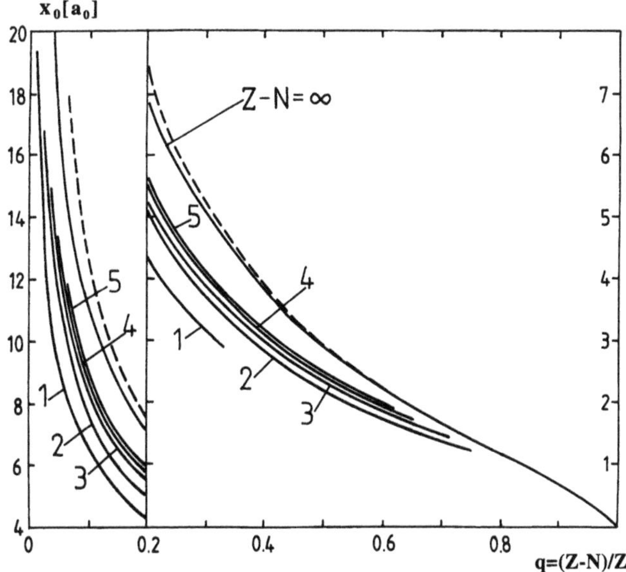

Fig. 3.18. Reduced radius x_0 vs degree of ionization $q = (Z - N)/Z$ for different values of $Z - N$ in the TFD model [3.52]; the curve $Z - N = \infty$ corresponds to the pure TF model (neglecting exchange)

Static multipole polarizabilities $\beta_\varkappa(0)$ are important atomic constants required for obtaining the refraction index, shift and broadening of spectral lines of atoms and ions in an external electric field, the identification of transitions between excited states in atoms and many other characteristics. The accurate values for $\beta_1(0)$ and $\beta_2(0)$ calculated in the relativistic approximations are given in [3.48, 51].

A semiempirical method for obtaining static multipole polarizabilities $\beta_\varkappa(0)$ based on the Thomas–Fermi (TF) model for the atomic electron density is described in [3.47, 52]. The values of $\beta_\varkappa(0)$ can be estimated from Figs. 3.18, 19, where the dependence of $\beta_\varkappa(0)$ on the boundary atomic radius r_0 is shown for $\varkappa = 1, 2$ and 3. Knowing r_0, one can obtain the values of $\beta_\varkappa$. For neutral atoms, the reasonable dependence of r_0 on the nuclear charge Z was obtained in [3.53] using the Thomas–Fermi–Dirac (TFD) model with account for exchange and correlation corrections. This dependence is can be presented in the form

$$x_0 = 4Z^{0.4}[a_0]\,, \quad Z = N\,, \quad x_0 = Z^{1/3}r_0/0.8853\,, \qquad (3.3.24)$$

where Z is the nuclear charge, N the total number of atomic electrons and x_0 is the reduced boundary radius.

The reduced radius x_0 of ions can be found from Fig. 3.18, where the dependence of x_0 on the ionization degree $q = (Z - N)/Z$ calculated in the TF model is shown.

Quadrupole polarizabilities of neutral atoms are listed in Table 3.3.

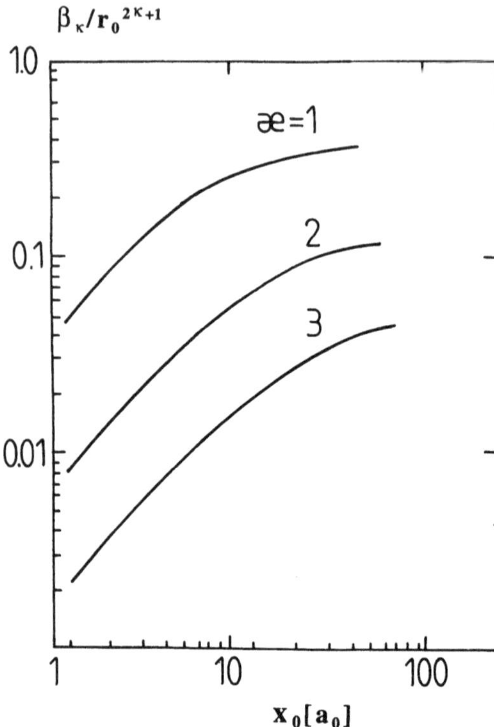

Fig. 3.19. $\beta_\varkappa/r_0^{2\varkappa+1}$ vs reduced radius x_0 for $\varkappa = 1$, 2 and 3 in the TF model [3.52]

3.4 Bremsstrahlung

BremsStrahlung (BS) (or *decelaration radiation*) is the process in which one photon is radiated in the scattering of an electron from an atom or ion whose internal state remains unchanged

$$X^{z+} + e(\mathbf{p}_0) \rightarrow X^{z+} + e(\mathbf{p}_1) + \hbar\omega\,, \tag{3.4.1}$$

where $\mathbf{p}_{0,1}$ are the electron momenta before and after collision.

In the non-relativistic approximation, the photon and electron energies are related by

$$\hbar\omega = \frac{1}{2m}(\mathbf{p}_0^2 - \mathbf{p}_1^2) = \hbar kc\,, \tag{3.4.2}$$

where k is the momentum of radiated photon.

BS is closely related to elastic electron scattering, radiative recombination (Sect. 3.1) and pair production. BS is also a fundamental process for a variety of applications in astrophysics, biology, fusion-plasma problems, transport and energy loss in charged particle beams, etc.

In general, BS is a quite complicated theoretical problem because it requires bound conditions and the use of the continuum Coulomb wave functions. The

Table 3.3. Static quadrupole β_2 and octupole β_3 polarizabilies of neutral atoms [3.54]

Atom	$\beta_2[a_0^5]$	$\beta_3[a_0^7]$
He	2.443(0)	1.060(1)
Ne	6.422(0)	3.427(1)
Ar	5.021(1)	5.313(2)
Kr	9.555(1)	1.260(3)
Xe	2.126(2)	3.602(3)
H	1.500(1)	1.313(2)
Li	1.383(3)	3.680(4)
Na	1.799(3)	5.117(4)
K	4.597(3)	1.502(5)
Rb	5.979(3)	2.127(5)
Ca	9.478(3)	3.399(5)
Be	3.026(2)	4.126(3)
Mg	8.280(2)	1.474(4)
Ca	2.717(3)	6.151(4)

* 2.443(0) means 2.443×10^0

problem is solved only in the non-relativistic limit and for some special cases, where numerical calculations were performed. Experiments are also limited in number, especially for positive ions. The general theory of BS is described in [3.55–63]. The BS spectra for electron energies from 100 eV to 10 MeV are tabulated in [3.64], the angular distribution in [3.65].

3.4.1 Basic Formulas

There are several analytical expressions for BS cross sections in the pure Coloumb case.

The BS cross section in the frequency interval ω, $\omega + \mathrm{d}\omega$ is usually expressed as

$$\mathrm{d}\sigma = g(\eta_0, \eta_1)\,\mathrm{d}\sigma^{\mathrm{Kr}}\,, \tag{3.4.3}$$

$$\mathrm{d}\sigma^{\mathrm{Kr}} = \frac{16}{3\sqrt{3}}\alpha^3\eta_0^2\frac{\mathrm{d}\omega}{\omega}[\pi a_0^2]\,, \tag{3.4.4}$$

$$\eta_0 = \frac{Ze^2}{\hbar v_0}\,, \quad \eta_1 = \frac{Ze^2}{\hbar v_1}\,, \tag{3.4.5}$$

where $\mathrm{d}\sigma^{\mathrm{Kr}}$ is the Kramers BS cross section [3.66], g the Gaunt factor and Z the nuclear charge. So, in general, the BS cross section is a function of the two variables, η_0 and η_1.

Nonrelativistic quantum mechanics in dipole approximation gives the Sommerfeld formula [3.57]:

$$g(\eta_0, \eta_1) = \frac{\pi\sqrt{3}}{(e^{2\pi\eta_0} - 1)(1 - e^{-2\pi\eta_1})} x_0 \frac{d}{dx_0} |{}_1F_2(i\eta_0, i\eta_1, 1; x_0)|^2 \tag{3.4.6}$$

$$x_0 = -4\eta_0\eta_1/(\eta_0 - \eta_1)^2 , \tag{3.4.7}$$

where F is the hypergeometric function.

In the quasiclassical limit, $\eta_1 > \eta_0 \gg 1$, (3.4.6) yields [3.58, 66]:

$$g = \frac{\pi\sqrt{3}}{4} i\nu H_{i\nu}^{(1)}(i\nu) H_{i\nu}^{(1)\prime}(i\nu) , \quad \nu = \frac{Ze^2}{mv_0^3}\omega , \tag{3.4.8}$$

where $H^{(1)}$ is the Hankel function of complex argument and index, $H_{i\nu}^{(1)\prime}$ is its derivative with respect to the argument.

The function (3.4.8) is monotonic; its limiting expressions are

$$g = \begin{cases} 1 , & \nu \gg 1 \\ \frac{\sqrt{3}}{\pi} \ln\left(\frac{2}{\gamma_0 \nu}\right) , & \nu \ll 1 , \end{cases} \tag{3.4.9}$$

where $\gamma_0 = 1.781$ is the Euler constant.

The BS spectrum for a point Coulomb potential has a logarithmic divergence in the soft-photon region ($\nu \ll 1$) and it becomes flat for harder photons ($\nu \gg 1$) (Fig. 3.20).

In the case of large initial electron velocity ($\eta_0 \ll 1$) one derives from (3.4.6) the Born–Elwert approximation

$$g = \frac{\sqrt{3}}{\pi} f_E \ln \frac{\eta_1 + \eta_0}{\eta_1 - \eta_0} , \quad \eta_0 \ll 1 , \tag{3.4.10}$$

where f_E is the Elwert factor

$$f_E = \frac{\eta_1}{\eta_0} \frac{1 - e^{-2\pi\eta_0}}{1 - e^{-2\pi\eta_1}} . \tag{3.4.11}$$

If the final electron energy is also large, $2\pi\eta_1 \ll 1$, (3.4.10) gives the nonrelativistic Born approximation:

$$g = \frac{\sqrt{3}}{\pi} \ln \frac{\eta_1 + \eta_0}{\eta_1 - \eta_0} , \quad f_E = 1 , \quad 2\pi\eta_0 < 2\pi\eta_1 \ll 1 . \tag{3.4.12}$$

The general approximate formula for the Gaunt factor g can be obtained with the semiclassical method [3.60]:

$$g = \frac{\pi\sqrt{3}}{4} i\nu \left(1 + \frac{1}{\gamma\eta_0}\right) H_{i\nu}^{(1)}\left[i\nu\left(1 + \frac{1}{\gamma\eta_0}\right)\right] H_{i\nu}^{(1)\prime}\left[i\nu\left(1 + \frac{1}{\gamma\eta_0}\right)\right] , \tag{3.4.13}$$

which gives both limiting cases (the Born approximation and the classical limit). For $\eta_0/\nu \gg 1$, (3.4.13) becomes (3.4.8). For the limiting case $\omega \to 0$ ($\nu \ll 1$) and arbitrary η_0 one has from (3.4.13)

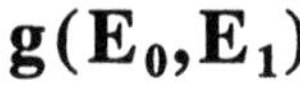

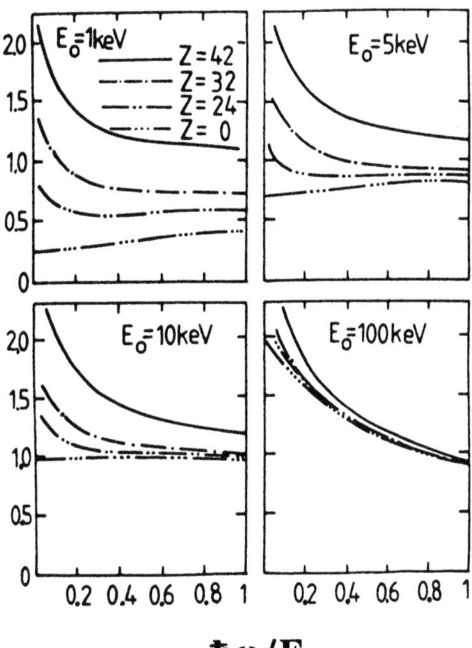

Fig. 3.20. The calculated Gaunt factor g for BS from Mo^{z+} ions as a function of the incident electron energy E_0 [3.55]

$$g = (\sqrt{3}/\pi)\ln[2/(\nu(\gamma + 1/\eta_0)] , \tag{3.4.14}$$

which gives the classical limit (3.4.9). In the Born approximation and for $\omega \to 0$ one can obtain from (3.4.12) or (3.4.14)

$$g = (\sqrt{3}/\pi)\ln(2\eta_0/\nu) , \quad \eta_1 \approx \eta_0 . \tag{3.4.15}$$

In general, the semiclassical result (3.4.13) is a good approximation to the Sommerfeld formula (3.4.6) for all the parameters, except the so-called short wavelength limit

$$\nu \to 0 , \quad \eta_0 \to 0 , \quad \nu/\eta_0 \to 1 .$$

This case corresponds to large electron velocities, for which the Sommerfeld formula (3.4.6) is not valid and one must take into account relativistic and retardation effects.

4 Electron–Atom Collisions

The fundamental processes of excitation and ionization of atoms and ions by electron impact are considered in this Chapter including multiple ionization. Analytical approximation formulas and tables for cross sections and Maxwellian rate coefficients are given.

4.1 Excitation

4.1.1 Basic Relations

The excitation of atoms and ions by electrons

$$X_z + e(E) \to X_z^* + e(E') \tag{4.1.1}$$

is described by three main processes:

(*i*) excitation of outer-shell electrons

$$X_z(\gamma nl^q) + e \to X_z^*(\gamma nl^{q-1}n'l') + e\,;$$

(*ii*) excitation of inner-shell electrons

$$X_z(\gamma nl^q\beta) + e \to X_z^*(\gamma nl^{q-1}\beta n'l') + e\,;$$

(*iii*) resonant excitation

$$X_z(\gamma_0) + e \to X_{z-1}^{**}(\gamma' nl) \to X_z^*(\gamma_1) + e\,.$$

Here, γ denotes a set of quantum numbers, q is the number of equivalent electrons of the nl shell and E and E' are the energies of the incident and scattered electrons, respectively. Reactions (*iii*) are especially important for the excitation of highly charged ions.

At threshold ($E = \Delta E$), the excitation cross section of a neutral atom is zero: $\sigma_{\text{th}} = 0$, $z = 1$. In the case of positive ions, because of the long-range attractive Coulomb force, the excitation cross sections for processes (*i–iii*) are finite at threshold $\sigma_{\text{th}} = \text{const.}$, $z > 1$ and, as a rule, have their maximum there.

The cross sections of the dipole (optically allowed) transitions ($\Delta l = \pm 1$, $\Delta S = 0$) fall off according to the Bethe formula:

$$\sigma^{\mathrm{dip}} \approx \frac{A}{E} + B\frac{\ln E}{E}, \quad E \gg \Delta E, \tag{4.1.2}$$

where A and B are constants. B is related to the oscillator strength ($B \propto f$) and A is obtained from numerical calculations. For other types of transitions

$$\sigma \propto \begin{cases} E^{-1} \text{ for transitions with } |\Delta l| \neq 1, \quad \Delta S = 0 & (4.1.3) \\ E^{-3} \text{ for intercombination transitions}, \quad \Delta S = 1. & (4.1.4) \end{cases}$$

The proportionality (4.1.4) is realized at very high energies $E \gg \Delta E$.

The excitation cross section σ of ions may be presented in the form

$$\sigma_{01}[\pi a_0^2] = \Omega(x)/(g_0 E), \quad x = E/\Delta E, \tag{4.1.5}$$

where g_0 is the statistical weight of the initial state and E the incident electron energy in Ry units. The quantity Ω is termed the *collision strength* which is symmetrical for direct 0–1 and inverse 1–0 transitions:

$$\Omega_{01} = \Omega_{10}. \tag{4.1.6}$$

The excitation cross sections σ_{01} and σ_{10} are related by the Klein–Rosseland formula:

$$\begin{aligned} &(E + \Delta E) g_0 \sigma_{01}(E + \Delta E) = E g_1 \sigma_{10}(E), \\ &g_0 \sigma_{01} \approx g_1 \sigma_{10}, \quad E \gg \Delta E; \\ &\sigma_{10} \to \infty, \quad E \to 0. \end{aligned} \tag{4.1.7}$$

The excitation rate coefficient $\langle v\sigma \rangle$ is defined by

$$\langle v\sigma \rangle = \int v\sigma(v) f(v)\, \mathrm{d}^3 v \ [\mathrm{cm}^3\, \mathrm{s}^{-1}], \tag{4.1.8}$$

where v is the relative velocity of colliding particles and $f(v)$ is the velocity distribution function of the incident particles. For the isotropic distribution function $f(v)$ one has

$$\begin{aligned} \langle v\sigma \rangle &= \int_{v_{\min}}^{\infty} v\sigma(v) f(v)\, \mathrm{d}v = \int_{\Delta E}^{\infty} v\sigma(E) f(E)\, \mathrm{d}E \\ &= K \int_{\Delta E}^{\infty} \frac{\sigma}{\pi a_0^2} \mathrm{e}^{-E/T} \frac{E\, \mathrm{d}E}{\mathrm{Ry}^{1/2} T^{3/2}}, \end{aligned} \tag{4.1.9}$$

$$\begin{aligned} &E = \mu v^2/2, \quad v_{\min} = (2\Delta E/\mu)^{1/2}, \\ &K = 2\sqrt{\pi a_0^2} c/137 = 2.17 \times 10^{-8}\ \mathrm{cm}^3\, \mathrm{s}^{-1}, \end{aligned} \tag{4.1.10}$$

where μ is the reduced mass of colliding particles, ΔE is the threshold energy of a given process.

The Maxwellian (isotropic) function has the form:

$$f(v) = (2/\pi)^{1/2} (\mu/T)^{3/2} \mathrm{e}^{-\mu v^2/2T} \tag{4.1.11}$$

or

$$f(E) = 2(E/\pi)^{1/2}T^{-3/2}e^{-E/T}, \quad E = \mu v^2/2, \tag{4.1.12}$$

where T is the plasma temperature. The functions $f(v)$ and $f(E)$ are normalized to unity

$$\int_0^\infty f(v)\,dv = \int_0^\infty f(E)\,dE = 1.$$

Maxwellian excitation rate coefficients corresponding to the cross sections (4.1.5) are often presented in the form

$$\langle v\sigma\rangle = K\cdot\frac{G(\beta)e^{-\beta}}{g_0}\,[\mathrm{cm^3\,s^{-1}}], \quad G(\beta) = \beta e^{\beta}\int_1^\infty e^{-\beta x}\Omega(x)\,dx, \tag{4.1.13}$$

with

$$x = E/\Delta E, \quad \beta = \Delta E/T,$$

where the electron temperature T is in Ry units and the constant K is defined by (4.1.10); $G(\beta)$ is called the *effective collision strength.*

According to the detailed balance principle, the excitation and de-excitation (quenching) rates are related by

$$g_0\langle v\sigma_{01}\rangle = g_1\langle v\sigma_{10}\rangle e^{-\Delta E/T}. \tag{4.1.14}$$

A survey of available experimental data on excitation cross sections and otpical excitation functions in neutral atoms from H to W is given in [4.1]; experimental techniques for measuring electron–atom collision cross sections are described in [4.1–3]. General aspects of excitation processes in collisions of atoms and ions by electrons are considered in [4.4–9]; theoretical and experimental data on excitation cross sections are given in [4.10–15].

4.1.2 Transitions in Hydrogen

Experimental excitation cross sections in hydrogen are available only for the excitation of the $n = 2$ and $n = 3$ levels [4.16–24]. Accurate theoretical calculations of the excitation cross sections in hydrogen are presented in [4.25–32]. Recommended data obtained on the basis of available experimental data and sophisticated calculations are given in [4.13]; these data are reproduced in Tables 4.1–4.

Allowed and intercombination transitions between Rydberg states are considered in Sect. 4.1.6.

4.1.3 Transitions in Helium

The excitation cross sections and rate coefficients of He atoms from the ground and excited states are required for many physical applications such as gas discharges, rare-gas lasers, fusion and astrophysical plasmas, plasma diagnostics, and heating of plasmas by neutral He beams etc.

Table 4.1. Excitation cross sections (in cm^2) in hydrogen for transitions $1s$–$2s$, $2p$ and $n = 2$ vs incident electron energy E (in eV)

E [eV]	$1s$–$2s$	$1s$–$2p$	$1s \to n = 2$
1.02 E + 01	1.14 E − 17	1.41 E − 17	2.55 E − 17
2.00 E + 01	8.46 E − 18	4.49 E − 17	5.33 E − 17
4.00 E + 01	6.73 E − 18	6.44 E − 17	7.12 E − 17
6.00 E + 01	5.46 E − 18	6.46 E − 17	7.01 E − 17
8.00 E + 01	4.57 E − 18	6.05 E − 17	6.51 E − 17
1.00 E + 02	3.92 E − 18	5.59 E − 17	6.00 E − 17
2.00 E + 02	2.27 E − 18	3.92 E − 17	4.15 E − 17
4.00 E + 02	1.22 E − 18	2.48 E − 17	2.61 E − 17
6.00 E + 02	8.38 E − 19	1.85 E − 17	1.94 E − 17
8.00 E + 02	6.37 E − 19	1.49 E − 17	1.55 E − 17
1.00 E + 03	5.13 E − 19	1.25 E − 17	1.31 E − 17
2.00 E + 03	2.61 E − 19	7.22 E − 18	7.50 E − 18
4.00 E + 03	1.31 E − 19	4.08 E − 18	4.22 E − 18
6.00 E + 03	8.79 E − 20	2.90 E − 18	3.00 E − 18
8.00 E + 03	6.60 E − 20	2.27 E − 18	2.35 E − 18
1.00 E + 04	5.28 E − 20	1.88 E − 18	1.94 E − 18

Table 4.2. As in Table 4.1 for transitions $1s$–$3s$, $3p$, $3d$ and $n = 3$

E [eV]	$1s$–$3s$	$1s$–$3p$	$1s$–$3d$	$1s \to n = 3$
1.24 E + 01	2.25 E − 18	4.71 E − 18	2.92 E − 18	1.08 E − 17
2.00 E + 01	3.12 E − 18	9.91 E − 18	4.58 E − 18	1.79 E − 17
4.00 E + 01	1.04 E − 18	1.24 E − 17	2.42 E − 18	1.58 E − 17
6.00 E + 01	8.93 E − 19	1.17 E − 17	1.58 E − 18	1.43 E − 17
8.00 E + 01	7.99 E − 19	1.07 E − 17	1.17 E − 18	1.27 E − 17
1.00 E + 02	7.12 E − 19	9.68 E − 18	9.23 E − 19	1.13 E − 17
2.00 E + 02	4.41 E − 19	6.55 E − 18	4.32 E − 19	7.42 E − 18
4.00 E + 02	2.44 E − 19	4.08 E − 18	2.06 E − 19	4.55 E − 18
6.00 E + 02	1.68 E − 19	3.03 E − 18	1.34 E − 19	3.34 E − 18
8.00 E + 02	1.27 E − 19	2.44 E − 18	1.00 E − 19	2.67 E − 18
1.00 E + 03	1.03 E − 19	2.05 E − 18	8.12 E − 20	2.23 E − 18
2.00 E + 03	5.17 E − 20	1.19 E − 18	4.12 E − 20	1.27 E − 18
4.00 E + 03	2.58 E − 20	6.76 E − 19	2.04 E − 20	7.18 E − 19
6.00 E + 03	1.72 E − 20	4.79 E − 19	1.36 E − 20	5.12 E − 19
8.00 E + 03	1.29 E − 20	3.75 E − 19	1.02 E − 20	3.98 E − 19
1.00 E + 04	1.03 E − 20	3.10 E − 19	8.22 E − 21	3.31 E − 19

Experimental data for electron impact excitation of neutral helium from its ground state to the $n^{1,3}L$ ($n \leqslant 6$) states have been reviewed in [4.33] and also combined with close-coupling calculations and Born-approximation data. In [4.34], these data have been presented by analytical expressions containing up to five fitting parameters. Theoretical data on excitation cross sections and Maxwellian rate coefficients are given in [4.35]. Experimental collision strengths

Table 4.3 As in Table 4.1 for transitions $1s$–n = 4, 5, 6, 8 and 10

E [eV]	$1s \to n = 4$	$1s \to n = 5$	$1s \to n = 6$	$1s \to n = 8$	$1s \to n = 10$
1.35 E + 01	–	–	5.52 E – 19	2.13 E – 19	1.05 E – 19
1.40 E + 01	7.02 E – 18	4.62 E – 18	–	–	–
2.00 E + 01	9.44 E – 18	5.99 E – 18	1.13 E – 18	4.55 E – 19	2.28 E – 19
4.00 E + 01	7.46 E – 18	4.36 E – 18	1.55 E – 18	6.32 E – 19	3.18 E – 19
6.00 E + 01	5.98 E – 18	3.21 E – 18	1.50 E – 18	6.09 E – 19	3.07 E – 19
8.00 E + 01	5.03 E – 18	2.58 E – 18	1.36 E – 18	5.53 E – 19	2.79 E – 19
1.00 E + 02	4.37 E – 18	2.20 E – 18	1.22 E – 18	4.97 E – 19	2.50 E – 19
2.00 E + 02	2.72 E – 18	1.33 E – 18	7.77 E – 19	3.16 E – 19	1.59 E – 19
4.00 E + 02	1.62 E – 18	7.77 E – 19	4.58 E – 19	1.86 E – 19	9.38 E – 20
6.00 E + 02	1.17 E – 18	5.69 E – 19	3.32 E – 19	1.35 E – 19	6.79 E – 20
8.00 E + 02	9.36 E – 19	4.56 E – 19	2.63 E – 19	1.07 E – 19	5.38 E – 20
1.00 E + 03	7.83 E – 19	3.79 E – 19	2.19 E – 19	8.90 E – 20	4.48 E – 20
2.00 E + 03	4.41 E – 19	2.12 E – 19	1.23 E – 19	4.99 E – 20	2.51 E – 20
4.00 E + 03	2.46 E – 19	1.19 E – 19	6.82 E – 20	2.77 E – 20	1.39 E – 20
6.00 E + 03	1.73 E – 19	8.37 E – 20	4.81 E – 20	1.95 E – 20	9.79 E – 21
8.00 E + 03	1.35 E – 19	6.52 E – 20	3.74 E – 20	1.52 E – 20	7.63 E – 21
1.00 E + 04	1.11 E – 19	5.37 E – 20	3.08 E – 20	1.25 E – 20	6.27 E – 21

Table 4.4. As in Table 4.1 for transitions $n = 2 \to n' = 3$, 4, 5 and 6

E [eV]	$n = 2 \to n = 3$	$n = 2 \to n = 4$	$n = 2 \to n = 5$	$n = 2 \to n = 6$
2.00 E + 00	2.73 E – 15	–	–	–
4.00 E + 00	2.91 E – 15	5.81 E – 16	1.90 E – 16	8.56 E – 17
6.00 E + 00	3.03 E – 15	7.31 E – 16	2.62 E – 16	1.26 E – 16
8.00 E + 00	3.40 E – 15	7.56 E – 16	2.78 E – 16	1.36 E – 16
1.00 E + 01	3.57 E – 15	7.38 E – 16	2.73 E – 16	1.34 E – 16
2.00 E + 01	2.87 E – 15	5.51 E – 16	2.05 E – 16	1.01 E – 16
4.00 E + 01	1.78 E – 15	3.40 E – 16	1.26 E – 16	6.19 E – 17
6.00 E + 01	1.34 E – 15	2.49 E – 16	9.15 E – 17	4.50 E – 17
8.00 E + 01	1.10 E – 15	1.98 E – 16	7.26 E – 17	3.56 E – 17
1.00 E + 02	9.32 E – 16	1.66 E – 16	6.05 E – 17	2.96 E – 17
2.00 E + 02	5.45 E – 16	9.36 E – 17	3.39 E – 17	1.65 E – 17
4.00 E + 02	3.17 E – 16	5.22 E – 17	1.88 E – 17	9.12 E – 18
6.00 E + 02	2.28 E – 16	3.68 E – 17	1.32 E – 17	6.40 E – 18
8.00 E + 02	1.80 E – 16	2.87 E – 17	1.03 E – 17	4.98 E – 18
1.00 E + 03	1.50 E – 16	2.37 E – 17	8.44 E – 18	4.09 E – 18
2.00 E + 03	8.32 E – 17	1.29 E – 17	4.58 E – 18	2.21 E – 18
4.00 E + 03	4.55 E – 17	6.97 E – 18	2.46 E – 18	1.19 E – 18
6.00 E + 03	3.21 E – 17	4.86 E – 18	1.71 E – 18	8.24 E – 19
8.00 E + 03	2.50 E – 17	3.75 E – 18	1.32 E – 18	6.35 E – 19
1.00 E + 04	2.04 E – 17	3.07 E – 18	1.08 E – 18	5.19 E – 19

Table 4.5. Experimental $1\,^1S$–$n\,^1S$ collision strengths (4.1.5) in He vs incident electron energy E (in eV) [4.33]

E [eV]	$2\,^1S$	$3\,^1S$	$4\,^1S$	$5\,^1S$	$6\,^1S$
2.30^{1}	3.36^{-2}				
2.50^{1}		6.06^{-3}	2.28^{-3}	8.56^{-4}	3.97^{-4}
3.00^{1}	6.02^{-2}	9.80^{-3}	4.06^{-3}	1.76^{-3}	9.02^{-4}
3.50^{1}		1.20^{-2}	4.83^{-3}	2.21^{-3}	1.16^{-3}
4.00^{1}	7.05^{-2}	1.32^{-2}	5.08^{-3}	2.40^{-3}	1.27^{-3}
4.50^{1}			5.38^{-3}	2.56^{-3}	
5.00^{1}	8.10^{-2}	1.41^{-2}	5.56^{-3}	2.68^{-3}	1.42^{-3}
6.00^{1}	8.87^{-2}		5.66^{-3}	2.65^{-3}	1.42^{-3}
8.00^{1}	1.00^{-1}	1.74^{-2}	6.36^{-3}	2.86^{-3}	1.55^{-3}
1.00^{2}	1.09^{-1}	1.95^{-2}	7.57^{-3}	3.50^{-3}	1.86^{-3}
1.50^{2}	1.29^{-1}	2.16^{-2}	8.76^{-3}	4.14^{-3}	2.08^{-3}
2.00^{2}	1.40^{-1}	2.44^{-2}	9.89^{-3}	4.70^{-3}	2.34^{-3}
2.50^{2}		2.74^{-2}	1.09^{-2}	5.22^{-3}	2.74^{-3}
3.00^{2}	1.59^{-1}	2.98^{-2}	1.13^{-2}	5.79^{-3}	2.93^{-3}
3.50^{2}		3.01^{-2}	1.18^{-2}	5.99^{-3}	3.07^{-3}
4.00^{2}	1.64^{-1}	3.06^{-2}	1.21^{-2}	6.15^{-3}	3.18^{-3}
5.00^{2}	1.77^{-1}	3.53^{-2}	1.36^{-2}	6.60^{-3}	3.55^{-3}
6.00^{2}	1.62^{-1}	3.39^{-2}	1.32^{-2}	6.82^{-3}	3.74^{-3}
8.00^{2}	1.79^{-1}	3.82^{-2}	1.51^{-2}	7.02^{-3}	3.74^{-3}
1.00^{3}	1.81^{-1}	3.87^{-2}	1.47^{-2}	7.20^{-3}	4.00^{-3}
1.50^{3}		3.94^{-2}	1.48^{-2}	7.56^{-3}	4.16^{-3}
2.00^{3}		4.09^{-2}	1.51^{-2}	7.47^{-3}	4.03^{-3}

and effective collision strengths for excitation of helium from its ground state are given in Tables 4.5–8.

Experimental data on excitation cross sections between excited states in He and very scarce [4.36, 37]. Theoretical calculations of excitation cross sections for n–n' transitions for the states n, $n' = 2$, 3, 4 and 5 have been recently reviewed in [4.38]. In [4.39], the results of the Coulomb–Born approximation with exchange are presented together with the fitting parameters for the cross sections and rates of spin-allowed transitions which are approximated by

$$\sigma = \frac{\pi a_0^2}{2l_0 + 1}\left(\frac{\mathrm{Ry}}{DE}\right)^2 \left(\frac{E_1}{E_0}\right)^{3/2} \frac{C}{u + \varphi}\left(\frac{u}{u + a}\right)^{1/2}, \tag{4.1.15}$$

$$\langle v\sigma \rangle = \frac{10^{-8}\,\mathrm{cm^3\,s^{-1}}}{2l_0 + 1}\left(\frac{\mathrm{Ry}E_1}{DEE_0}\right)^{3/2} \mathrm{e}^{-\Delta E/T} \frac{A\beta^{1/2}(\beta + 1)}{(\beta + \chi)(a\beta + 1)^{1/2}}, \tag{4.1.16}$$

with

$$a = \Delta E/DE\,, \quad u = (E - \Delta E)/DE\,, \quad \beta = DE/T\,, \tag{4.1.17}$$

where l_0 is the orbital quantum number of the electron in its initial state, $E_{0,1}$ are the atomic energies of the initial and final states counted from the ionization limit, and ΔE is the excitation energy. In [4.39], a scaling value $DE = 5$ Ry was

Table 4.6. As in Table 4.5 for $1\,{}^1S$–$n\,{}^1P$ transitions

E [eV]	$2\,{}^1P$	$3\,{}^1P$	$4\,{}^1P$
3.00^1	9.40^{-2}	1.85^{-2}	7.77^{-3}
3.20^1	1.18^{-1}	2.54^{-2}	
3.50^1	1.59^{-1}	3.25^{-2}	
4.00^1	2.15^{-1}	4.61^{-2}	1.96^{-2}
4.50^1	2.90^{-1}	6.24^{-2}	
5.00^1	3.43^{-1}	7.69^{-2}	3.48^{-2}
6.00^1	4.77^{-1}	1.08^{-1}	4.73^{-2}
7.00^1	5.71^{-1}	1.37^{-1}	
8.00^1	6.78^{-1}	1.62^{-1}	7.69^{-2}
9.00^1	7.66^{-1}	1.84^{-1}	
1.00^2	8.44^{-1}	1.04^{-1}	8.94^{-2}
1.20^2	9.66^{-1}	2.42^{-1}	
1.50^2	1.15^0	2.83^{-1}	1.28^{-1}
1.80^2	1.32^0	3.31^{-1}	
2.00^2	1.39^0	3.48^{-1}	1.49^{-1}
2.50^2	1.59^0	3.93^{-1}	
3.00^2	1.74^0	4.44^{-1}	1.81^{-1}
3.50^2	1.87^0	4.77^{-1}	
4.00^2	1.99^0	5.05^{-1}	2.05^{-1}
5.00^2	2.37^0	5.39^{-1}	2.21^{-1}
6.00^2	2.32^0	5.56^{-1}	2.39^{-1}
8.00^2	2.51^0	6.12^{-1}	2.53^{-1}
1.00^3	2.62^0	6.63^{-1}	2.67^{-1}
1.50^3	2.93^0	7.37^{-1}	2.91^{-1}
2.00^3	3.09^0	7.82^{-1}	3.18^{-1}

used. The fitting parameters C, φ, A and χ for triplet–triplet and singlet–singlet transitions are given in Tables 4.9–10. Maximum errors (in %) of fits to the excitation cross sections and rates by (4.1.15–17) are also given. The reduced electron energy u and the electron temperature β vary in the limits:

$$0.02 \leqslant u \leqslant 16.0\,, \quad 0.25 \leqslant \beta \leqslant 8.0\,. \tag{4.1.18}$$

For some transitions between excited states in He, the calculated excitation cross sections are shown in Figs. 4.1–7 in comparison with available experimental data.

4.1.4 Dipole Transitions. Model Potentials

For dipole (optically allowed) transitions with $\Delta l = \pm 1$, $\Delta S = 0$ induced by electron impact, the Bethe formula is often used

$$\sigma = \frac{8f}{E\,\Delta E} \ln \frac{q_0}{k_0 - k_1} [\pi a_0^2]\,, \quad k_0 \pm k_1 = \sqrt{2E}(1 \pm \sqrt{1 - \Delta E/E})\,, \tag{4.1.19}$$

Table 4.7. As in Table 4.5 for $1\,^1S$–$n\,^1D$ transitions

E [eV]	$3\,^1D$	$4\,^1D$	$5\,^1D$	$6\,^1D$
2.50^{1}	4.01^{-3}	2.19^{-3}	1.28^{-3}	7.44^{-4}
3.00^{1}	5.72^{-3}	3.06^{-3}	1.80^{-3}	10.5^{-3}
3.50^{1}	7.25^{-3}	3.92^{-3}	2.39^{-3}	1.39^{-3}
4.00^{1}	8.66^{-3}	4.65^{-3}	2.76^{-3}	1.60^{-3}
4.50^{1}	9.93^{-3}	5.23^{-3}	3.02^{-3}	1.76^{-3}
5.00^{1}	1.09^{-2}	5.77^{-3}	3.10^{-3}	1.80^{-3}
6.00^{1}	1.14^{-2}	6.02^{-3}	3.23^{-3}	1.87^{-3}
8.00^{1}	1.24^{-2}	6.58^{-3}	3.54^{-3}	2.06^{-3}
1.00^{2}	1.20^{-2}	6.61^{-3}	3.53^{-3}	2.05^{-3}
1.50^{2}	1.17^{-2}	6.45^{-3}	3.46^{-3}	2.10^{-3}
2.00^{2}	1.18^{-2}	6.48^{-3}	3.48^{-3}	2.02^{-3}
2.50^{2}	1.21^{-2}	6.64^{-3}		
3.00^{2}	1.12^{-2}	6.17^{-3}	3.38^{-3}	1.97^{-3}
4.00^{2}	1.07^{-2}	5.88^{-3}	3.33^{-3}	1.94^{-3}
5.00^{2}	1.01^{-2}	5.56^{-3}		
6.00^{2}	1.00^{-2}	5.51^{-3}	3.17^{-3}	1.80^{-3}
8.00^{2}	1.04^{-2}	5.72^{-3}	3.10^{-3}	1.80^{-3}
1.00^{3}	1.04^{-2}	5.73^{-3}	3.12^{-3}	1.81^{-3}
1.50^{3}	9.78^{-3}	5.38^{-3}	2.93^{-3}	1.70^{-3}
2.00^{3}	9.73^{-3}	5.35^{-3}	2.91^{-3}	1.69^{-3}

where f is the oscillator strength. The resonable estimation for the cutting-off parameter q_0 is given by [4.8]

$$q_0 = \min\{k_0 + k_1, \sqrt{2E_0}\}\,,$$

where E_0 is the atomic energy of the initial state counted from the ionization limit.

Expressions for the dipole excitation cross sections and Maxwellian rate coefficients can be obtained in a closed analytical form if a model potential of the type [4.48]

$$V^{\mathrm{M}}(r) = \frac{\lambda r}{(r^2 + r_0^2)}\,, \quad \lambda^2 \propto f/\Delta E \tag{4.1.20}$$

is applied, where r_0 is the effective radius. The model potential (4.1.20) serves the correct asymptotics at $r \to 0$ and $r \to \infty$. A reasonable fit of the model potential to the exact one can be achieved if one sets

$$r_0 = \frac{n_0^* n_1^*}{z(\Delta + 0.5)}\,, \quad \Delta = n_1^* - n_0^*\,, \tag{4.1.21}$$

$$n^* = n - \Delta = (E_{nl}/z^2\mathrm{Ry})^{-1/2}\,, \tag{4.1.22}$$

where Δ is the quantum defect (1.1.3) and z is the spectroscopic symbol. With the model potential (4.1.20–22), the expression for the Born excitation cross section

Table 4.8. Maxwellian-averaged collision strengths (4.1.13) for singlet–singlet transitions in He [4.33]

T [eV]	$2\,^1S$	$3\,^1S$	$4\,^1S$	$3\,^1D$	$4\,^1D$	$2\,^1P$	$3\,^1P$	$4\,^1P$
1.00^{0}	3.30^{-2}	6.69^{-3}	2.02^{-3}	4.11^{-3}	1.61^{-3}	1.64^{-2}	3.79^{-3}	1.48^{-3}
2.00^{0}	3.86^{-2}	7.01^{-3}	2.42^{-3}	4.32^{-3}	2.00^{-3}	2.61^{-2}	6.47^{-3}	2.70^{-3}
3.00^{0}	4.19^{-2}	7.13^{-3}	2.73^{-3}	4.58^{-3}	2.26^{-3}	3.62^{-2}	9.09^{-3}	3.90^{-3}
4.00^{0}	4.45^{-2}	7.21^{-3}	2.99^{-3}	4.86^{-3}	2.48^{-3}	4.68^{-2}	1.17^{-2}	5.11^{-3}
5.00^{0}	4.67^{-2}	7.32^{-3}	3.21^{-3}	5.13^{-3}	2.67^{-3}	5.78^{-2}	1.44^{-2}	6.33^{-3}
7.00^{0}	5.04^{-2}	7.64^{-3}	3.55^{-3}	5.67^{-3}	3.00^{-3}	8.07^{-2}	1.99^{-2}	8.84^{-3}
1.00^{1}	5.05^{-2}	8.30^{-3}	3.92^{-3}	6.38^{-3}	3.41^{-3}	1.16^{-1}	2.84^{-2}	1.27^{-2}
1.50^{1}	6.11^{-2}	9.43^{-3}	4.35^{-3}	7.30^{-3}	3.92^{-3}	1.74^{-1}	4.24^{-2}	1.91^{-2}
2.00^{1}	6.61^{-2}	1.04^{-2}	4.67^{-3}	7.99^{-3}	4.29^{-3}	2.30^{-1}	5.60^{-2}	2.53^{-2}
3.00^{1}	7.40^{-2}	1.21^{-2}	5.18^{-3}	8.91^{-3}	4.80^{-3}	3.32^{-1}	8.12^{-2}	3.67^{-2}
4.00^{1}	8.04^{-2}	1.34^{-2}	5.60^{-3}	9.49^{-3}	5.12^{-3}	4.22^{-1}	1.04^{-1}	4.66^{-2}
5.00^{1}	8.59^{-2}	1.45^{-2}	5.97^{-3}	9.88^{-3}	5.33^{-3}	5.03^{-1}	1.24^{-1}	5.54^{-2}
7.00^{1}	9.48^{-2}	1.62^{-2}	6.60^{-3}	1.04^{-2}	5.61^{-3}	6.43^{-1}	1.60^{-1}	7.04^{-2}
1.00^{2}	1.05^{-1}	1.83^{-2}	7.35^{-3}	1.07^{-2}	5.82^{-3}	8.16^{-1}	2.03^{-1}	8.83^{-2}
1.50^{2}	1.17^{-1}	2.08^{-2}	8.31^{-3}	1.09^{-2}	5.95^{-3}	1.04^{0}	2.59^{-1}	1.11^{-1}
2.00^{2}	1.25^{-1}	2.28^{-2}	9.05^{-3}	1.10^{-2}	5.98^{-3}	1.22^{0}	3.02^{-1}	1.28^{-1}
3.00^{2}	1.37^{-1}	2.57^{-2}	1.01^{-2}	1.09^{-2}	5.97^{-3}	1.48^{0}	3.65^{-1}	1.53^{-1}
4.00^{2}	1.44^{-1}	2.77^{-2}	1.08^{-2}	1.08^{-2}	5.93^{-3}	1.67^{0}	4.12^{-1}	1.72^{-1}
5.00^{2}	1.49^{-1}	2.92^{-2}	1.14^{-2}	1.07^{-2}	5.88^{-3}	1.81^{0}	4.48^{-1}	1.86^{-1}
7.00^{2}	1.56^{-1}	3.13^{-2}	1.22^{-2}	1.06^{-2}	5.80^{-3}	2.04^{0}	5.05^{-1}	2.08^{-1}
1.00^{3}	1.62^{-1}	3.33^{-2}	1.29^{-2}	10.4^{-2}	5.71^{-3}	2.28^{0}	5.66^{-1}	2.31^{-1}
2.00^{3}	1.71^{-1}	3.65^{-2}	1.40^{-2}	1.01^{-2}	5.51^{-3}	2.75^{0}	6.83^{-1}	2.77^{-1}
3.00^{3}	1.74^{-1}	3.77^{-2}	1.44^{-2}	9.98^{-3}	5.41^{-3}	3.03^{0}	7.50^{-1}	3.30^{-1}
4.00^{3}	1.76^{-1}	3.84^{-2}	1.47^{-2}	9.89^{-3}	5.35^{-3}	3.22^{0}	7.96^{-1}	3.21^{-1}
5.00^{3}	1.77^{-1}	3.89^{-2}	1.48^{-2}	9.84^{-3}	5.31^{-3}	3.36^{0}	8.30^{-1}	3.34^{-1}
7.00^{3}	1.78^{-1}	3.94^{-2}	1.50^{-2}	9.77^{-3}	5.25^{-3}	3.56^{0}	8.80^{-1}	3.54^{-1}
1.00^{4}	1.79^{-1}	3.98^{-2}	1.52^{-2}	9.72^{-3}	5.21^{-3}	3.77^{0}	9.30^{-1}	3.74^{-1}

can be obtained in the form

$$\sigma = \frac{8f}{E\Delta E}[\Phi(x_{\min}) - \Phi(x_{\max})][\pi a_0^2]\,,$$

$$\Phi(x) = (x^2/2)[K_0(x)K_2(x) - K_1^2(x)] \qquad (4.1.23)$$

$$x_{\max,\min} = r_0(k_0 \pm k_1)\,,$$

where $K(x)$ is the McDonald function. The function $\Phi(x)$ to within 1.5% is fitted by the formula

$$\Phi(x) \approx e^{-2x}\ln\left(2.193 + 0.681\frac{1 + \ln(1 + 0.8\sqrt{x})}{x}\right).$$

At high electron-impact energies $E \gg \Delta E$, (4.1.23) yields

$$\sigma = \frac{8f}{E\Delta E}\ln\frac{1.36\,E^{1/2}}{r_0\Delta E}[\pi a_0^2]\,, \quad E \gg \Delta E\,, \qquad (4.1.24)$$

Table 4.9. Transition energies ΔE (in eV), oscillator strengths f and fitting parameters (4.1.15–18) for triplet–triplet transitions [4.39]

Transition	ΔE [eV]	f	C	φ	Error [%]	A	χ	Error [%]
$2s$–$2p$	1.14	0.539	1.46×10^3	0.224	12	1.13×10^3	0.24	6
–$3s$	2.89		54.2	0.0532	19	87.0	0.624	5
–$3p$	3.19	0.0636	78.8	0.0702	50	57.8	0.0983	13
–$3d$	3.25		143	0.0337	25	258	0.712	2
–$4s$	3.77		25.5	0.0264	35	48.8	0.726	2
–$4p$	3.89	0.0253	59.5	0.0596	70	45.7	0.106	11
–$4d$	3.91		79.5	0.0182	30	169	0.869	3
–$4f$	3.92		12.3	3.59×10^{-3}	25	38.1	1.50	12
$2p$–$3s$	1.75	0.069	482	0.475	12	260	0.125	20
–$3p$	2.04		259	0.0575	20	385	0.572	7
–$3d$	2.11	0.61	4.85×10^3	0.391	25	2.77×10^3	0.120	20
–$4s$	2.63	0.0105	105	0.181	40	76.4	0.171	15
–$4p$	2.74		119	0.0256	20	230	0.801	1
–$4d$	2.77	0.123	1.59×10^3	0.157	40	1.28×10^3	0.206	10
–$4f$	2.77		119	8.38×10^{-3}	40	249	0.786	1
$3s$–$3p$	0.288	0.89	9.11×10^3	0.162	30	7.93×10^3	0.285	15
–$3d$	0.355		343	0.011	2	638	0.793	3
–$4s$	0.875		202	0.0416	2	320	0.657	10
–$4p$	0.989	0.050	176	0.0312	60	180	0.252	10
–$4d$	1.02		160	0.0223	20	285	0.716	5
–$4f$	1.02		203	4.54×10^{-3}	20	43	0.919	1
$3p$–$3d$	0.0664	0.112	1.59×10^4	5.44×10^{-2}	30	1.89×10^4	0.441	10
–$4s$	0.587	0.145	2.44×10^3	0.269	26	1.62×10^3	0.181	18
–$4p$	0.700		909	0.0527	3	1.30×10^3	0.556	9
–$4d$	0.729	0.48	7.86×10^3	0.313	17	4.80×10^3	0.147	20
–$4f$	0.730		1.78×10^3	0.0269	10	3.10×10^3	0.748	5
$3d$–$4s$	0.520		84.8	6.69×10^{-3}	12	172	0.890	1
–$4p$	0.634	0.022	516	0.0465	60	498	0.252	12
–$4d$	0.662		1.25×10^3	0.0698	7	1.58×10^3	0.453	12
–$4f$	0.663	1.01	3.52×10^4	0.502	18	1.96×10^4	0.121	25
$4s$–$4p$	0.114	1.21	2.90×10^4	0.108	25	3.09×10^4	0.356	12
–$4d$	0.142		1.31×10^3	8.86×10^{-3}	2	2.49×10^3	0.822	4
–$4f$	0.143		211	2.46×10^{-3}	7	439	0.916	1
$4p$–$4d$	0.0281	0.20	6.65×10^4	0.0286	18	8.79×10^4	0.494	8
–$4f$	0.0292		1.17×10^3	5.80×10^{-3}	4	2.26×10^3	0.842	3
$4d$–$4f$	0.00105	3.88×10^{-3}	8.51×10^4	5.74×10^{-3}	17	1.44×10^5	0.670	4

Table 4.10. As in Table 4.9 for singlet–singlet transitions

Transition	ΔE [eV]	f	C	φ	Error [%]	A	χ	Error [%]
$2s$–$2p$	0.565	0.346	1.79×10^3	0.141	13	1.61×10^3	0.302	14
–$3s$	2.30		70.7	0.0690	25	105	0.571	8
–$3p$	2.46	0.167	309	0.579	60	144	0.0214	30
–$3d$	2.46		209	0.0656	30	322	0.572	10
–$4s$	3.05		30.6	0.0387	50	52.6	0.602	7
–$4p$	3.12	0.0526	160	0.393	70	88.9	0.0320	30
–$4d$	3.12		76.7	0.0250	70	141	0.603	7
–$4f$	3.12		21.7	0.0254	60	41.9	0.676	5
$2p$–$3s$	1.74	0.044	340	0.690	20	156	0.0835	25
–$3p$	1.89		299	0.0996	25	397	0.523	10
–$3d$	1.89	0.695	6.32×10^3	0.598	25	3.26×10^3	0.111	20
–$4s$	2.49	7.90×10^{-3}	82.7	0.286	50	50.7	0.0971	20
–$4p$	2.56		139	0.0741	25	218	0.662	5
–$4d$	2.56	0.121	1.96×10^3	0.356	15	1.30×10^3	0.195	15
–$4f$	2.56		172	0.0528	35	273	0.591	7
$3s$–$3p$	0.154	0.571	1.10×10^4	0.106	20	1.11×10^4	0.359	12
–$3d$	0.154		291	0.011	5	531	0.772	4
–$4s$	0.752		238	0.0512	40	375	0.650	7
–$4p$	0.816	0.165	775	0.447	40	413	0.0976	25
–$4d$	0.816		341	0.0443	60	583	0.712	6
–$4f$	0.816		262	0.0305	20	462	0.770	5
$3p$–$4s$	0.60	0.0943	1.50×10^3	0.332	60	977	0.154	20
–$4p$	0.66		989	0.0655	2	1.37×10^3	0.546	10
–$4d$	0.66	0.62	1.09×10^4	0.344	40	6.87×10^3	0.159	20
–$4f$	0.66		1.94×10^3	0.0391	20	3.22×10^3	0.722	6
$3d$–$3p$	1.97×10^{-4}	1.95×10^{-4}	3.00×10^4	4.52×10^{-3}	20	5.06×10^4	0.683	3
–$4s$	0.599		51.1	0.0256	10	88.6	0.742	5
–$4p$	0.66	0.011	306	0.0455	40	326	0.323	10
–$4d$	0.66		1.28×10^3	0.0856	1	1.56×10^3	0.441	13
–$4f$	0.66	1.014	3.46×10^4	0.506	30	1.96×10^4	0.120	25
$4s$–$4p$	0.0627	0.783	3.59×10^4	0.0620	40	4.20×10^4	0.428	5
–$4d$	0.0627		1.21×10^3	7.76×10^{-3}	6	2.30×10^3	0.821	4
–$4f$	0.0636		168	5.31×10^{-3}	4	332	0.865	3
$4d$–$4p$	9.93×10^{-6}	4.20×10^{-5}	7.85×10^5	0.0418	100	2.65×10^5	0.162	10
–$4f$	8.51×10^{-4}	3.16×10^{-3}	8.73×10^4	5.48×10^{-3}	17	1.48×10^5	0.676	3
$4f$–$4p$	8.18×10^{-4}		1.03×10^3	6.64×10^{-3}	1	1.98×10^3	0.870	3

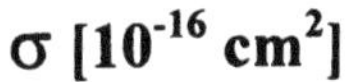

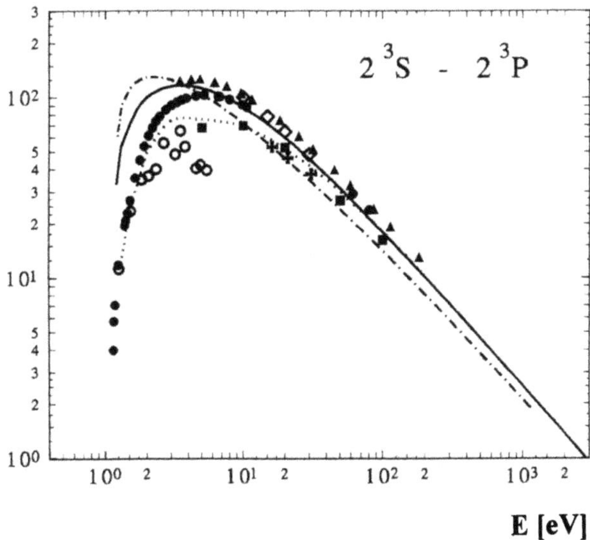

Fig. 4.1. Calculated excitation cross section for transition $2\,^3S$–$2\,^3P$ in He: *squares* ten-channel eikonal approximation [4.40]; *full circles* five-state *R*-matrix calculations [4.41]; *full triangles* Hartree–Fock approximation [4.42]; *open triangles* *R*-matrix calculation with eleven lowest target states included [4.43]; *diamonds* Distorted Wave Approximation (DWA) [4.44]; *open circles* *R*-matrix calculations [4.45]; *crosses* DWA [4.46]; *dotted curve* preferred data obtained from compilation [4.38]; *solid curve* CBE results (ATOM code) [4.39]; *dot-dashed curve* dipole model (4.1.23).

which corresponds to the Maxwellian excitation rate

$$\langle v\sigma\rangle = 1.74\frac{f\beta^{1/2}}{z\Delta E}\mathrm{e}^{-\Delta E/T}\ln\frac{1.26\,z}{\Delta E r_0\beta^{1/2}}\,, \quad \beta = z^2\mathrm{Ry}/T \ll 1\,. \tag{4.1.25}$$

A comparison of the model dipole cross sections (4.1.23) with experimental and theoretical data are given in Fig.s 4.1, 2 and 5.

The Born, Coulomb–Born and classical excitation cross sections are considered in [4.8, 50], the dipole approximation for transitions between highly-excited states of ions in [4.54].

The dipole transitions induced by electron impact in ions have been considered in [4.55] using the Coulomb–Born approximation by regularization of the interaction potential. The result for the excitation cross section for transitions n_0–n_1 has the form

$$\sigma = \frac{[\pi a_0^2]}{E}f\left(\frac{\mathrm{Ry}}{\Delta E}\right)^2\sigma_0\,,$$

with

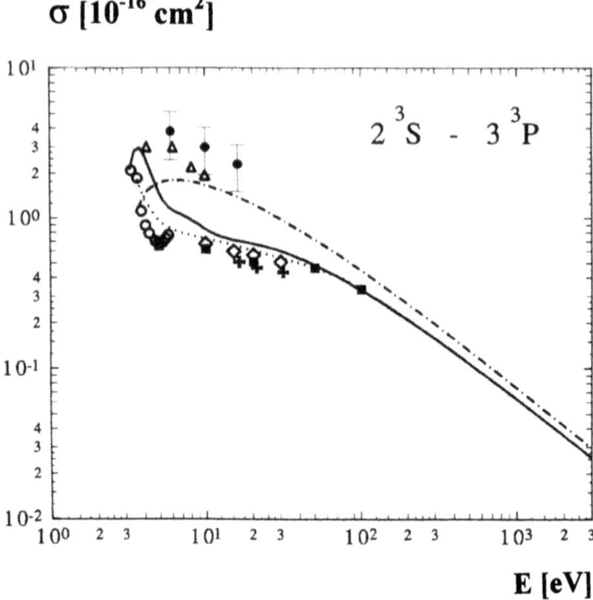

Fig. 4.2. Same as in Fig. 4.1 for $2\,^3S$–$3\,^3P$ transition. *Full circles* experiment [4.37]

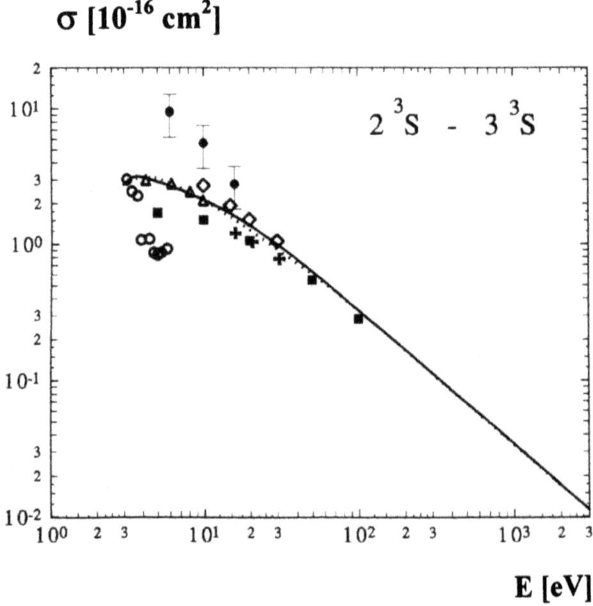

Fig. 4.3. Same as in Figs. 4.1, 2 for $2\,^3S$–$3\,^3S$ transition

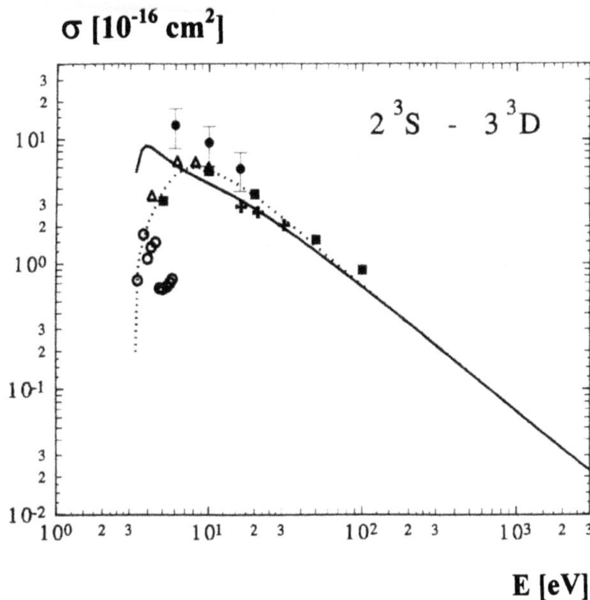

Fig. 4.4. Same as in Figs. 4.1, 2 for $2\,^3S$–$3\,^3D$ transition

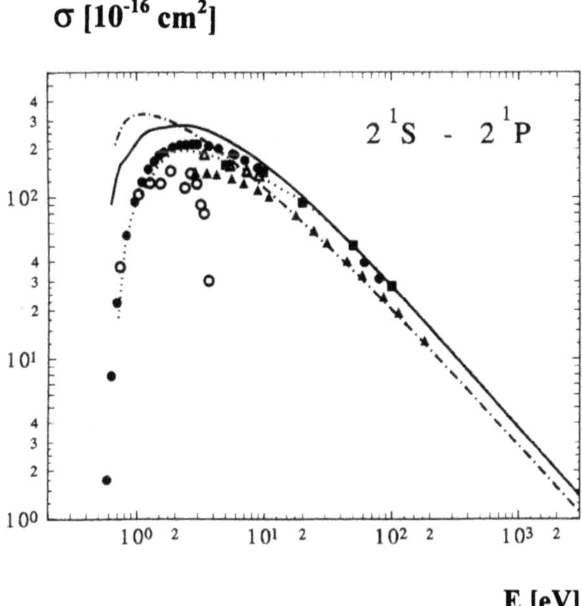

Fig. 4.5. Same as in Figs. 4.1, 2 for $2\,^1S$–$2\,^1P$ transition

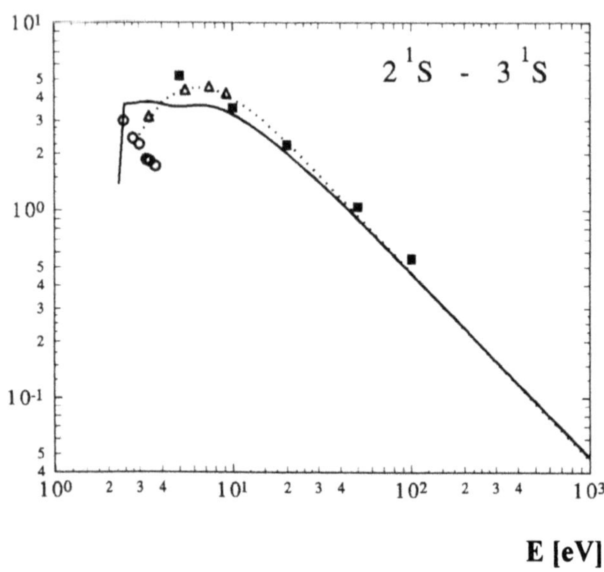

Fig. 4.6. Same as in Figs. 4.1, 2 for $2\,^1S$–$3\,^1S$ transition

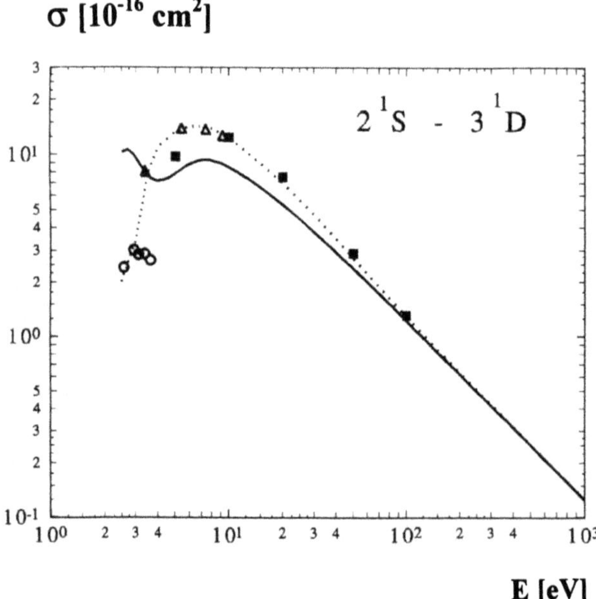

Fig. 4.7. Same as in Figs. 4.1, 2 for $2\,^1S$–$3\,^1D$ transition

$$\sigma_0 = \frac{\Delta E}{E} \frac{2\pi^2 \exp(2\pi\eta_0)}{[\exp(2\pi\eta_0) - 1][\exp(2\pi\eta_1]} \frac{\mathrm{d}}{\mathrm{d}x} |F(\mathrm{i}\eta_0, \mathrm{i}\eta_1, 1; x)|^2 \tag{4.1.26}$$

$$q_m = q_0 + (k_0 + k_1 - q_0)[1 + 2n_0 n_1 (E - \Delta E)/z^2 \mathrm{Ry}]^{-1}; \quad q_0 = z\Delta n/n_0^2 ,$$

where $F(x)$ is a hypergeometric function, f is the oscillator strength for transitions n_0–n_1 (see Sect. 2.5).

Equations (4.1.26) describe the dipole Coulomb–Born cross sections near threshold with an accuracy of a factor of two. At high electron energies, (4.1.26) give the same result as the Bethe–Born formula (4.1.19).

4.1.5 Transitions Between Rydberg States

Transitions between highly excited (Rydberg) states with the principal quantum numbers $n \gg 1$ in atoms and ions induced by electrons and heavy particles are of interest for many problems connected with astrophysical and laboratory plasmas. The recent advances in experimental astrophysical research made it, possible to detect the excited states of atoms with principal quantum numbers n up to $n_{\max} \approx 800$.

If the relative velocity v of colliding particles $v \gg 1$ (in atomic units), the Born approximation is used for transitions with $\Delta S = 0$ and the Ochkur approximation for intercombination transitions $\Delta S = 1$ (see [4.8] and Sect. 4.1.6). For highly excited states, the cross sections $\sigma(n_0$–$n_1)$, i.e., averaged over the orbital quantum numbers $l_0 l_1$, are of interest for transitions with $\Delta n \neq 0$. For transitions with $\Delta n = 0$, the partial cross sections $\sigma(nl_0$–$nl_1)$ are important.

For transitions with $\Delta n \neq 0$, the analytical formulas for the excitation cross sections $\sigma^{\mathrm{B}}(n_0$–$n_1)$ and corresponding rates $\langle v\sigma^{\mathrm{B}}(n_0$–$n_1)\rangle$ were obtained in the Born approximation [4.9]:

$$\sigma^{\mathrm{B}}(n_0\text{–}n_1) = \frac{8}{n_0^2} \frac{z_p^2}{z^4} \left(\frac{z}{v}\right)^2 \left[\left(1 - \frac{0.25}{\Delta n}\right) \frac{(E_0 E_1)^{3/2}}{(\Delta E)^4} \ln(1 + x^2) \right.$$
$$\left. + (1 - 0.6/\Delta n) - \frac{E_1^{3/2}}{(\Delta E)^2} \left(\frac{4}{3\Delta E} + \frac{1}{E_0}\right) \frac{x^2}{1 + x^2} \right] [\pi a_0^2] , \tag{4.1.27}$$

$$x = vn/z , \quad E_0 = 1/n^2 , \quad E_1 = 1/n_1^2 , \quad \Delta E = |E_0 - E_1| , \quad \Delta n = n_1 - n_0 ,$$

where z_{p} is the charge of projectile, z is the spectroscopic symbol of the target. Equation (4.1.27) was obtained under the assumption $n \gg \Delta n \gg 1$. The additional terms with fitting parameters 0.25 and 0.6 are added to describe the transitions with small Δn (including $\Delta n = 1$).

The first term in (4.1.27), proportional to $1/\Delta E^4 \approx 1/\Delta n^4$, corresponds to the dipole interaction (*distant* encounters). The second one ($\propto 1/\Delta E^3 \approx 1/\Delta n^3$) corresponds to the *close* collisions.

The Born rate coefficient corresponding to the electron-induced cross section (4.1.27) is given by

$$\langle v\sigma^{\mathrm{B}}(n_0-n_1)\rangle = \frac{16}{\sqrt{\pi}}\left(\frac{\pi a_0^2 v_0}{z^3}\right)\left(\frac{z^2\mathrm{Ry}}{T}\right)^{1/2}\frac{\mathrm{e}^{-\Delta E/T}}{n_0^2}$$
$$\times\left\{\left(1-\frac{0.25}{\Delta n}\right)\frac{(E_0E_1)^{3/2}}{(\Delta E)^4}\varphi(E_{n0}/T)\right.$$
$$\left.+\left(1-\frac{0.6}{\Delta n}\right)\frac{(E_1)^{3/2}}{(\Delta E)^2}\left(\frac{4}{3\Delta E}+\frac{1}{E_0}\right)[1-(E_0/\theta)\varphi(E_0/\theta)]\right\}, \tag{4.1.28}$$

where $E_{n0} = z^2\mathrm{Ry}/n_0^2$, $\theta = T/z^2\mathrm{Ry}$ and $\varphi(x) = \mathrm{e}^x|\mathrm{Ei}(-x)|$.

In some cases, the Born approximation overestimates the excitation cross sections up to 5 to 10 times for the incident electron energies $E \ll z^2\mathrm{Ry}$. More accurate results can be obtained by the semiclassical method. Since at large energies both approximations give the same results it is convenient to write the cross section in the form [4.9]

$$\sigma^{\mathrm{cl}}(n_0-n_1) = \sigma^{\mathrm{B}}(n_0-n_1)f(x,y)$$
$$f(x,y) = \ln[1 + x/y(1 + 1.25n_0/(xy)^{1.5}][\ln(1 + x/y)]^{-1}, \quad y = z\Delta n/z_{\mathrm{p}} \tag{4.1.29}$$

and, respectively,

$$\langle v\sigma^{\mathrm{cl}}(n_0-n_1)\rangle = \langle v\sigma^{\mathrm{B}}(n_0-n_1)\rangle g(\beta),$$
$$g(\beta) = \ln[(1 + n_0\theta^{1/2}/z\Delta n)/(1 + 2.5/z\Delta n\sqrt{\theta})][\ln[1 + (n_0\theta^{1/2}/z\Delta n)]^{-1}. \tag{4.1.30}$$

For transitions nl_0–nl_1 with no change of the principal quantum number, $\Delta n = 0$ and $n \gg 1$, the Born cross section can be written in the form [4.53]

$$\sigma^{\mathrm{B}}(l_0,l_0\pm 1) = \frac{n^4z_{\mathrm{p}}^2}{z^4}\frac{6\max\{l_0,l_0\pm 1\}}{(2l_0+1)}\frac{\gamma_1}{v^2}\mathrm{Ei}\left(-\frac{0.5a_{\Delta_l}n^2\Delta\varepsilon}{zv}\right)[\pi a_0^2],$$
$$\Delta l = |l_0 - l_1| = 1, \tag{4.1.31}$$

$$\sigma^{\mathrm{B}}(l_0,l_0\pm\Delta_l) = \frac{n^4z_{\mathrm{p}}^2}{z^4}8(2\Delta l+1)[2(l_0\pm\Delta l)+1]$$
$$\times\begin{pmatrix} l_0 & l_0\pm\Delta l & \Delta l \\ 0 & 0 & 0\end{pmatrix}^2\frac{\gamma_{\Delta_l}}{v^2\Delta l^4}[\pi a_0^2],$$
$$\gamma_k^2 = \prod_{i=0}^{k-1}\left[1-\left(\frac{l-i}{\pi}\right)^2\right], \quad \Delta l \geqslant 2. \tag{4.1.32}$$

The parameters a_{Δ_l} are found from the numerical calculations:

$\Delta l =$	1	2	3	4	5
$a_{\Delta_l} =$	0.586	0.454	0.422	0.441	0.405

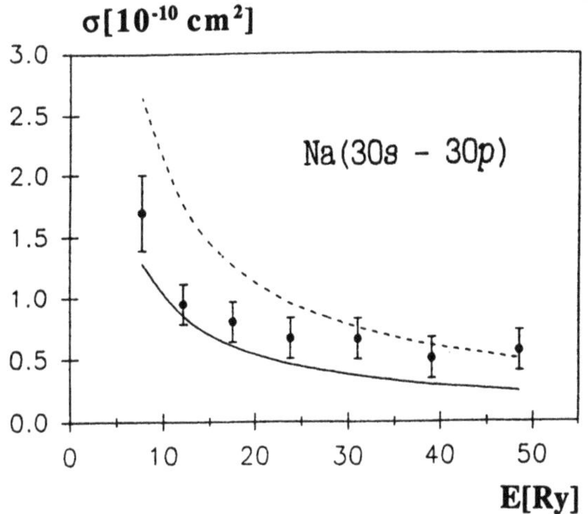

Fig. 4.8. Excitation cross sections for $30s$–$30p$ transition between Rydberg states in Na. *Solid circles* experiment [4.55]; *dashed curve* theory [4.56]; *solid curve* dipole model (4.1.24)

As can be seen from (4.1.31, 32), the cross section falls off rapidly with increasing Δl, and the total excitation cross section from the nl_0 level is dominated by dipole transitions $\Delta l = 1$.

A comparison of the classical and the Born excitation cross sections in collisions of electrons with Rydberg atoms is considered in [4.54]. Measurements of the dipole excitation cross sections in Na induced by electrons between levels with the principal quantum numbers $n \approx 30$ have been carried out in [4.55] (Fig. 4.8).

General theory and experiments on elementary processes involving Rydberg states are considered in [4.8, 9, 50, 57, 58].

4.1.6 Intercombination Transitions

Excitation intercombination transitions (i.e., with a change of atomic spin $\Delta S = 1$) have been considered in detail in various review papers [4.6–8]. Such transitions take place due to exchange effects between the incident and bound electrons, and are very important in solving plasma kinetics problems, e.g., for definition of the relative concentrations of atoms, line intensities, etc.

Calculations of the intercombination cross sections are usually performed using the close-coupling method or modifications of the Born–Oppenheimer approach: the Ochkur method [4.49] and that of the orthogonalised wave functions of the initial and final states [4.57].

The Ochkur approximation for intercombination transitions is similar to the Born approximation for excitation of neutral atoms. Due to the additional factor q^2/E in the exchange amplitude as compared to the Born approximation, the intercombination cross section decreases at large incident electron energies E as

$$\sigma \propto E^{-3}, \quad E \to \infty, \quad \Delta S = 1, \tag{4.1.33}$$

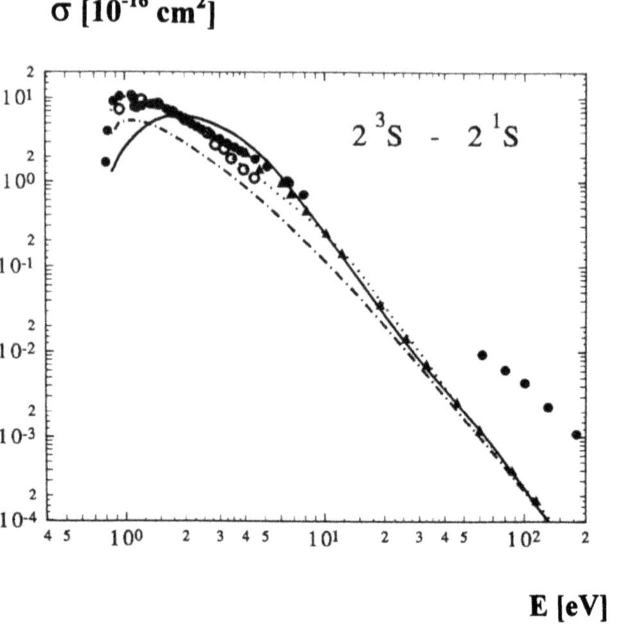

Fig. 4.9. Excitation cross section for intercombination $2\,^3S$–$2\,^1S$ in He (see Fig. 4.1 for notations). *Dot-dashed curve* CBE calculations [4.39] using Ochkur approximation [4.49]

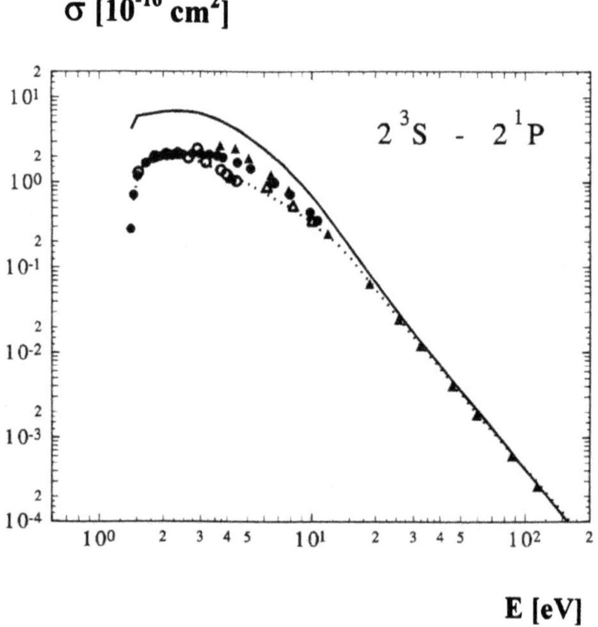

Fig. 4.10. Same as in Fig. 4.7 for $2\,^3S$–$2\,^1P$ transition

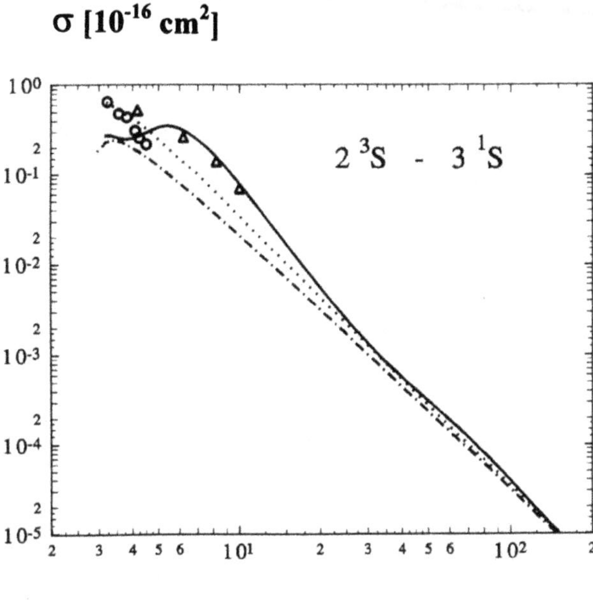

Fig. 4.11. Same as in Fig. 4.7 for $2\,^3S$–$3\,^1S$ transition

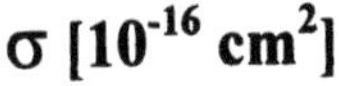

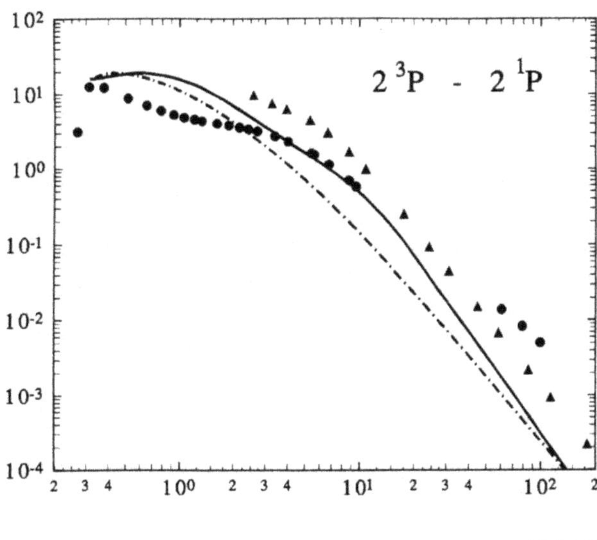

Fig. 4.12. Same as in Fig. 4.7 for $2\,^3P$–$2\,^1P$ transition

Table 4.11. Maxwellian-averaged collision strengths for singlet-triplet transitions in He from its ground state [4.33]

T [eV]	$2\,^3S$	$3\,^3S$	$4\,^3S$	$3\,^3D$	$4\,^3D$	$2\,^3P$	$3\,^3P$
1.00^0	6.27^{-2}	1.20^{-2}	5.09^{-3}	2.12^{-3}	7.13^{-4}	2.31^{-2}	7.87^{-3}
2.00^0	6.25^{-2}	1.34^{-2}	5.85^{-3}	2.31^{-3}	9.39^{-4}	3.18^{-2}	1.04^{-2}
3.00^0	6.06^{-2}	1.41^{-2}	6.25^{-3}	2.52^{-3}	1.11^{-3}	3.75^{-2}	1.20^{-2}
4.00^0	5.87^{-2}	1.45^{-2}	6.47^{-3}	2.68^{-3}	1.23^{-3}	4.15^{-2}	1.30^{-2}
5.00^0	5.70^{-2}	1.48^{-2}	6.58^{-3}	2.80^{-3}	1.32^{-3}	4.45^{-2}	1.38^{-2}
7.00^0	5.42^{-2}	1.50^{-2}	6.65^{-3}	2.93^{-3}	1.40^{-3}	4.84^{-2}	1.47^{-2}
1.00^1	5.07^{-2}	1.48^{-2}	6.54^{-3}	2.95^{-3}	1.43^{-3}	5.15^{-2}	1.52^{-2}
1.50^1	4.62^{-2}	1.40^{-2}	6.19^{-3}	2.84^{-3}	1.38^{-3}	5.30^{-2}	1.53^{-2}
2.00^1	4.25^{-1}	1.32^{-2}	5.79^{-3}	2.67^{-3}	1.31^{-3}	5.25^{-2}	1.50^{-2}
3.00^1	3.69^{-2}	1.16^{-2}	5.09^{-3}	2.36^{-3}	1.16^{-3}	4.95^{-2}	1.39^{-2}
4.00^1	3.27^{-2}	1.03^{-2}	4.53^{-3}	2.11^{-3}	1.03^{-3}	4.59^{-2}	1.28^{-2}
5.00^1	2.95^{-2}	9.27^{-3}	4.08^{-3}	1.90^{-3}	9.33^{-4}	4.25^{-2}	1.19^{-2}
7.00^1	2.47^{-2}	7.76^{-3}	3.42^{-3}	1.60^{-3}	7.85^{-4}	3.68^{-2}	1.02^{-2}
1.00^2	2.01^{-2}	6.29^{-3}	2.77^{-3}	1.30^{-3}	6.38^{-4}	3.05^{-2}	8.47^{-3}
1.50^2	1.54^{-2}	4.86^{-3}	2.13^{-3}	9.98^{-4}	4.90^{-4}	2.38^{-2}	6.60^{-3}
2.00^2	1.26^{-2}	4.03^{-3}	1.74^{-3}	8.14^{-4}	4.00^{-4}	1.95^{-2}	5.42^{-3}
3.00^2	9.35^{-3}	3.11^{-3}	1.29^{-3}	5.99^{-4}	2.94^{-4}	1.45^{-2}	4.01^{-3}
4.00^2	7.47^{-3}	2.62^{-3}	1.03^{-3}	4.76^{-4}	2.34^{-4}	1.15^{-2}	3.19^{-3}
5.00^2	6.23^{-3}	2.30^{-3}	8.57^{-4}	3.95^{-4}	1.94^{-4}	9.57^{-3}	2.66^{-3}
7.00^2	4.70^{-3}	1.92^{-3}	6.46^{-4}	2.96^{-4}	1.45^{-4}	7.18^{-3}	1.99^{-3}
1.00^3	3.45^{-3}	1.60^{-3}	4.74^{-4}	2.16^{-4}	1.06^{-4}	5.24^{-3}	1.46^{-3}
2.00^3	1.84^{-3}	1.11^{-3}	2.54^{-4}	1.14^{-4}	5.61^{-5}	2.77^{-3}	7.71^{-4}
3.00^3	1.26^{-3}	8.84^{-4}	1.74^{-4}	7.80^{-5}	3.83^{-5}	1.89^{-3}	5.26^{-4}
4.00^3	9.62^{-4}	7.41^{-4}	1.32^{-4}	5.92^{-5}	2.91^{-5}	1.43^{-3}	4.00^{-4}
5.00^3	7.77^{-4}	6.41^{-4}	1.07^{-4}	4.77^{-5}	2.34^{-3}	1.16^{-3}	3.22^{-4}
7.00^3	5.62^{-4}	5.08^{-4}	7.72^{-5}	3.44^{-5}	1.69^{-5}	8.33^{-4}	2.32^{-4}
1.00^4	3.97^{-4}	3.90^{-4}	5.46^{-5}	2.43^{-5}	1.19^{-5}	5.88^{-4}	1.64^{-4}

i.e., much faster than the spin-allowed transitons (see Sect. 4.1.1). Calculated cross sections for transitions between excited states in He are shown in Figs. 4.9–12. The Maxwellian rate coefficients for intercombination transitions form the ground state in He are given in Table 4.11.

In some cases, the intercombination cross sections averaged over the orbital quantum numbers l are of interest. The numerical calculations of the Ochkur amplitudes are given in [4.58, 59]. For transition $n_0S_0 - n_1S_1$ with $\Delta S = 1$ the excitation intercombination cross section has the form [4.59]:

$$\sigma = \frac{(2S_1 + 1)}{2(2S_\mathrm{p} + 1)} 8(v_0/v)^2 (E_{\min}/E_{n_0})^{1/2} n_1^{-3} \left[\mathrm{Ry}/\Delta E + \frac{4E_{\min}}{3\Delta E} (\mathrm{Ry}/\Delta E)^2 \right] \frac{\pi a_0^2}{z^2}, \tag{4.1.34}$$

where S_p is the spin of atomic core v_0 is atomic unit of velocity and $E_{\min}$ is a minimum energy of the four energies – the incident and bound electrons in the initial and final states. The cross sections (4.1.34) agree quite well with calculations in Ochkur approximation (Fig. 4.13).

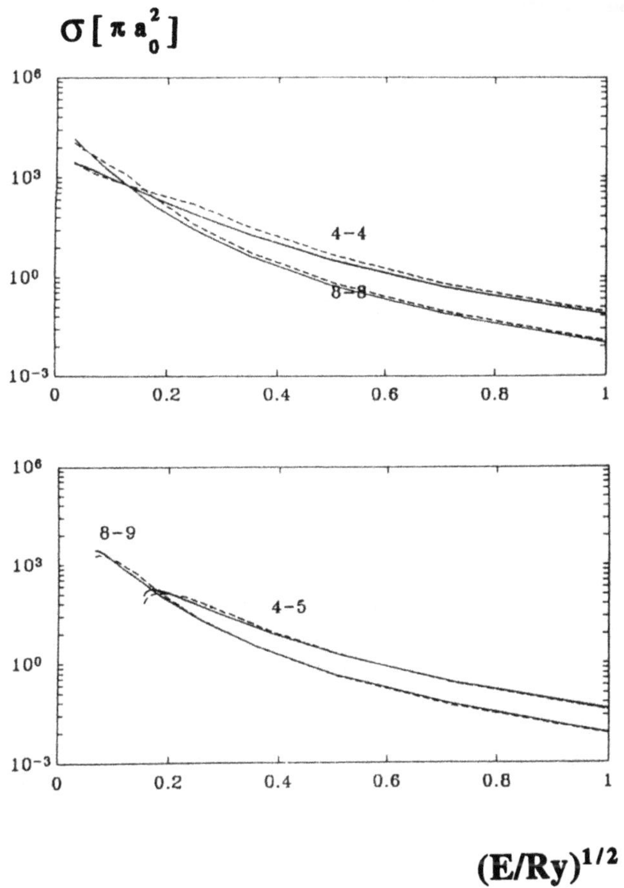

Fig. 4.13. The l-averaged intercombination singlet–triplet transitions n–n' in He. *Solid curves* (4.1.34); *dashed curves* Ochkur approximation [4.49]

The l-resolved intercombination transitions in Rydberg atoms in collisions with electrons are considered in [4.60].

4.2 Single Ionization

The most important elementary processes in electron–atom collisions is ionization. The general problems of single ionization processes are considered in [4.8, 50, 63, 64], experimental data on ionization cross sections are presented in [4.65–67] and on ionization rate coefficients in [4.12]. Bibliography on excitation and ionization cross sections are given in [4.68]. Recommended data on ionization cross sections and rates as well as corresponding fitting parameters are presented in [4.69–72].

4.2.1 General Properties

Three main processes are distinguished in single ionization of atoms and ions by electron impact

$$X_z + e \to X_{z+1} + 2e\,,$$

namely:

(*i*) Direct Ionization (DI),

(*ii*) Excitation-Autoionization (EA),

(*iii*) resonant processes (REDA, READI). The process (*iii*) is of a special importance in ionization of positive ions by electron impact [4.73].

The cross section σ for DI

$$X_z(nl^q) + e \to X_{z+1}(nl^{q-1}) + 2e \tag{4.2.1}$$

is a smooth function of the incident electron energy E and has the following asymptotic behavior:

$$\sigma \propto \begin{cases} u^{3/2}\,, & u \to 0\,, \quad z = 1 \\ u\,, & u \to 0\,, \quad z > 1 \\ \dfrac{A}{u} + B\dfrac{\ln u}{u}\,, & u \to \infty \end{cases} \tag{4.2.2}$$

$$\sigma_{\max} \propto I^{-2} \propto z^{-4}\,, \quad u_{\max} \approx 1.5\,,$$

where $u = E/I - 1$ is the scaled incident electron energy, I is the binding energy of the outer or inner shell nl^q of the target and q is number of equivalent electrons. The constant A is associated with the classical ionization cross section and B (the Bethe constant) with the corresponding photoionization cross section, respectively (Sect. 4.2.3).

Classical and quantum theories give the following scaling laws for σ and $\langle v\sigma \rangle$

$$F(u) = I^2\sigma\,, \quad u = E/I - 1\,, \tag{4.2.3}$$

$$G(\beta) = I^{3/2} e^{\beta} \langle v\sigma \rangle / q\,, \quad \beta = I/T\,. \tag{4.2.4}$$

The functions $F(u)$ and $G(\beta)$ should be universal for given quantum numbers nl and ions of a given isoelectronic sequence.

Almost all experimental data on ionization cross sections of atoms and ions by electron impact have been obtained for atomic targets being in their ground state except ionization of excited H(2*s*) and He(2*s*) atoms. Ionization cross sections of H and He in their ground and excited states are shown in Figs. 4.14–17.

For many-electron atoms, a contribution from ionization of inner-shell electrons can be significant and reach up to a factor of 2. Among inner-shell ionization processes, the *K*-shell ionization processes have been investigated in more detail [4.74–78]. The scaled *K*-shell ionization cross section of light atoms is shown in Fig. 4.18; the full curve corresponds to the universal cross section which is described by

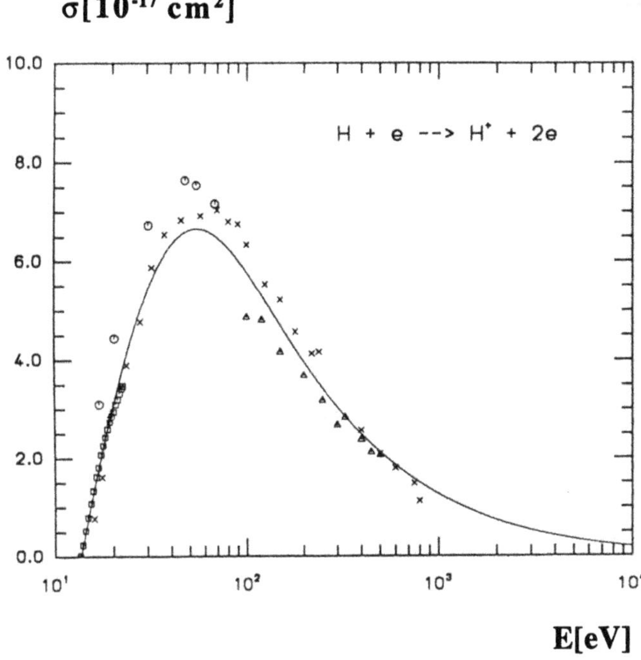

Fig. 4.14. Electron-impact ionization cross sections of H atoms by electron impact (from [4.66]). *Symbols* experiment; *solid curve* Lotz formula (4.2.8)

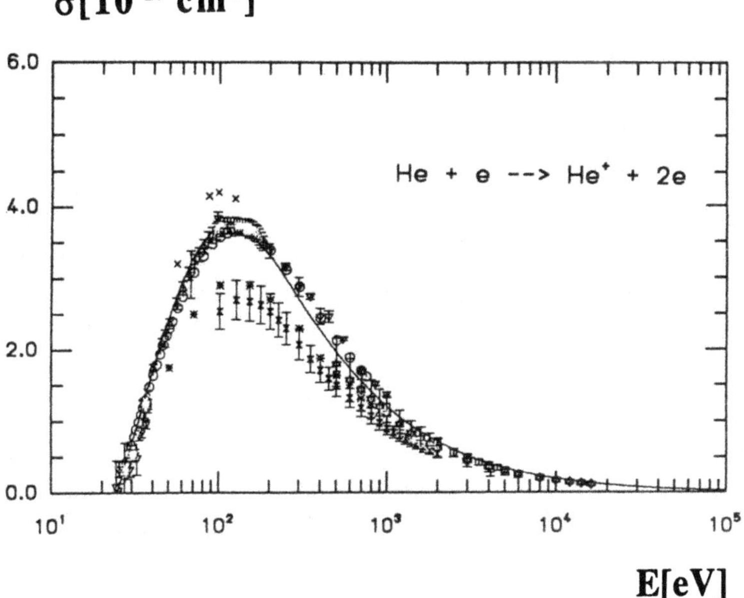

Fig. 4.15. Same as in Fig. 4.14 for He atoms

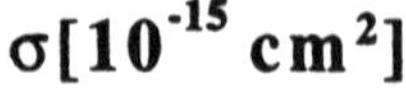

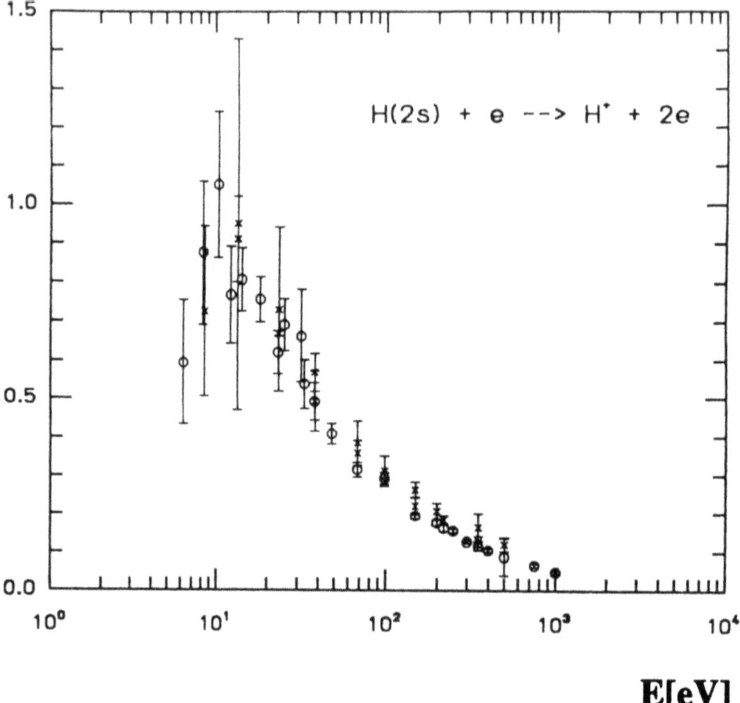

Fig. 4.16. Same as in Fig. 4.14 for excited H(2s) atoms

$$\sigma_{1s} I_{1s}^2 = \frac{44.3 \times 10^{-14}\,\text{cm}^2\,\text{eV}^2}{u + 6.0} \left(\frac{u}{u+1} \right)^{3/2}, \quad u = E/I_{1s} - 1\,, \tag{4.2.5}$$

where I_{1s} is the K-shell binding energy.

For heavy atoms with a nuclear charge $Z_n \geqslant 30$ (and ions with a charge $z \geqslant 50$), the inner-shell ionization cross sections do not follow the scaling relation (4.2.3) due to the influence of the relativistic effects on the wave functions of the free and bound electrons [4.73]. Figure 4.19 shows the K-shell ionization cross sections for heavy atomic targets where relativistic effects strongly enhance the shape of the ionization cross sections.

For targets having more than one electron shell, the total cross section can contain a structure on top of DI cross section caused by indirect (multi-step) ionization mechanisms. One of these mechanisms which significantly contribute to the total cross section is excitation of the inner-shell electrons into autoionizing states followed by autoionization decay

$$X_z + e \rightarrow X_z^{**} + e \rightarrow X_{z+1} + 2e\,,$$

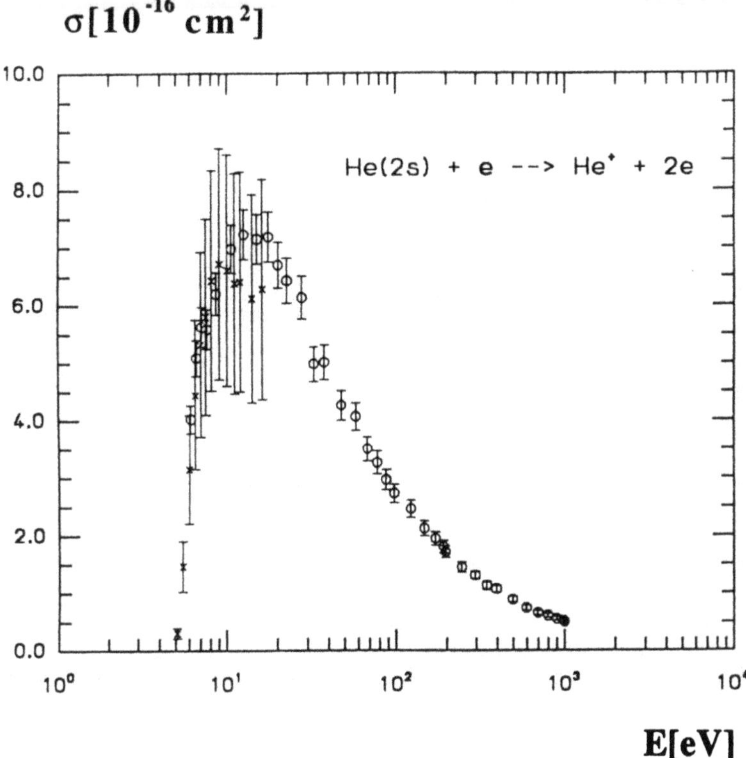

Fig. 4.17. Same as in Fig. 4.14 for excited He(2s) atoms

which is termed Excitation-Autoionization (EA). For example, for Li-like ions, EA processes are

$$X_z(1s^2 2s) + e \rightarrow [X_z(1s2snl)]^{**} + e \rightarrow X_{z+1}(1s^2) + 2e\,, \quad n \geqslant 2\,.$$

The sum of DI and EA cross sections is presented in the form

$$\sigma = \sigma_{\mathrm{DI}} + \sum_j B_j \sigma_{\mathrm{ex}}(j)\,, \tag{4.2.6}$$

$$B_j = \sum_m W_{jm} \bigg/ \left(\sum_m W_{jm} + \sum_l A_{jl} \right), \tag{4.2.7}$$

where B_j is the branching ratio coefficient of the autoionizing j state of the target, A and W are radiative and autoionization probabilities and σ_{ex} is the electron-impact excitation cross section of inner-shell electrons. Equation (4.2.6) is valid if DI and EA processes are treated as independent; an interference of these two processes is also possible.

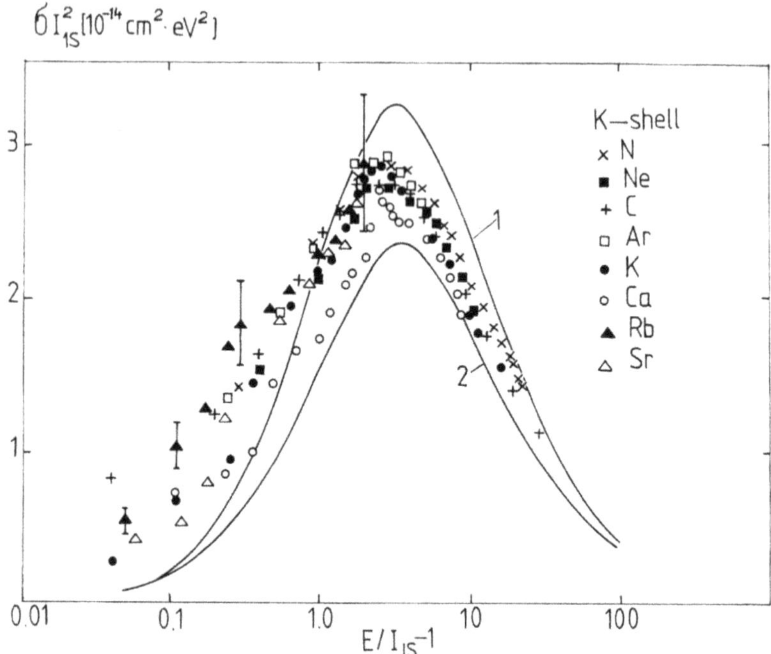

Fig. 4.18. Scaled K-shell ionization cross sections in the non-relativistic energy region (from [4.78]). *Symbols* experimental data; *curve 1* the Lotz formula (4.2.8); *curve 2* CBE calculations [4.78]

4.2.2 Approximation Formulas

Ionization cross sections σ of neutral atoms by electron impact including inner-shell ionization are usually estimated by the semiempirical Lotz formula [4.79] obtained on the basis of experimental data:

$$\sigma = 0.614q\,(\mathrm{Ry}/I)^2 \frac{a\ln(u+1)}{u+1}(1 - b\,\mathrm{e}^{-cu})[\pi a_0^2]\,, \quad u = E/I - 1\,, \qquad (4.2.8)$$

where I is the binding energy of the shell nl^q, and q is the number of equivalent electrons. The fitting parameters a, b and c for different atomic subshells are given in Table 4.12.

In the case of ionization of positive ions by electrons another Lotz formula is used [4.80, 81]:

$$\sigma = 2.76q\,(\mathrm{Ry}/I)^2 \frac{\ln(u+1)}{u+1}[\pi a_0^2]\,. \qquad (4.2.9)$$

Equation (4.2.9) has been deduced on the basis of numerical calculations of H-like ions in the Coulomb–Born approximation with exchange. The Maxwellian ionization rate coefficient $\langle v\sigma\rangle$ corresponding to the cross section (4.2.9) has the form [4.80]

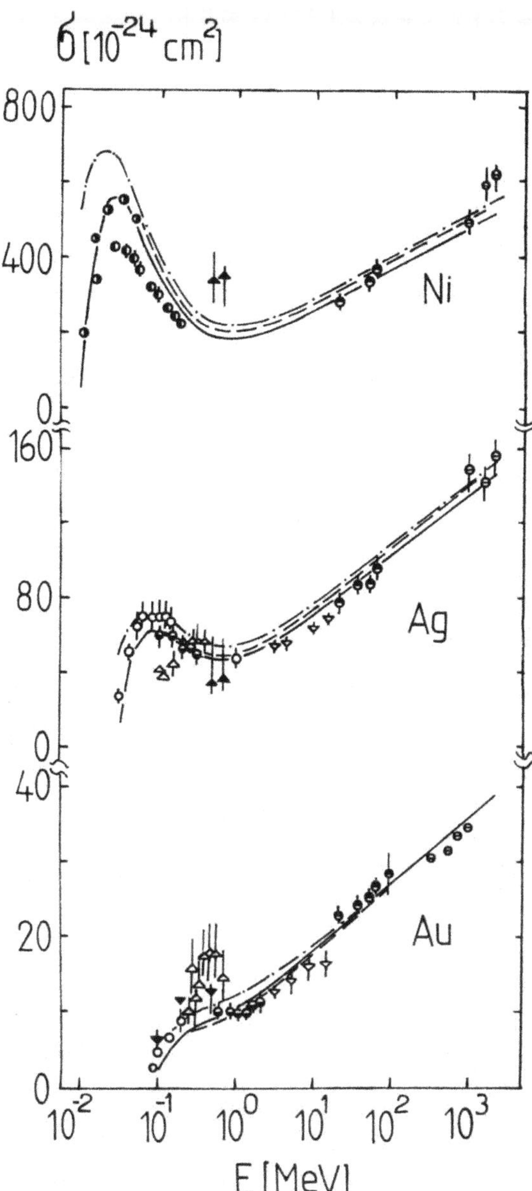

Fig. 4.19. *K*-shell ionization cross section of Ni, Ag, and Au atoms by electron impact (from [4.73]). *Symbols* experimental data; *curves* relativistic calculations (see [4.73] for details)

Table 4.12. Constants a, b, and c of the empirical formula (4.2.8) for different shell in neutral atoms. The quantity n is equal to 4, 5, 6, or 7 for the s shell; equal to 4, 5, or 6 for the p shell; equal to 5 or 6 for the d shell; equal to 4 or 5 for the f shell. The constant a is given in 10^{-14} cm^2 (eV)^2

	$1s$	$1s^2$	$2p$	$2p^2$	$2p^3$	$2p^4$	$2p^5$	$2p^6$	$3d$	$3d^2$	$3d^3$	$3d^4$	$3d^5$	$3d^6$	$3d^7$	$3d^8$	$3d^9$	$3d^{10}$
a	4	4	3.8	3.5	3.2	3.0	2.8	2.6	3.7	3.4	3.1	2.8	2.5	2.2	2.0	1.8	1.6	1.4
b	0.60	0.75	0.6	0.7	0.8	0.85	0.90	0.92	0.6	0.7	0.8	0.85	0.90	0.92	0.93	0.94	0.95	0.96
c	0.56	0.50	0.4	0.3	0.25	0.22	0.20	0,19	0.4	0.3	0.25	0.20	0.18	0.17	0.16	0.15	0.14	0.13
	$2s$	$2s^2$	$3p$	$3p^2$	$3p^3$	$3p^4$	$3p^5$	$3p^6$	$4d$	$4d^2$	$4d^3$	$4d^4$	$4d^5$	$4d^6$	$4d^7$	$4d^8$	$4d^9$	$4d^{10}$
a	4	4	4	4	4	4	4	4	4	3.8	3.5	3.2	3.0	2.8	2.6	2.4	2.2	2.0
b	0.3	0.5	0.35	0.40	0.45	0.50	0.55	0.6	0.3	0.45	0.6	0.7	0.8	0.85	0.90	0.92	0.93	0.94
c	0.6	0.6	0.6	0.6	0.6	0.5	0.45	0.4	0.6	0.5	0.4	0.3	0.25	0.20	0.18	0.17	0.16	0.15
	$3s$	$3s^2$	np	np^2	np^3	np^4	np^5	np^6	nd	nd^2	nd^3	nd^4	nd^5	nd^6	nd^7	nd^8	nd^9	nd^{10}
a	4	4	4	4	4	4	4	4	4	4	3.8	3.6	3.4	3.2	3.0	2.8	2.6	2.4
b	0	0.3	0	0	0.2	0.3	0.4	0.5	0	0.2	0.3	0.45	0.6	0.7	0.8	0.85	0.90	0.92
c	0	0.6	0	0	0.6	0.6	0.6	0.5	0	0.6	0.6	0.5	0.4	0.3	0.25	0.20	0.18	0.17
	ns	ns^2	nf	nf^2	nf^3	nf^4	nf^5	nf^6	nf^7	nf^8	nf^9	nf^{10}	nf^{11}	nf^{12}	nf^{13}	nf^{14}		
a	4	4	3.7	3.4	3.1	2.8	2.5	2.2	2.0	1.8	1.6	1.4	1.3	1.2	1.1	1.0		
b	0	0	0.6	0.7	0.8	0.95	0.90	0.92	0.93	0.94	0.95	0.96	0.96	0.97	0.97	0.96		
c	0	0	0.4	0.3	0.25	0.20	0.18	0.17	0.16	0.15	0.14	0.13	0.12	0.12	0.11	0.11		

Table 4.13. Fitting parameters for ionization cross sections and rates of multicharged ions for nl states (4.2.12–15); $u \leqslant 14$, $1/8 \leqslant \beta \leqslant 8.0$

nl-state	C	φ	A	χ
1s	7.96	2.70	5.65	0.40
2s	6.69	2.03	6.23	0.52
2p	6.93	1.47	9.05	0.73
3s	6.00	1.59	7.37	0.70
3p	6.24	1.31	9.11	0.82
3d	6.57	1.08	11.7	1.00
4s	5.77	1.43	7.76	0.76
4p	6.00	1.26	9.11	0.86
4d	6.23	1.11	10.8	0.97
4f	7.06	1.00	13.5	1.07
5s	5.66	1.36	7.96	0.79
5p	5.88	1.23	9.13	0.87
5d	6.08	1.12	10.4	0.96
5f	6.26	1.08	11.1	1.00
5g	6.47	1.04	11.9	1.03
6s	5.60	1.32	8.07	0.80
6p	5.82	1.22	9.13	0.88
6d	6.00	1.13	10.2	0.95
6f	6.24	1.07	11.2	1.00
6g	6.33	1.04	11.7	1.03
6h	6.44	1.01	12.2	1.06

$$\begin{aligned} \langle v\sigma \rangle &= 6q\sqrt{\beta}(\mathrm{Ry}/I)^{3/2}\mathrm{e}^{-\beta}f(\beta) \times 10^{-8}\ \mathrm{cm^3\,s^{-1}}\,, \\ f(\beta) &= \mathrm{e}^{\beta}|\mathrm{Ei}(-\beta)|\,, \quad \beta = I/T\,, \end{aligned} \tag{4.2.10}$$

where T is the electron temperature and $\mathrm{Ei}(x)$ is the integral exponent. The function $f(x)$ is fitted to within 3% by

$$f(x) = \mathrm{e}^{x}|\mathrm{Ei}(-x)| \approx \ln\left(1 + \frac{0.562 + 1.4x}{x(1 + 1.4x)}\right), \quad x > 0\,. \tag{4.2.11}$$

The Lotz formulas (4.2.8–10) are convenient and useful for estimating the ionization cross sections and rates accurate to within a factor of 2.

The fitting parameters for ionization cross sections and rates obtained on the basis of experimental data and sophisticated calculations are given in [4.69, 70, 77]. Approximation parameters for ionization cross sections of H-like atoms are given in Sect. 4.2.3.

4.2.3 Fitting Parameters for H-like Ions. High-Energy Behavior

Fitting parameters for ionization cross section σ and Maxwellian rate coefficients $\langle v\sigma \rangle$ of H-like ions are presented in Table 4.13. They have been obtained in [4.84] by the least-squares method on the basis of accurate calculations [4.72] in the CBE approximation. The approximation formulas have the form:

$$\sigma = q(\mathrm{Ry}/I)^2 \frac{Cu}{(u+1)(u+\varphi)} [\pi a_0^2] , \quad u = E/I - 1 , \tag{4.2.12}$$

$$\langle v\sigma \rangle = 10^{-8}\ \mathrm{cm^3\,s^{-1}} q(\mathrm{Ry}/I)^{3/2} \mathrm{e}^{-\beta} G(\beta) , \tag{4.2.13}$$

$$G(\beta) = \frac{A\sqrt{\beta}}{\beta + \chi} , \quad \beta = I/T , \tag{4.2.14}$$

$$0 \leqslant u \leqslant 16 , \quad 1/8 \leqslant \beta \leqslant 8 , \tag{4.2.15}$$

where C, φ, A and χ are fitting parameters.

Cross section (4.2.12) has a maximum at

$$\sigma_{\max} = q(\mathrm{Ry}/I)^2 \frac{C}{(\sqrt{\varphi}+1)^2} [\pi a_0^2] , \quad u_{\max} = \sqrt{\varphi} .$$

The accuracy of (4.2.12–15) with parameters given in Table 4.13 is to within 15% as compared to the CBE numerical calculations [4.72].

For physical applications, the l-averaged ionization cross sections

$$\sigma_n = n^{-2} \sum_{l=0}^{n-1} (2l+1)\sigma_{nl}$$

are also required. Using the parameters given in Table 4.13 and applying the LSM method, one can obtain the corresponding fitting parameters for σ_n and $\langle v\sigma_n \rangle$ (Table 4.14).

Since the fitting parameters in Tables 4.13, 14 are smooth functions of the quantum numbers n and l, they can be extrapolated into the range of the higher numbers $n \geqslant 7$. Expansion of the fitting parameters on the $1/n^2$ parameter gives:

$$b = b_0 + b_1(1/n^2) + \cdots + b_k(1/n^{2k}) , \quad n \geqslant 7 , \quad b = C, \varphi, A, \chi . \tag{4.2.16}$$

The two first expansion coefficients b_0 na b_1 obtained by the LSM method for parameters C, φ, A, χ are presented in Table 4.15 that makes it possible to obtain ionization cross sections and rates for the states *ns*, *np*, *nd* and l-averaged values for $n \geqslant 7$.

Unfortunately, it is not possible to obtain expansion coefficients for the states *nl* with $n \geqslant 7$, $l \geqslant 3$ because of the lack of the CBE ionization cross sections for this range of quantum numbers.

Table 4.14. Fitting parameters C, φ and A, χ for the l-averaged cross sections σ_n and rates $\langle v\sigma_n \rangle$ (4.2.12–15); $u \leqslant 14$, $1/8 \leqslant \beta \leqslant 8$

n-state	C	φ	A	χ
$n = 1$	7.96	2.70	5.65	0.40
2	6.82	1.55	8.33	0.68
3	6.44	1.23	10.2	0.90
4	6.30	1.11	10.9	0.97
5	6.24	1.06	11.2	1.01
6	6.21	1.03	11.4	1.03

Table 4.15. Expansion coefficients (4.2.12–16) for ionization from the states with the principal quantum numbers $n \geqslant 7$; $u \leqslant 16$, $1/8 \leqslant \beta \leqslant 8$

Coeff.	State *ns*	*np*	*nd*	*l*-averaged
C_0	5.46	5.68	5.81	6.13
C_1	4.92	5.02	6.85	2.75
φ_0	1.23	1.19	1.14	0.97
φ_1	3.23	1.11	−0.59	2.33
A_0	8.28	9.13	9.70	11.8
A_1	−8.15	−0.37	17.8	−13.6
χ_0	0.84	0.90	0.94	1.07
χ_1	−1.28	−0.66	0.52	−1.54

The behavior of the ionization cross section for the process

$$X_z(nl^q) + e \rightarrow X_{z+1}(nl^{q-1}) + 2e$$

at high incident electron energies $E \gg I$ is of interest for many applications, e.g., for the investigation of the relation between the classical and quantum-mechanical cross sections.

The electron-impact ionization processes in the classical approximation and its modifications have been considered in various publications [4.50, 82–84]. There are three main expressions for the *classical* ionization cross sections derived particle from the Stabler formula [4.85] for the energy transfer from the incident electron to the target. These formulas do not comprise the Bethe logarithm (4.2.2) and have the form:

$$\sigma = q(\mathrm{Ry}/I)^2 \Phi(u)[\pi a_0^2]\,, \quad u = E/I - 1\,, \tag{4.2.17}$$

where

$$\Phi(u) = \frac{4u}{u+1} \tag{4.2.18}$$

in the classical Thomson formula [4.50],

$$\Phi(u) = \frac{4}{3} u \frac{5u+7}{(u+1)^3}\,, \tag{4.2.19}$$

and

$$\Phi(u) = \frac{4}{3}(u+1)^{-3} \begin{cases} u(5u+3)\,, & u \leqslant 1 \\ 4u^2 + 5u - 1\,, & u > 1 \end{cases} \tag{4.2.20}$$

in [4.84]. Classical ionization cross sections (4.17–20) have the asymptotic law $\sigma \propto A/u$ at $u \rightarrow \infty$, where the constant $A = 4.0$, 20/3 and 16/3, respectively.

The accurate calculations of the ionization cross sections in the Born approximation are given in [4.84, 86–88]. In [4.84], the expression for the ionization cross section at high electron energy has been deduced as

Table 4.16. Bethe constant $B(nl)$, $1s \geqslant nl \geqslant 6h$, for H-like systems [4.84]. The constants $B(n)$ correspond to the l-averaged values

State nl	$B(nl)$	State nl	$B(nl)$	State nl	$B(nl)$
1s	1.12	5s	2.37	$n = 7$	1.25
$n = 1$	1.12	5p	2.30	$n = 8$	1.25
2s	1.62	5d	1.94	$n = 9$	1.23
2p	1.05	5f	1.20	$n = 10$	1.26
$n = 2$	1.19	5g	0.40	$n = 11$	1.26
3s	1.93	$n = 5$	1.24	$n = 12$	1.26
3p	1.63	6s	2.55	$n = 13$	1.26
3d	0.83	6p	2.55	$n = 14$	1.26
$n = 3$	1.22	6d	2.31	$n = 15$	1.27
4s	2.16	6f	1.72		
4p	2.01	6g	0.92		
4d	1.47	6h	0.26		
4f	0.60	$n = 6$	1.24		
$n = 4$	1.23				

$$\sigma(nl) = (\mathrm{Ry}/I)^2 \frac{[\pi a_0^2]}{u}\left[qA + \frac{B(nl)}{n}\ln(4u)\right], \quad u \gg 1\,, \tag{4.2.21}$$

where the constant B (Bethe constant) is given by

$$B(nl) = n\frac{1}{n\alpha}\int_x^\infty \frac{\sigma_{nl}(\omega)}{\pi a_0^2}\frac{I}{\hbar\omega}\,\mathrm{d}\,\hbar\omega\,. \tag{4.2.22}$$

Here α is the fine-structure constant and $\sigma_{nl}(\omega)$ is the photoionization cross section from the nl state at frequency ω.

In the case of the ionization of H-like systems ($q = 1$),

$$\mathrm{H}_z(nl) + e \to \mathrm{H}_z^+ + 2e$$

from the states $1s \leqslant nl \leqslant 6h$, th constant $B(nl)$ in (4.2.16) has been found analytically [4.84] (Table 4.16).

The use of the Kramers (l-averaged) photoionization cross sections $\sigma_n^{\mathrm{Kr}}(\omega)$ (3.2.2) yields for all n states a single value of $B(n)$:

$$B(n) = \frac{64}{9\sqrt{3}\pi} \approx 1.31\,, \tag{4.2.23}$$

which is close to those given in Table 4.16 for the states $n \geqslant 2$.

It is seen from (4.2.22) that the logarithmic term is important if the scaled electron energy u satisfies the inequality

$$u > \mathrm{e}^{-4n}/4\,, \tag{4.2.24}$$

where n is the principal quantum number of the target state. For example, if $n = 1$ the dipole term prevails at energies $u > 10$; if $u < 10^3$ the logarithmic term has to be taken into account for the states with $n = 1$, 2 and 3. This is evident from Fig. 4.20, where ionization cross sections from the states $n = 1, 2$ and 4 of hydrogen ions calculated in the Coulomb–Born approximation with exchange (CBE) by the ATOM code [4.47] are shown.

The scaled Born ionization cross sections for the states from 1*s* up to 5*g* are given in Tables 4.17–19. A comparison of the asymptotic formula (4.2.15) with available *l*-averaged Born cross sections for H-like systems is given in Tables 4.20, 21.

4.3 Multiple Ionization

Multiple Ionization (MI) of atoms and ions by electron impact

$$e + A^{z+} \to e + A^{(z+n)+} + ne\,, \quad n > 1 \tag{4.3.1}$$

is an important elementary process which has to be taken into account in many physical applications: the charge-state evaluation of atoms exposed to an electron beam [4.89, 90], calibration of ion-beam probes used for plasma diagnostics in thermonuclear research reactors [4.91], formation of highly charged recoil ions, etc. Theoretical aspects of multiple electron transitions in atomic collisions, including multiple ionization, has been developed by *J.H. McGuire* (see [4.92–96]).

At present, MI processes are investigated mostly experimentally using the crossed-beam technique or its modifications for neutral atomic targets from Ne up to U and ejection up to 13 electrons, and with those for positive ions from Ar^+ up to W^{4+} with ejection of up to four electrons in the energy range from threshod to 10 keV (see, e.g., [4.66, 90, 97–116]). The typical relation between single- and multiple-ionization cross sections in e + Fe collisions is shown in Fig. 4.21.

The measured threshold energy E_{th} for MI cross sections σ_n corresponds to the minimal ionization energy I_n required to remove n outmost electrons, i.e.,

$$E_{\text{th}} = I_n = \sum_{i=0}^{n-1} I_{i,i+1}\,, \tag{4.3.2}$$

where $I_{i,i+1}$ is the one-electron ionization energy from the charge i to $i+1$. The values of I_n can be obtained from Table 4.22 [4.117, 118]. For example, the minimal energy I_3 required for ionization of three electrons in Ar is estimated to be: $I_3 = I(\text{Ar}) + I(\text{Ar}^+) + I(\text{Ar}^{2+}) = 15.76 + 27.6 + 40.9 = 84.26$ eV. The minimal ionization energies of the neutral atomic targets are given in Table 4.22.

The quantum mechanical calculations of MI cross sections, even for $n = 2$, are still unknown. Therefore, analytical semiempirical formulas are of special interest.

The double-ionization cross sections σ_2 are often estimated by the Gryzinski binary-encounter approximation formula [4.119] which reproduces the shape of the cross section reasonably well, but can greatly overestimates its size.

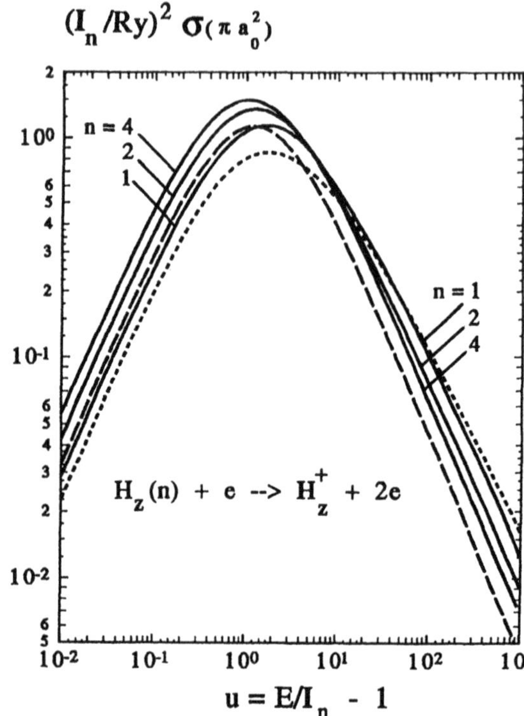

Fig. 4.20. Scaled ionization cross sections for H-like ions ($z = 10$, $n = 1$, 2 and 4) vs reduced incident electron energy $u = E/I - 1$: *solid curves* Coulomb–Born-Exchange cross sections (ATOM code [4.47]); *dashed curve* classical ionization cross section (4.2.15), corresponding to the constant $A = 16/3$; *dot-dashed curve* Lotz formula (4.2.9)

Table 4.17. Scaled Born ionization cross sections $z^4\sigma(nl)$ (in cm^2) of H-like systems [4.84] from 1*s* up to 3*d* states vs $u = E/I_{nl} - 1$*

u \ nl	1*s*	2*s*	2*p*	3*s*	3*p*	3*d*
0.065	3.08 − 18	6.92 − 17	5.81 − 17	4.10 − 16	3.70 − 16	2.73 − 16
0.25	1.85 − 17	3.70 − 16	3.43 − 16	2.00 − 15	1.97 − 15	1.74 − 15
1.00	6.15 − 17	9.42 − 16	1.08 − 15	4.67 − 15	5.13 − 15	5.88 − 15
4.00	7.39 − 17	9.77 − 16	1.12 − 15	4.70 − 15	5.02 − 15	5.86 − 15
13.0	4.39 − 17	5.67 − 16	5.93 − 16	2.64 − 15	2.71 − 15	2.92 − 15
16.0	3.74 − 17	4.83 − 16	4.93 − 16	2.24 − 15	2.26 − 15	2.41 − 15
50.0	1.58 − 17	2.01 − 16	1.92 − 16	9.06 − 16	8.89 − 16	9.00 − 16
200	4.80 − 18	6.05 − 17	5.49 − 17	2.69 − 16	2.56 − 16	2.46 − 16
800	1.39 − 18	1.73 − 17	1.51 − 17	7.61 − 17	7.10 − 17	6.55 − 17
3200	3.92 − 19	4.84 − 18	4.08 − 18	2.11 − 17	1.93 − 17	1.72 − 17

* 3.08 − 18 means 3.08×10^{-18}

Table 4.18. As in Table 4.77 for 4 l states

u \ nl	$4s$	$4p$	$4d$	$4f$
0.065	1.40 − 15	1.30 − 15	1.05 − 15	7.33 − 16
0.25	6.32 − 15	6.48 − 15	6.20 − 15	5.34 − 15
1.00	1.50 − 14	1.59 − 14	1.72 − 14	1.95 − 14
4.00	1.45 − 14	1.50 − 14	1.66 − 14	1.89 − 14
13.0	7.80 − 15	7.86 − 15	8.48 − 15	9.04 − 15
16.0	6.56 − 15	6.54 − 15	6.99 − 15	7.35 − 15
50.0	2.61 − 15	2.56 − 15	2.64 − 15	2.67 − 15
200	7.65 − 16	7.36 − 16	7.41 − 16	7.13 − 15
800	2.15 − 16	2.04 − 16	2.01 − 16	1.86 − 16
3200	5.90 − 17	5.57 − 17	5.35 − 17	4.78 − 17

Table 4.19. As in Table 4.17 for 5 l states

u \ nl	$5s$	$5p$	$5d$	$5f$	$5g$
0.065	3.53 − 15	3.37 − 15	2.89 − 15	2.26 − 15	1.60 − 15
0.25	1.53 − 14	1.60 − 14	1.59 − 14	1.48 − 14	1.26 − 14
1.00	3.65 − 14	3.84 − 14	4.05 − 14	4.39 − 14	4.92 − 14
4.00	3.37 − 14	3.43 − 14	3.73 − 14	4.15 − 14	4.64 − 14
13.0	1.76 − 14	1.75 − 14	1.87 − 14	2.01 − 14	2.13 − 14
16.0	1.48 − 14	1.46 − 14	1.55 − 14	1.65 − 14	1.72 − 14
50.0	5.82 − 15	5.66 − 15	5.91 − 15	6.07 − 16	6.11 − 15
200	1.69 − 15	1.63 − 15	1.67 − 15	1.66 − 15	1.61 − 15
800	4.73 − 16	4.52 − 16	4.54 − 16	4.40 − 16	4.14 − 16
3200	1.29 − 16	1.24 − 16	1.23 − 16	1.16 − 16	1.06 − 16

Table 4.20. Ionization cross sections σ (in cm^2) from $1s$ states in hydrogen vs incident electron energy E

E [eV]	[4.13] Recomm.	[4.84] Born approx.	(4.2.22) $A = 20/3$ $B = 1.12$	(4.2.22) $A = 16/3$ $B = 1.12$
20	8.59 − 16	3.59 − 17	1.38 − 15	1.13 − 15
100	2.99 − 16	6.40 − 17	1.43 − 16	1.24 − 16
400	2.41 − 17	2.46 − 17	3.71 − 17	3.32 − 17
800	1.39 − 17	1.40 − 17	1.94 − 17	1.74 − 17
2000	6.26 − 18	6.31 − 17	8.32 − 17	7.51 − 17
4000	3.51 − 18	3.42 − 18	4.38 − 18	3.98 − 18
10000	1.53 − 18	1.50 − 18	1.87 − 17	1.71 − 17

Table 4.21. The l-averaged ionization cross sections (in cm^2) for the states $n = 2, 4$ and 10 in neutral hydrogen

$n = 2$				
E [eV]	[4.13] Recomm.	[4.84] Born approx.	(4.2.22) $A = 20/3$ $B = 1.28$	(4.2.22) $A = 16/3$ $B = 1.28$
20	9.92 − 16	1.00 − 15	2.47 − 15	2.09 − 15
100	3.20 − 16	3.12 − 16	4.81 − 16	4.14 − 16
400	1.00 − 16	9.20 − 16	1.28 − 16	1.11 − 16
800	5.28 − 17	4.88 − 17	6.64 − 17	5.84 − 17
2000	2.32 − 17	2.08 − 17	2.79 − 17	2.47 − 17
4000	1.24 − 17	1.10 − 17	1.45 − 17	1.29 − 17
10000	5.42 − 18	4.63 − 18	6.07 − 18	5.43 − 18

$n = 4$				
E [eV]	[4.13] Recomm.	[4.88] Born approx.	(4.2.22) $A = 20/3$ $B = 1.31$	(4.2.22) $A = 16/3$ $B = 1.31$
20	6.15 − 15	5.30 − 15	8.14 − 15	6.80 − 15
100	1.41 − 15	1.21 − 15	1.68 − 15	1.41 − 15
400	3.77 − 16	3.24 − 16	4.38 − 16	3.73 − 16
800	1.94 − 16	1.67 − 16	2.24 − 16	1.92 − 16
2000	8.03 − 17	6.96 − 17	9.26 − 17	7.98 − 17
4000	4.12 − 17	3.57 − 17	4.74 − 17	4.09 − 17
10000	1.70 − 17	1.48 − 17	1.95 − 17	1.70 − 17

$n = 10$				
E [eV]	[4.13] Recomm.	[4.88] Born approx.	(4.2.22) $A = 20/3$ $B = 1.31$	(4.2.22) $A = 16/3$ $B = 1.31$
20	4.11 − 14	4.46 − 14	4.52 − 14	3.72 − 14
100	8.55 − 15	9.30 − 15	9.25 − 15	7.65 − 15
400	2.19 − 15	2.41 − 15	2.36 − 15	1.96 − 15
800	1.11 − 15	1.22 − 15	1.20 − 15	9.95 − 16
2000	4.51 − 16	5.00 − 16	4.85 − 16	4.05 − 16
4000	2.28 − 16	2.54 − 16	2.45 − 15	2.05 − 15
10000	9.27 − 17	1.04 − 16	9.96 − 17	8.36 − 17

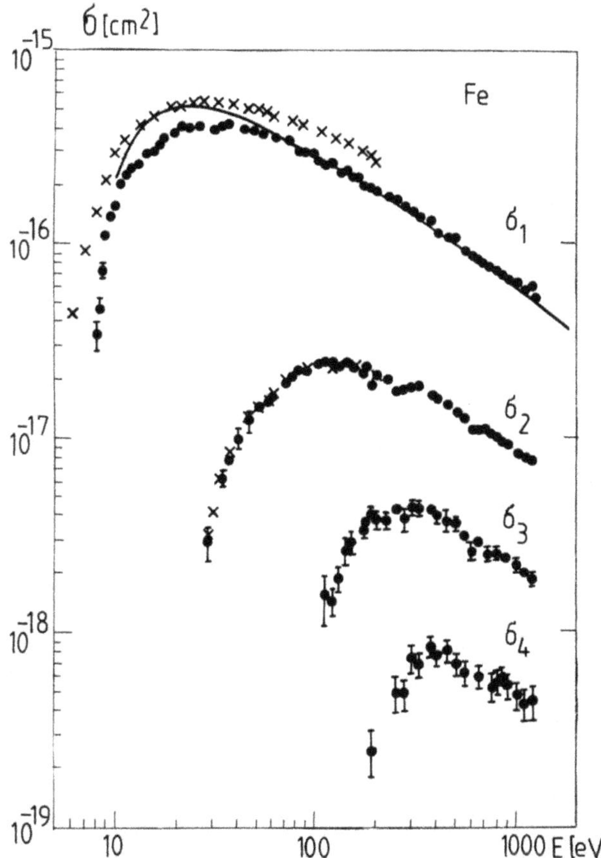

Fig. 4.21. Single- and multiple-ionization cross sections of Fe by electron impact. *Crosses* experiment [4.102]; *solid circles* [4.106]; *solid curve* single-ionization cross section calculated in the Born approximation [4.117]

Semiempirical formulas to predict double- and triple-ionization cross sections of some specific atomic targets are considered in [4.120, 121]. A scaling of multiple-ionization cross sections and semiempirical formulas for σ_n are considered in [4.122].

In [4.123–125], on the basis of available experimental data and the assumption of the Born–Bethe dependence of σ_n on the incident electron energy E, the semiempirical formulas for MI cross sections of atoms and ions by electron impact for ejection of three or more electrons was suggested

$$\sigma_n(u) = \frac{a(n)N^{b(n)}}{(I_n/\mathrm{Ry})^2}\left(\frac{u}{u+1}\right)^c \frac{\ln(u+1)}{u+1}\,[10^{-18}\,\mathrm{cm}^2]\,, \quad n \geqslant 2\,, \tag{4.3.3}$$

$$u = E/I_n - 1\,, \tag{4.3.4}$$

Table 4.22. The minimal ionization energies I_n, $n \leqslant 7$ (in eV) for neutral atoms (estimated from [4.117, 118])

Atom	Number of the target electrons N	I_1	I_2	I_3	I_4	I_5	I_6	I_7
He	2	24.6	79.0					
Ne	10	21.6	62.7	126	223	349	513	734
Mg	12	7.65	22.6	103	212	353	539	788
Ar	18	15.8	43.4	84.3	144	219	310	450
Fe	26	7.87	23.5	56.2	114	197	305	438
Cu	29	7.73	29.0	77.3	153	256	386	543
Ga	31	6.00	25.8	56.7	118	214	344	508
Ge	32	7.90	24.3	58.6	105	192	317	479
Se	34	9.75	32.1	66.3	112	183	269	416
Kr	36	14.0	41.9	83.7	139	209	294	410
Ag	47	7.58	29.1	70.1	131	211	310	429
In	49	5.79	23.9	51.9	106	184	284	408
Sn	50	7.34	22.4	53.1	94.8	169	267	392
Sb	51	8.64	26.0	52.8	98.0	155	250	372
Te	52	9.01	28.3	58.1	98.4	160	233	353
Xe	54	12.13	35.7	70.8	117	177	249	347
Pb	82	7.42	23.0	55.3	98.9	166	254	363
Bi	83	7.29	24.4	51.3	97.4	156	241	350
U	92	6.0	17.6	35.7	66.6	116	185	275

Table 4.23 Fitting parameters $a(n)$ and $b(n)$ in (4.3.3) for removal of $2 \leqslant n \leqslant 10$ electrons from atoms or ions

n	$a(n)$	$b(n)$
2	14.0	1.08
3	6.30	1.20
4	0.50	1.73
5	0.14	1.85
6	0.049	1.96
7	0.021	2.00
8	0.0096	2.00
9	0.0049	2.00
10	0.0027	2.00

where the constant $c = 1$ for neutral targets and $c = 0.75$ for positive and negative ions; E is the incident electron energy, I_n is the minimal ionization energy required to eject n electrons from the target. The approximation parameters $a(n)$ and $b(n)$ were obtained by fitting (4.3.3) to reliable experimental data for electron–atom and electron–ion collisions at low as well as high electron energies. The results for ejection of $2 \leqslant n \leqslant 10$ electrons are given in Table 4.23; for $n > 10$ one can use the asymptotic values:

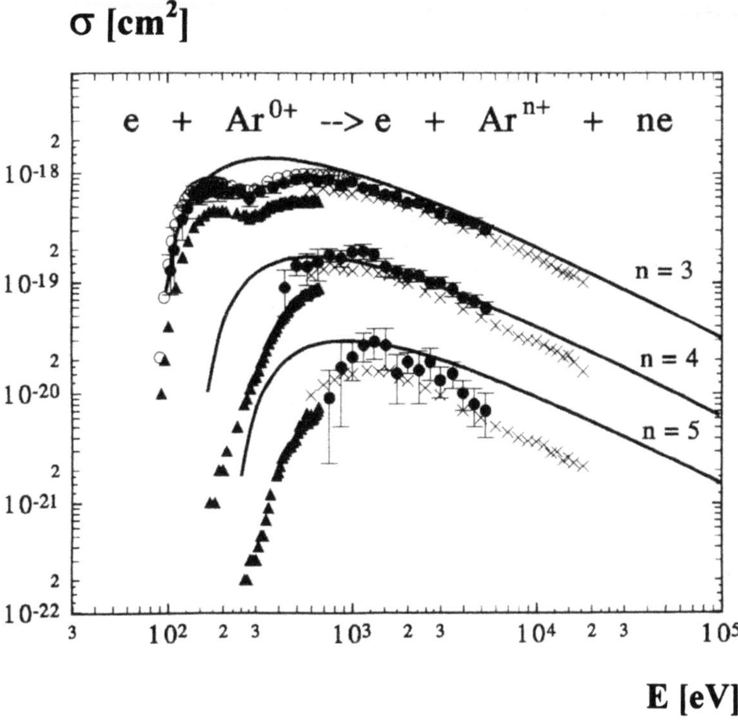

Fig. 4.22. Triple-, quadruple- and quintuple-ionization cross sections of Ar by electron impact. *Experiment*: (×) [4.97]; (o) [4.100]; • [4.104]; ▵ [4.105]. *Solid curves*: semiempirical formula (4.3.3)

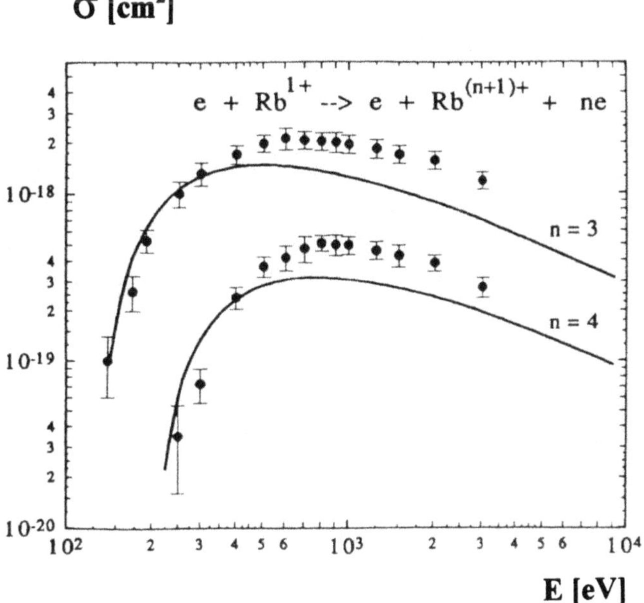

Fig. 4.23. Triple- and quadruple-ionization cross sections of Rb^+ ions by electrons. *Solid circles* experiment [4.110]; *solid curves* (4.3.3)

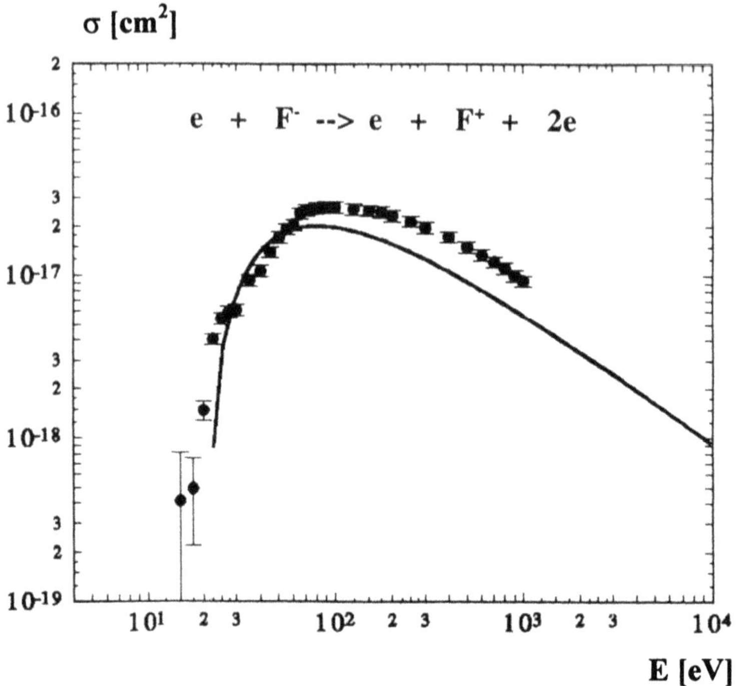

Fig. 4.24. Double ionization cross sections of F^- ions by electron impact. *Solid circles* experiment [4.125]; *solid curve* (4.3.3)

$$a(n) \approx 1350n^{-5.7}, \quad b(n) = \text{const} = 2.00, \quad n > 10. \tag{4.3.5}$$

According to (4.3.5), the increasing number of the ejected electrons n, the multiple-ionization cross section approximately decreases as $\sigma_n \propto n^{-6}$.

The cross section σ_n (4.3.3) reaches its maximum at $u_n^{\max} \approx 3.2$, which corresponds to $E_n^{\max} \approx 4.2I_n$, i.e.,

$$\sigma_n^{\max} \approx 0.27a(n)N^{b(n)}(I_n/\text{Ry})^{-2}[10^{-18}\,\text{cm}^2] \tag{4.3.6}$$

Equation (4.3.3) gives a quite good description of the MI cross-section behavior in a wide range of the incident electron energies. Figures 4.22–24 show typical MI cross sections of neutral metallic and inert-gas targets by electron impact in comparison with experimental data. The accuracy of the semiempirical formula (4.3.3) is about a factor of 2 (see [4.123–125] for details).

In general, the MI processes of atoms and ions are much richer and of greater variety than the single-electron ionization. They must be taken into account in considering the kinetics of a plasma and its ionization states. An adequate theory is still required for such applications.

5 Ion–Atom Collisions

Collisions of ions with neutral species are currently a subject of fundamental interest. Processes as excitation, ionization and electron capture are important in many applied areas, e.g., confined plasma research, beam transport and neutral beam injection for plasma heating, production of slow highly charged recoil ions, diagnostics of the edge plasma and others. The general experimental and theoretical aspects of ion–atom, collisions are considered in various reviews and monographs [5.1–12]. In this Chapter, excitation, ionization and electron capture processes arising in ion–atom collisions are considered. Special attention is paid to the processes involving negative hydrogen atoms.

5.1 Excitation

Excitation processes of atoms by ion impact

$$X^{z+} + A \rightarrow X^{z+} + A^* \tag{5.1.1}$$

have been investigated mostly for hydrogen and helium targets [5.13–27]; for heavy atomic targets, the data on the excitation cross sections are given in [5.28–30].

Recommended excitation cross sections for proton–hydrogen and proton–helium collisions [5.14, 16] are given in Tables 5.1–3. It is seen that the cross sections for excitation from the ground state into the $n = 2$ states comprise two distinguished maxima.

Several theoretical approaches beyond the first Born approximation are applied for calculations of excitation cross sections: Dipole – Approximation – Close – Coupling (DACC), the Classical – Trajectory – Monte – Carlo approximation (CTMC), the Unitarized – Distorted – Wave – Approximation (UDWA) and the Atomic – Orbital Close-Coupling method (AO–CC) (see [5.10] for details).

For H and He targets, the excitation cross sections σ are scaled by the law derived in [5.31] within the DACC approximation valid for relatively high energies ($u > 15$)

$$F(u) = \sigma(E)/z\,, \quad u = E/z\ [\mathrm{keV/amu}]\,, \quad u > 15\,, \tag{5.1.2}$$

Table 5.1. Proton-impact excitation cross sections (in cm^2) of H atoms from the ground state into the $2s$, $2p$ and $n = 2$ states [5.16]

Energy [eV/amu]	H (2s)	H (2p)	H (n = 2)
5.00E + 02	1.16E − 19	–	–
6.00E + 02	–	5.60E − 18	5.77E − 18
1.00E + 03	4.97E − 19	2.30E − 17	2.34E − 17
2.00E + 03	1.65E − 18	2.89E − 17	3.04E − 17
5.00E + 03	5.81E − 18	3.05E − 17	3.63E − 17
1.00E + 04	4.48E − 18	2.35E − 17	2.80E − 17
2.00E + 04	9.81E − 18	4.12E − 17	5.10E − 17
5.00E + 04	1.45E − 17	8.57E − 17	1.00E − 16
1.00E + 05	8.53E − 18	8.15E − 17	9.00E − 17
2.00E + 05	4.47E − 18	5.94E − 17	6.38E − 17
5.00E + 05	1.88E − 18	3.36E − 17	3.55E − 17
1.00E + 06	9.65E − 19	1.99E − 17	2.09E − 17
2.00E + 06	4.88E − 19	1.15E − 17	1.20E − 17
5.00E + 06	1.96E − 19	5.30E − 18	5.50E − 18

Table 5.2. Proton-impact excitation cross sections (in cm^2) for the processes $H^+ + H(n = 2) \to H^+ + H(m)$ [5.16]

Energy [eV/amu]	$n = 2 \to m = 3$	$n = 2 \to m = 4$	$n = 2 \to m = 5$
5.00E + 02	2.73E − 16	2.31E − 17	7.89E − 18
1.00E + 03	4.54E − 16	5.20E − 17	1.78E − 17
2.00E + 03	6.95E − 16	1.04E − 16	3.56E − 17
5.00E + 03	1.18E − 15	2.33E − 16	7.99E − 17
1.00E + 04	1.68E − 15	3.55E − 16	1.22E − 16
2.00E + 04	2.03E − 15	4.35E − 16	1.49E − 16
5.00E + 04	1.77E − 15	3.71E − 16	1.26E − 16
1.00E + 05	1.33E − 15	2.54E − 16	8.56E − 17
2.00E + 05	8.59E − 16	1.52E − 16	5.22E − 17
5.00E + 05	4.33E − 16	7.52E − 17	2.60E − 17
1.00E + 06	2.49E − 16	4.09E − 17	1.45E − 17
2.00E + 06	1.39E − 16	2.23E − 17	7.90E − 18
5.00E + 06	6.27E − 17	1.02E − 17	3.38E − 18

where z is the charge of the incident ion. Recommended, scaled excitation cross sections (5.1.2) for He from the ground state are given in Table 5.4.

In the work [5.32], using the adiabatic theory, DAAC and classical impulse approximation, the scaling relations for dipole-allowed ($1s$–np) and dipole-forbidden ($1s$–nl, $\Delta l \neq 1$) transitions in atoms from the ground state have been suggested in the form:

Table 5.3. Recommended excitation cross sections (in units of 10^{-20} cm^2) for $H^+ + He(1\,^1S) \to H^+ + He(n\,^1L)$ reactions [5.14]

E [keV]	$2\,^1S$	$3\,^1S$	$4\,^1S$	$2\,^1P$	$3\,^1P$	$4\,^1P$	$3\,^1D$	$4\,^1D$
6	71.1			194.5			19.5	
8							27.0	
10	241	27.4	8.82	331.3	45.5	14.8	29.9	12.6
12.5		36.1	10.4				37.1	16.7
14	460			311.4				
15		43.1	11.5		89.3	32.5	38.7	16.5
17.5								13.5
20	598	84.6	20.6	295.3	101	40.0	29.9	11.6
25		112	33.8		100	40.0	21.8	10.2
28	536			428				
30		116	40.8	425	104	45.6	23.4	10.4
35			44.8	512	125	52.5	26.7	11.9
40	567	128	48.9	556	136	56.3	32.0	13.1
50	547	131	47.1	793	194	75.0	29.3	13.7
60	520	112	44.8	954	234	87.5	29.1	14.2
70	523	94.7	45.1	958	234	98.7	27.9	13.6
80	483	102.6	41.7	1152	282	117	25.8	13.3
90	451	89.3	38.9	1183	289	114	23.3	12.8
100	407	90.8	35.1	1273	311	119	21.6	11.4
110	372	83.0	32.1	1340	328	144	20.0	11.2
120	343	77.8	29.6	1306	320	131	18.9	10.3
130	317	67.6	27.3	1333	326	135	18.1	9.81
140	287	69.8	24.7	1350	330	129	17.1	8.91
150	256	63.7	22.1	1287	323	130	16.1	8.19
200	219	49.3	18.9	1211	299	117	12.8	6.0
300	145	32.6	12.5	1006	256	101	7.1	3.8
400	110	24.8	9.5	864	220	93	6.0	3.2
500	87	19.6	7.5	787	197	77	4.5	2.4
600	69.6	15.7	6.0	682	178	72	3.9	2.1
700	59.2	13.3	5.1	647	165	63	3.2	1.7
800	53.4	12.0	4.6	568	150	58	2.8	1.5
1000	41.8	9.4	3.6	505	129	51	2.1	1.1
2000	19.78	4.46	1.71	304.2	75.1	30.1	1.037	0.552
3000	13.23	2.98	1.14	223.6	55.1	22.1	0.696	
4000	9.94	2.24	0.857	178.8	44.1	17.7	0.524	
5000	7.96	1.79	0.686	150.0	37.0	14.8	0.420	

$$F(u) = \sigma \frac{(\Delta E)^2}{zf}, \quad u = \frac{E}{z\Delta E}\,[\mathrm{keV/amu}], \quad \Delta l = \pm 1, \tag{5.1.3}$$

$$F(u) = \sigma \frac{(\Delta E)^4 n^3}{z}, \quad u = \frac{E}{z\Delta E}\,[\mathrm{keV/amu}], \quad \Delta l \neq \pm 1, \tag{5.1.4}$$

where ΔE (in a.u.) and f are the transition energy and the oscillator strength, respectively.

Table 5.4. Scaled excitation cross sections (5.1.2) for He from the ground state [5.15]

E/z [keV/amu]	σ/z [10^{-20} cm^2]						
	$3\,^1S$	$4\,^1S$	$5\,^1S$	$3\,^1P$	$3\,^1D$	$4\,^1D$	$5\,^1D$
15	65.7	14.2	6.99	106	25.8	10.9	6.30
20	74.7	24.2	11.9	132	38.6	16.3	8.89
30	93.6	33.4	17.9	181	56.4	23.8	12.1
40	103.2	36.1	19.0	227	62.1	26.2	12.6
50	102.9	34.2	18.9	269	64.0	27.0	12.4
60	96.6	31.1	18.2	302	62.8	26.5	11.8
70	88.9	28.1	17.0	327	59.3	25.0	11.0
80	82.1	25.3	15.8	345	54.5	23.0	10.3
90	76.3	22.8	14.6	359	49.1	20.7	9.55
100	71.2	20.6	13.5	369	44.1	18.6	8.90
110	66.6	18.7	12.4	377	40.1	16.9	8.30
120	62.4	17.0	11.4	381	37.0	15.6	7.76
130	58.4	15.7	10.6	382	34.1	14.4	7.27
140	54.5	14.6	9.85	379	31.8	13.4	6.84
150	50.7	13.7	9.25	374	29.2	12.3	6.45
160	46.8	12.8	8.67	367	26.3	11.1	6.07

According to [5.32], the scaled excitation cross sections for $1s$–np transitions in H and He are well described by the universal fitting formula

$$F(u) = \frac{A\exp(-a/u_{150}^{1/2})\ln(e + cu_{150})}{(1 + Cu_{150}^{-b})u_{150}}\,[10^{-17}\,\mathrm{cm}^2]\,, \qquad u_{150} = u/150\,, \tag{5.1.5}$$

where $F(u)$ and u are given by (5.1.3), and the fitting coefficients are: $A = 3.65$, $C = 0.26$, $a = 0.1$, $b = 2.15$ and $c = 3.5$; $e = 2.718282\ldots$.

For $1s$–ns and $1s$–nd transitions, one has

$$F(u) = \frac{AB}{Bu_{50}^{-b}\exp(a/u_{50}^{1/2}) + Au_{50}}\,[10^{-17}\,\mathrm{cm}^2]\,, \quad u_{50} = u/50\,, \tag{5.1.6}$$

where $F(u)$ and u are given by (5.1.4) and the fitting parameters are $A = 4.12$, $B = 2.45$, $a = 0.1$ and $b = 1.7$.

The scalings (5.1.3, 4) well describe the available experimental data for H and He (Fig. 5.1) for reduced energies $u > 25$ keV/amu. These scalings also show that, at a fixed energy E, the non-scaled excitation cross section as a function of z does not saturate at high z values but decreases after reaching a maximum.

At low energies, $E/z < 10$ keV/amu, AO–CC calculations of the excitation cross sections for the Li($2s$–$2p$) transition [5.29] and experimental and theoretical data on the Na($3s$–$3p$) transition [5.30] are nearly independent on the incident ion chanrge z (Fig. 5.2) (see [5.29–30] for details).

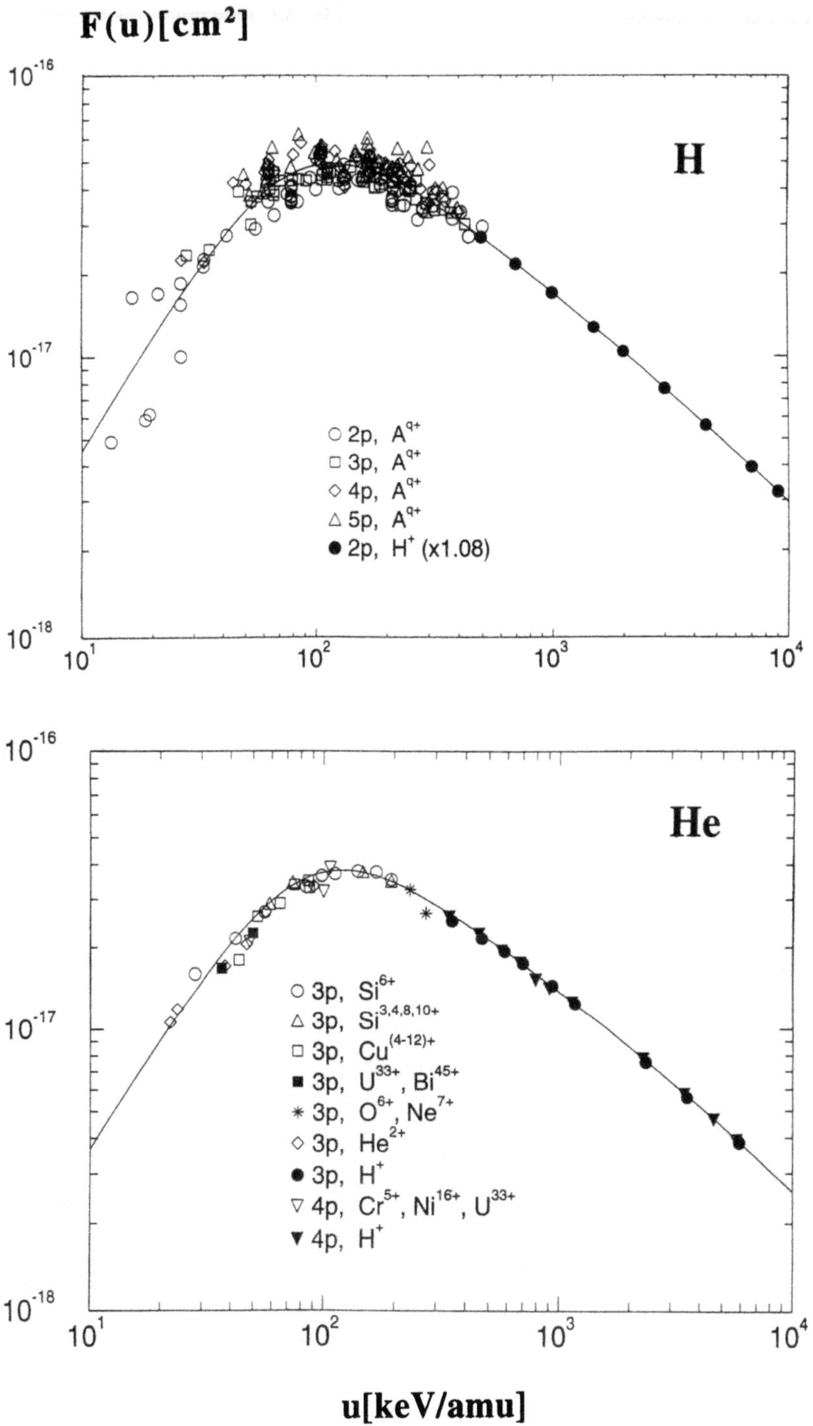

Fig. 5.1. Scaled dipole excitation cross sections (5.1.3) for $1s$–np transitions in H and He (n = 2–5) induced by protons and multicharged ions [5.32]. For H atoms: *symbols* experiment [5.15–17]; *solid curve* fitting (5.1.5). For He atoms: *symbols* experiment [5.14, 27]; *solid curve* fitting (5.1.5)

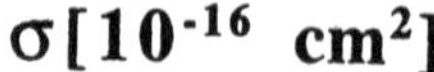

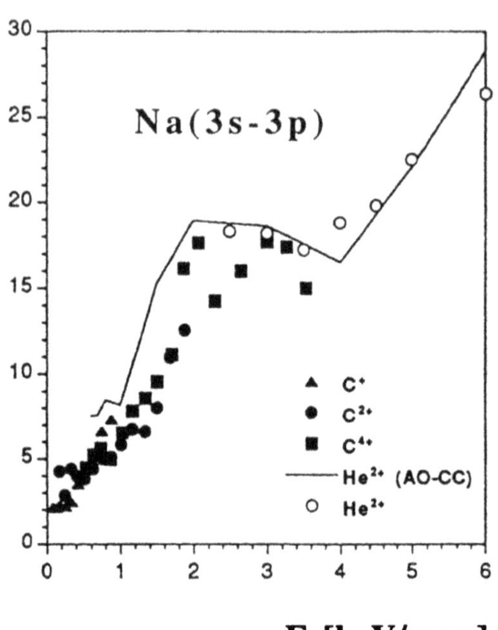

Fig. 5.2. Excitation cross section for transition Na($3s$–$3p$) induced by helium and carbon ions. Experimental data for incident helium ions are from [5.28] and for carbon ions from [5.30], respectively. *Solid curve* theory atomic-orbital close-coupling calculations [5.30]

5.2 Ionization

5.2.1 Single Ionization

Single ionization in ion–atom collisions

$$X^{z+} + A \rightarrow X^{z+} + A^{+} + e \qquad (5.2.1)$$

has been studied mainly for hydrogen and helium targets being in their ground state [5.4–6, 21, 31, 34–36]. Ionization of excited hydrogen atoms is considered in various theoretical works [5.16, 23, 37, 38]. The data on ionization of multi-electron atomic targets are given in [5.39, 40] and on the inner-shell ionization cross sections are presented in [5.41–49].

The recommended ionization cross section for proton–hydrogen collisions are given in Tables 5.5, 6. Figures 5.3, 4 show experimental and calculated ionization cross sections of H and He by protons, He^{2+} and Li^{3+} ions.

There are several scaling laws applying for ionization cross sections σ of H and He by charged ions. In the high velocity region ($v > 1$ a.u.) σ-values are scaled according to the law suggested in [5.31]:

$$F(u) = \sigma(E)/z\,, \quad u = E/z\ [\mathrm{keV/amu}]\,, \qquad (5.2.2)$$

where $F(u)$ is the universal function of the reduced energy u.

Table 5.5. Cross sections (in cm^2) for the ionization process $H^+ + H(n) \rightarrow H^+ + H^+ + e$ [5.15]

E [eV/amu]	$n = 1$	$n = 2$	$n = 3$
5.00E + 02	1.46E − 20	2.06E − 17	9.21E − 16
1.00E + 03	1.46E − 19	6.97E − 17	2.29E − 15
2.00E + 03	1.02E − 18	2.14E − 16	4.95E − 15
5.00E + 03	6.24E − 18	8.20E − 16	1.13E − 14
1.00E + 04	1.94E − 17	2.03E − 15	1.67E − 14
2.00E + 04	6.73E − 17	3.54E − 15	1.38E − 14
5.00E + 04	1.43E − 16	2.82E − 15	7.24E − 15
1.00E + 05	1.10E − 16	1.71E − 15	3.93E − 15
2.00E + 05	6.99E − 17	9.35E − 16	2.14E − 15
5.00E + 05	3.48E − 17	3.99E − 16	9.49E − 16
1.00E + 06	1.94E − 17	2.13E − 16	5.15E − 16
2.00E + 06	1.05E − 17	1.12E − 16	2.73E − 16
5.00E + 06	4.62E − 18	4.84E − 17	1.14E − 16

Table 5.6. Same as in Table 5.5, as a function of the scaled proton energy

$(3/n)^2E$ [eV/amu]	$n = 4$	$n = 5$	$n = 6$
5.00E + 02	6.34E − 15	2.60E − 14	7.81E − 14
1.00E + 03	1.37E − 14	5.22E − 14	1.56E − 13
2.00E + 03	2.74E − 14	9.86E − 14	2.47E − 13
5.00E + 03	5.03E − 14	1.23E − 13	2.25E − 13
1.00E + 04	4.69E − 14	8.91E − 14	1.38E − 13
2.00E + 04	3.02E − 14	4.98E − 14	7.50E − 14
5.00E + 04	1.35E − 14	2.28E − 14	3.47E − 14
1.00E + 05	7.58E − 15	1.26E − 14	1.89E − 14
2.00E + 05	4.14E − 15	6.76E − 15	1.00E − 14
5.00E + 05	1.80E − 15	2.91E − 15	4.31E − 15
1.00E + 06	9.49E − 16	1.53E − 15	2.25E − 15
2.00E + 06	4.97E − 16	7.98E − 16	1.18E − 15
5.00E + 06	2.10E − 16	3.37E − 16	4.95E − 16

Another scaling relation used was obtained within the Bethe–Born approximation [5.63, 64]:

$$\sigma(E) = z^2\sigma_B(E)e^{-Az/v} , \qquad (5.2.3)$$

where σ_B is the Bethe ionization cross section of H atoms by protons and A is the fitting parameter determined from experiment.

Analysis of the available experimental data on ionization cross sections of H and He at low energies ($v < 1$ a.u.) has shown [5.65] that in this energy range the scaling law

$$F(u) = \sigma(E)/z , \quad u = v/z^{1/4} \text{ [a.u.]} \qquad (5.2.4)$$

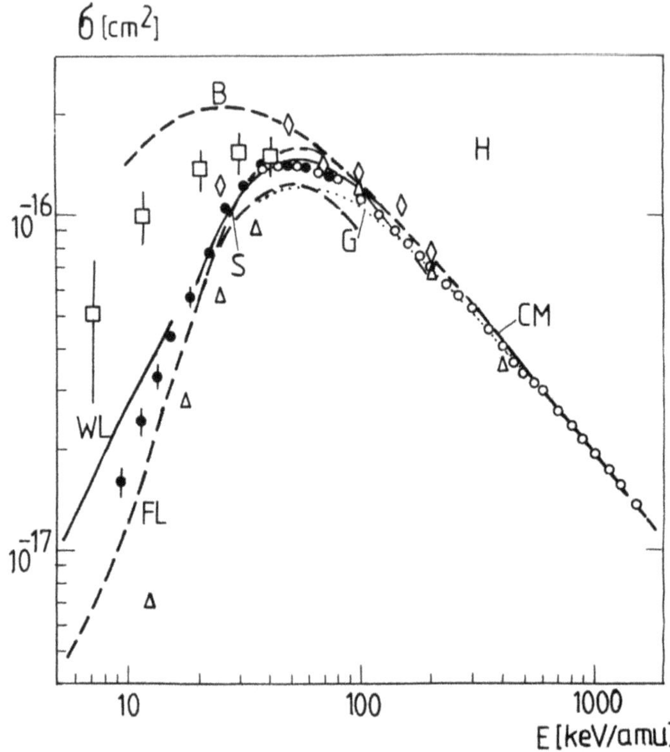

Fig. 5.3. Ionization cross sections of H by proton impact. *Experiment* • [5.50], ○ [5.51], □ [5.52]; *Theory B* Born approximation (5.53); △ CTMC calculations [5.54]; *S* double-centre atomic state method [5.55]; *G* Glauber approximation [5.56]; *CM – DW* approximation [5.57]; ◇ CTMC calculations [5.58]; *FL* double-centre atomic state method [5.59]; *WL* triple-centre atomic state method [5.60]

can be used (Fig. 5.5). The scaling (5.2.4) also agrees well with the results of the hidden-crossing model [5.66, 67] for ionization of H by highly charged ions that yields the function $F(u)$ in (5.2.4):

$$F(u) = Au\, e^{-c/u}, \quad u = v/z^{1/4}\ [\text{a.u.}]\,. \tag{5.2.5}$$

Here A and C are constants for a given target.

5.2.2 Double Ionization of He

Double ionization of He by heavy particles have been the subject of interest in atomic physics during the last years (see, e.g., [5.68, 74]). Its investigation is fundamental for the understanding of the role of the static and dynamic electron–electron correlations in a multiple ionization process.

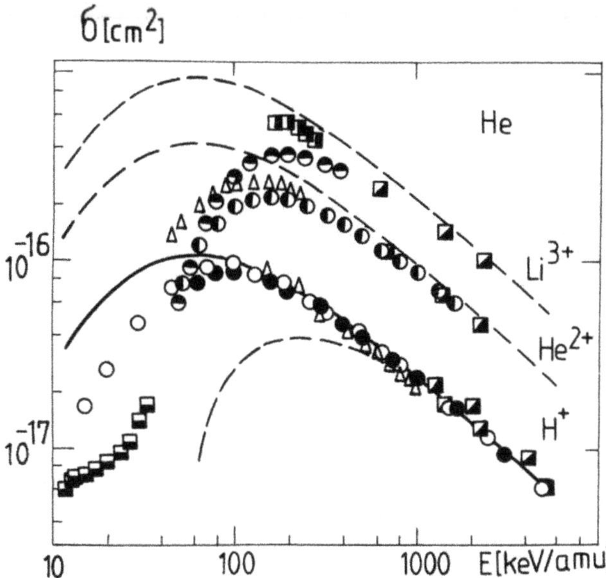

Fig. 5.4. Single-ionization cross sections of He by H^+, He^{2+} and Li^{3+} ions. *Symbols* experiment [5.61]; Theory: *solid curve* Born approximation [5.62]; *dashed curves* z^2 scaled Born approximation; *dot-dashed curve* ionization by electron impact [5.63]

From the theoretical point of view even in such a simple target, two general problems arise: a choice of the correlated two-electron wavefunction of a target such as helium, and a reasonable approximation for the continuum wavefunctions in the final channel. The first problem was analyzed in [5.68–73]. In [5.73], the first problem was reduced to the determination of the asymptotic value of the shake-off ratio of double-to-single ionization cross sections, and the second problem was considered using the generalization of the *Volkov–Keldych* states introduced in [5.36]. The double-ionization cross section σ_2 in [5.73] was obtained on the basis of the non-perturbation approach

$$\begin{aligned} &\sigma_2 = 2.6 \times 10^{-3}\sigma_1 + 0.864(z/v)^4 \exp(-4.1z/v^2 - 2.16v)\,[\pi a_0^2]\,, \\ &v^2/z > 1\,, \end{aligned} \tag{5.2.6}$$

where σ_1 is the single ionization cross section of He, z is the charge of the incident ion and v is the relative velocity (in a.u.) of colliding particles. Using the fit of *Gillespie* [5.64] for the single ionization cross section σ_1, the ratio $R = \sigma_2/\sigma_1$ was obtained as

$$R = \sigma_2/\sigma_1 = 2.6 \times 10^{-3} + (z/v)^2 F(z, v)\,, \tag{5.2.7}$$

$$F(z, v) = \frac{0.442 \exp(-2.1z/v^2 - 2.16/v)}{1.448 + \ln[v^2/(1 - v^2/c^2)] - v^2/c^2}\,, \tag{5.2.8}$$

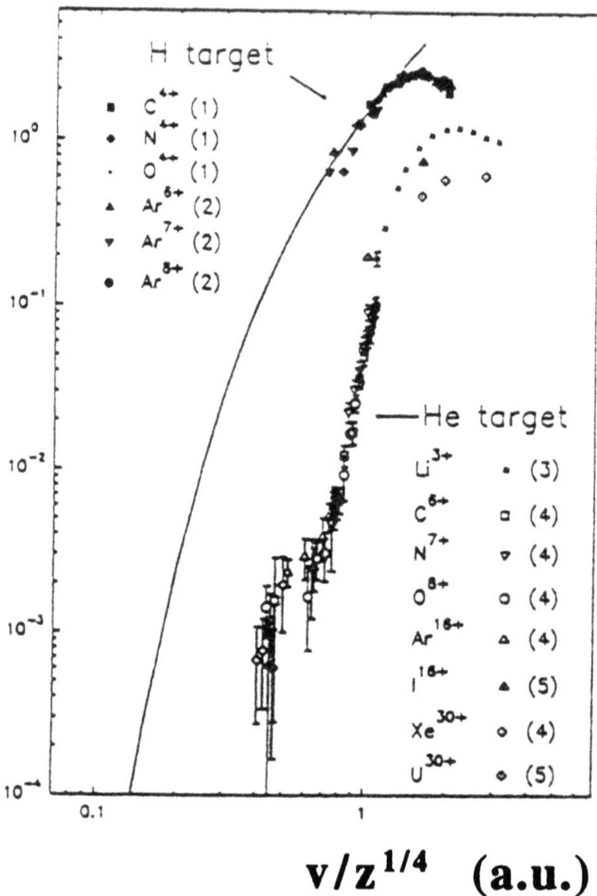

Fig. 5.5 Scaled ionization cross sections of H and He targets by positive ions. *Symbols* experiments: (from [5.65]); *Solid curve* Theory: ionization of H by multicharged ions [5.67]

where c is the velocity of light. According to (5.2.7, 8), the ratio R reaches its maximum at $v^2/z \approx 2$. Equations (5.2.7, 8) also show that there is no simple scaling law for the ratio R. Experimental values of R in collisions of He atoms with ions are given in Table 5.7 and Fig. 5.6 in comparison with theoretical results following from (5.2.6–8). In almost all cases, the agreement between theory and experiment is within to 50%.

5.2.3 Multiple Ionization

Analysis of the charge-state distributions of slow ions produced in ion–atom collisions has shown [5.12] that the main channels of reactions are: direct

Table 5.7. Ratio R of double-to-single ionization cross sections of He by multicharged ions [5.73]

Projectile	E [MeV u^{-1}]	v [a.u.]	R (experiment)	R (calculated)
C^{6+}	3.31	11.5	0.0175	0.0163
	4.58	13.4	0.0144	0.0127
	5.50	14.6	0.0114	0.0111
	6.33	15.7	0.0099	0.0099
N^{7+}	10	19.7	0.0015	0.0088
	15	24.2	0.0120	0.0066
	20	28.1	0.0090	0.0055
	30	34.0	0.0093	0.0045
	40	38.9	0.0059	0.0041
O^{8+}	2.84	10.7	0.0360	0.0295
	3.91	12.5	0.0247	0.0227
	4.61	13.4	0.0203	0.0202
	5.19	14.0	0.0197	0.0187
Ne^{10+}	90	55.9	0.0032	0.0040
	180	75.4	0.0030	0.0033
	400	97.5	0.0026	0.0030
	800	114	0.0028	0.0029
	1500	126.5	0.0026	0.0028
Si^{14+}	3.39	11.6	0.066	0.067
	3.86	12.4	0.055	0.060
	4.21	12.9	0.052	0.056
S^{16+}	3.11	11.2	0.077	0.089
	3.53	11.9	0.066	0.081
Ni^{28+}	80	53.2	0.0132	0.0143
	150	69.2	0.0092	0.0092
	240	82.9	0.0060	0.0071
	550	106.2	0.0056	0.0052
	1900	129.3	0.0041	0.0042
Kr^{36+}	500	103	0.018	0.007
	1000	120	0.015	0.006
Gd^{37+}	1.4	7.48	0.282	0.354
U^{44+}	1.4	7.48	0.319	0.384
U^{90+}	60	46.7	0.063	0.153
	120	64.7	0.048	0.079
	420	98.6	0.030	0.034

multiple outer- and inner-shell ionization, multiple electron capture, single- and multiple-electron capture with simultaneous ionization and shake-off processes.

Similar to ionization by electron impact (Sect. 4.2), Multiple Ionization (MI) of atoms in collisions with ions

$$X^{z+} + A \rightarrow X^{z+} + A^{k+} + ke\,, \quad k > 1 \tag{5.2.9}$$

has been investigated mostly experimentally [5.75–89].

Theoretical aspects of MI processes of atoms by ions are considered in [5.70, 75–79]. Usually, the theoretical approaches used for description of MI are based on the Independent Particle Model (IPM) when the target electrons

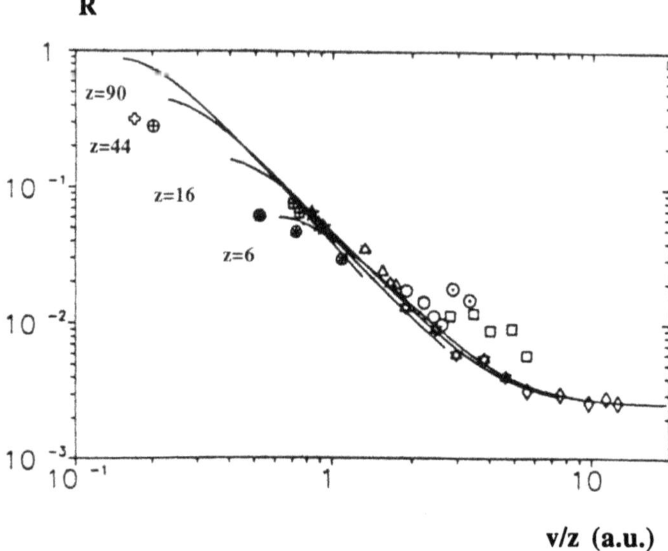

Fig. 5.6. Ratio R of double-to-single ionization of He by multicharged ions as a function of the reduced relative velocity v/z (in a.u.). Experiment: *symbols* from [5.73]. Theory: *solid curves* [5.73] for different z-values (indicated)

are treated independently from each other and shell structure effects are neglected. A review on classical, semiclassical and quantum-mechanical methods used in the theory of MI, excitation and capture in energetic ion–atom collisions is given in [5.76].

The typical behavior of MI cross sections in comparison with the loss, transfer and total ionization cross sections is shown in Fig. 5.7 for Ar^{10+} + Ne collisions. A contribution of the experimental K-electron ionization cross sections to the total MI is shown in Table 5.8. The total ionization cross section σ_{tot} is defined as the sum of the partial cross sections

$$\sigma_{tot} = \Sigma_k \sigma_k \,. \tag{5.2.10}$$

Usually, the cross section σ_k for producing a recoil ion a charge state k is obtained by normalizing the measured charge-state fractions to the *net* (or *gross*) ionization cross section σ_+

$$\sigma_+ = \Sigma_k k \sigma_k \,, \tag{5.2.11}$$

where σ_k is the k-fold ionization cross section. The CTMC calculations showed [5.80] that the calculated σ_+ values for a given rare-gas target can be described by the scaling relation

$$F(u) = \sigma_+/z \,, \quad u = E/z \text{ [keV/u]} \,, \tag{5.2.12}$$

where z is the charge of the incident ion.

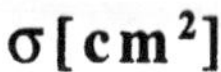

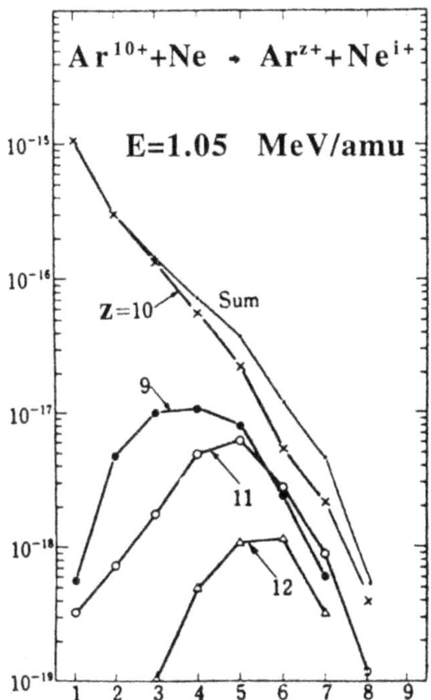

Fig. 5.7. Experimental cross sections for recoil Ne^{+i} ion production in collisions with Ar^{+10} ions of 1.05 MeV/amu energy [5.90]. Sum denotes the total cross section summed over pure (MI, $z = 10$), loss ($z = 11$ and 12) and electron capture with ionization ($z = 9$) cross sections, respectively

The use of the Thomas–Fermi model for the target–electron density gives the double scaling law of σ_{tot} on the incident ion charge z and the nuclear charge Z of the target for all targets and all projectiles (including electrons) in the form [5.91] (Fig. 5.8)

$$F(u) = \frac{\sigma_{\text{tot}}}{z} Z^{1/2}\,, \quad u = \frac{E}{zZ^{3/2}}\,[\text{keV/amu}]\,. \tag{5.2.13}$$

5.3 Electron Capture

Experimental and theoretical data on electron capture (or charge exchange) cross sections for collisions of neutral atoms with ions

$$X^{z+} + A - X^{(z-1)+} + A^{+} \tag{5.3.1}$$

are required for many fields of atomic physics.

Table 5.8. Experimental MI cross sections (in cm^2) of Ar atoms by heavy multicharged uranium, gold and bismuth ions (evaluated from [5.87–89])

$U^{z+} + Ar \rightarrow U^{z+} + Ar^{k+} + ke$

E [MeV/amu]	1.4	3.6	5.9	9.4	15.5	60	120	420
z	44	65	65	65	75	89	91	91
k = 1	2.4 − 14	1.5 − 14	–	7.7 − 15	5.4 − 15	1.0 − 14	–	8.6 − 15
2	1.0 − 14	6.1 − 15	4.6 − 15	3.3 − 15	2.4 − 15	2.0 − 15	1.6 − 15	6.7 − 16
3	5.2 − 15	3.6 − 15	2.6 − 15	1.8 − 15	9.7 − 16	4.1 − 16	4.3 − 16	2.0 − 16
4	3.4 − 15	1.8 − 15	1.3 − 15	8.0 − 16	3.8 − 16	3.0 − 16	2.4 − 16	8.7 − 17
5	1.8 − 15	1.0 − 15	7.7 − 16	4.9 − 16	2.6 − 16	1.2 − 16	1.5 − 16	5.0 − 17
6	1.2 − 15	5.4 − 16	4.4 − 16	2.5 − 16	1.9 − 16	9.0 − 17	8.0 − 17	2.6 − 17
7	5.5 − 16	3.6 − 16	2.3 − 16	1.8 − 16	1.4 − 16	7.0 − 17	6.0 − 17	2.4 − 17
8	1.2 − 16	2.3 − 16	2.0 − 16	1.5 − 16	1.3 − 16	8.0 − 17	3.9 − 17	1.3 − 17
9	–	7.7 − 17	1.1 − 16	1.0 − 16	7.7 − 17	1.0 − 17	2.2 − 17	5.4 − 18
10	–	4.9 − 17	8.2 − 17	8.2 − 17	5.6 − 17	2.5 − 17	1.4 − 17	3.5 − 18
11	–	3.3 − 17	4.1 − 17	5.1 − 17	4.4 − 17	5.6 − 18	9.0 − 18	2.7 − 18
12	–	2.6 − 17	2.6 − 17	3.6 − 17	2.6 − 17	–	5.0 − 18	2.2 − 18
13	–	2.0 − 17	1.5 − 17	2.8 − 17	1.8 − 17		2.8 − 18	–
14		–	8.2 − 18	1.3 − 17	9.0 − 18		–	–
15		–	–	–	7.7 − 18		4.6 − 18	–
17					3.0 − 18		3.1 − 18	
18					–		2.1 − 18	
							8.5 − 19	
σ_{tot}	4.6 − 14	2.9 − 14	–	1.5 − 14	1.0 − 14	1.3 − 14	–	9.7 − 15

	Au^{24+} + Ar (E = 5.9 MeV/amu)	Bi^{67+} + Ar (E = 300 MeV/amu)
k = 1	–	2.6 − 15
2	1.8 − 15	2.0 − 16
3	5.0 − 16	7.0 − 17
4	3.0 − 16	4.6 − 17
5	1.0 − 16	1.4 − 17
6	8.0 − 17	1.3 − 17
7	4.1 − 17	1.2 − 17
8	2.1 − 17	8.0 − 18
9	1.3 − 17	6.0 − 18
10	6.0 − 18	–
11	3.0 − 18	
σ_{tot}	–	3.0 − 15

2.4 − 14 means 2.4×10^{-14}

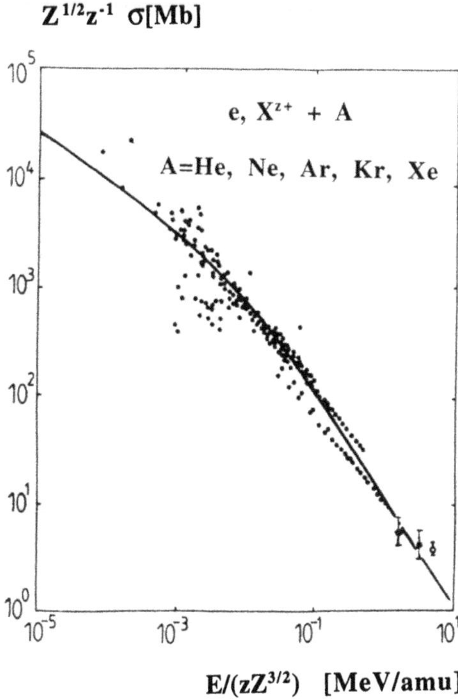

Fig. 5.8. Scaled total MI cross sections of rare-gas targets of the nuclear charge Z as a function of the reduced energy of the incident electrons and ions with the charge z (5.2.11). *Solid circles* experimental data from various papers; last three points on the energy correspond to the experimental data on double-electron ionization of He by highly charged ions [5.92]. *Solid curve* calculation in the Thomas–Fermi model for a target electron density [5.91]

The problems of Electron Capture (EC) in ion–aton collisions have been reviewed in [5.9, 93, 94] from the experimental point of view and in theoretical works [5.3–12]. Experimental data on EC cross sections for proton–atom collisions are given in [5.95, 96]. Mutliple-electron capture processes are considered in [5.97–99].

The behavior of electron capture cross sections σ are usually described in two energy ranges: adiabatic ($v \leqslant v_e$) and diabatic ($v \geqslant v_e$), where v is the relative velocity and v_e is the bound electron velocity. In the first region, the bound target electrons adjust to the changing field of the projectile, and quasi-molecular or close-coupling approaches can be used. In the diabatic (fast collision) region, the perturbation theory and its modifications are usually applied.

Some general features of σ for the reaction (5.3.1) are the following [5.100]. For a given target atom in the low-velocity region ($v \leqslant v_e$):

(1) for multicharged, partially stripped ions, σ values are large, nearly independent on v and $\sigma \propto z$. The EC cross section is constant at low energies because a lot of excited states of the resulting ion is involved. For a specific state the σ value (the partial cross section) may have a maximum but the total cross section is a sum over the partial cross sections with maxima at different energies and hence the total cross section is nearly constant;

(2) for almost fully stripped ions and ions in low charge states, the capture is highly selective, σ values are not scaled on z and are independent on v;

(3) the capture into a particular n state of the $X^{(z-1)+}$ ions with $n \approx z^{3/4}$ has the largest probability.

In the high energy region ($v \gtrsim v_e$):

(1) σ falls off rapidly with increasing v for one-shell targets (H or He) and decreases much slower for multielectron atoms because of the contribution from the capture of the inner-shell electrons;

(2) the total σ-values may contain the characteristic 'breaks' in the velocity dependence $\sigma(v)$, where the capture of electrons from neighbour shells gives nearly equal contribution,

(3) for partially stripped ions, σ depends only on z and v and is almost independent on ion species;

(4) the σ-value has a scaling law $\sigma \propto z^{\alpha(v)}$, where the function $\alpha(v)$ changes in the limits $1 \leqslant \alpha(v) \leqslant 5$;

(5) the distribution over n of the resulting ion $X^{(z-1)+}$ becomes broader with increasing v, and its maximum moves towards low n states.

The typical behavior of the EC cross sections is shown in Figs. 5.9, 10 for $H^+ + H(1s)$ and $C^{3+} + H$ collisions.

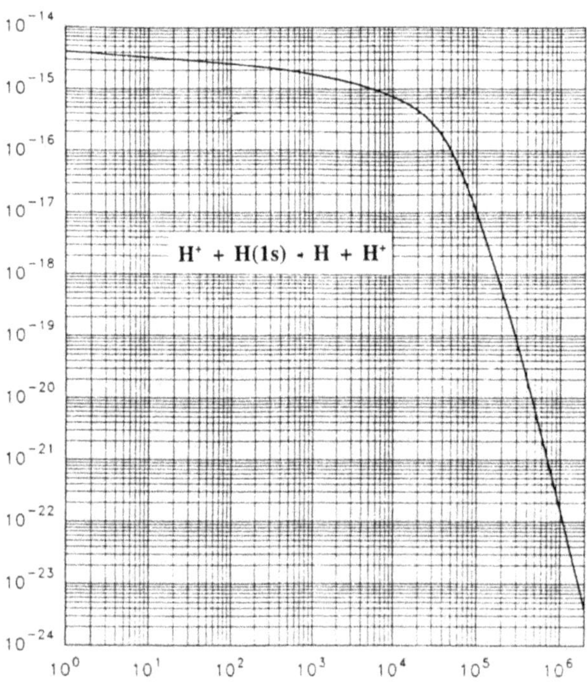

Fig. 5.9. Recommended total EC section in proton-H(1s) collisions [5.16]

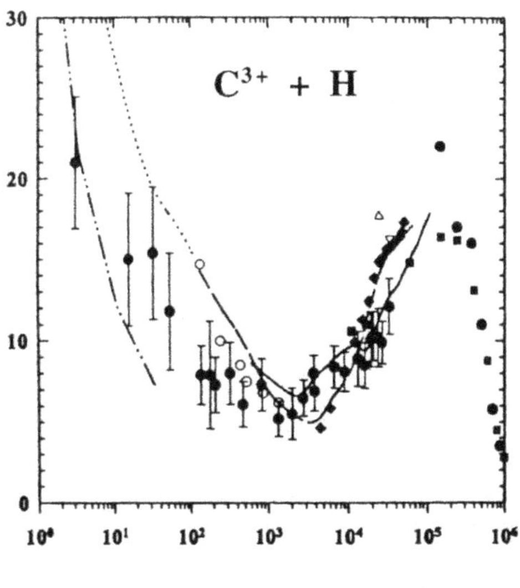

Fig. 5.10. Total EC cross sections in $C^{+3} + H$ collisions [5.101]. *Symbols* experimental data; *curves* theoretical predictions (see [5.101] for details)

Recommended data on EC cross sections in collisions of protons with excited hydrogen

$$H^+ + H^*(n) \to H + H^+ \tag{5.3.2}$$

for the states $n = 1–5$ are given in Table 5.9 in the reduced form. The cross sections are described by the scaling law [5.102]

$$F(u) = \sigma(E)/n^4 , \quad u = En^2 \text{ [keV/amu]} \tag{5.3.3}$$

accurate to within 20% for all n states including the ground state at incident proton energies $E > 200$ keV/amu. EC of ions on excited atoms is considered in [5.23, 102–108].

The partial and the total EC cross sections for reactions

$$H^+ + He(1s^2) \to H(nl) + He^+ \tag{5.3.4}$$

are shown in Fig. 5.11.

In the case of H and He atoms colliding with the positive ions, the total EC cross sections are described by the scaling law [5.100]

$$F(u) = \sigma(E)/z , \quad u = E/vz \text{ [keV/amu]} . \tag{5.3.5}$$

The universal function $F(u)$ has the form

$$\sigma(u) = \frac{A \ln(B/u) \text{ cm}^2}{1 + Cu^2 + Du^{4.5}} , \qquad u = E/\sqrt{z} \text{ [keV/amu]} \tag{5.3.6}$$

Here A, B, C and D are fitting parameters, given in Table 5.10.

Table 5.9. Scaled cross sections σ/n^4 (in cm^2) as a function of the reduced proton energy n^2E for electron capture $H^+ + H(n) \to H + H^+$ into n-states of the resulting H atom

n^2E [eV/amu]	$n = 1$	$n = 2$	$n = 3$	$n \geqslant 4$
1.20E – 01	4.96E – 15			
2.00E – 01	4.70E – 15			
5.00E – 01	4.33E – 15			
1.00E + 00	4.10E – 15			
2.00E + 00	3.83E – 15			
5.00E + 00	3.46E – 15			
1.00E + 01	3.17E – 15			
2.00E + 01	2.93E – 15			
5.00E + 01	2.65E – 15			
1.00E + 02	2.44E – 15			
2.00E + 02	2.22E – 15	–	–	–
5.00E + 02	1.97E – 15	1.26E – 15	6.91E – 16	5.45E – 16
1.00E + 03	1.71E – 15	1.19E – 15	6.70E – 16	5.36E – 16
2.00E + 03	1.44E – 15	1.09E – 15	6.43E – 16	5.27E – 16
5.00E + 03	1.10E – 15	1.03E – 15	6.23E – 16	5.23E – 16
1.00E + 04	7.75E – 16	9.59E – 16	6.01E – 16	5.16E – 16
2.00E + 04	4.45E – 16	8.62E – 16	5.74E – 16	5.08E – 16
5.00E + 04	9.93E – 17	7.92E – 16	5.54E – 16	5.00E – 16
1.00E + 05	1.01E – 17	7.25E – 16	5.34E – 16	4.91E – 16
2.00E + 05	6.09E – 19	6.32E – 16	5.07E – 16	4.81E – 16
5.00E + 05	6.03E – 21	5.45E – 16	4.77E – 16	4.69E – 16
1.00E + 06	1.57E – 22	4.03E – 16	3.85E – 16	3.82E – 16
2.00E + 06	3.78E – 24	9.93E – 17	9.93E – 17	9.93E – 17
5.00E + 06	2.56E – 26	1.01E – 17	1.01E – 17	1.01E – 17
1.00E + 07	5.99E – 28	6.09E – 19	6.09E – 19	6.09E – 19

One-electron State-Selective Electron Capture (SSEC)

$$X^{z+} + H, \quad He \to X^{(z-1)+}(nl) + H^+, He^+ \qquad (5.3.7)$$

is the relevant reaction the knowledge of which is required for studying radiation losses from multicharged ions due to EC recombination and for plasma diagnostics based on EC.

Most of the general theoretical results for SSEC have been obtained for collisions of bare ions with H atoms. In general, the distribution over nl states of the ion $X^{(z-1)+}$, given by theoretical models and confirmed by experiments with bare and closed-shell ions, is [5.100]:

(1) the principal quantum number of the dominantly populated ionic state is $n = n_m \propto z^{3/4}$;

(2) for the very-low-velocity limit $v \gg v_e$ and ions with $z \ll 1$, the most populated substate for a given n is the p state, if the initial electron state of H or He is n_0s. This results from the weak mixing of the sublevels in this velocity limit;

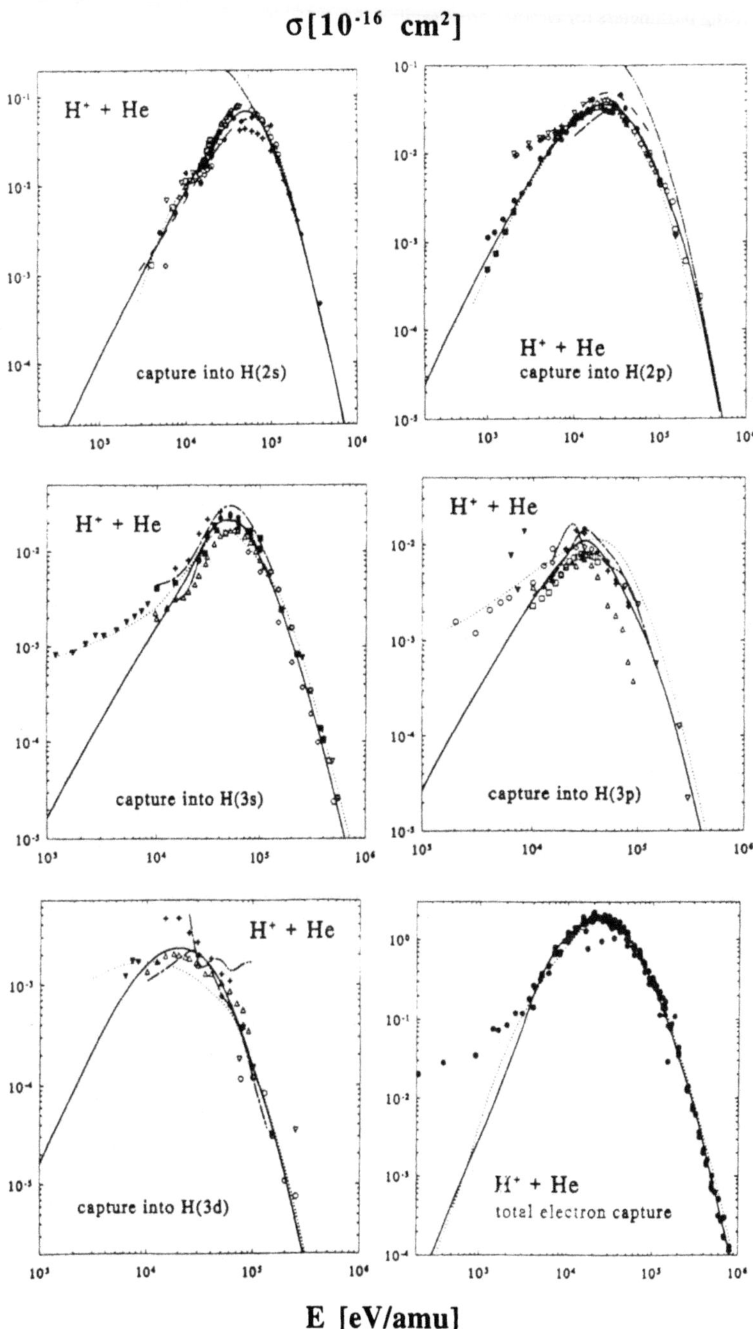

Fig. 5.11. The partial and the total EC cross sections in $H^+ + He$ collisions. Experiment: *symbols* (see [5.109] for details). Recommended data: *dotted curve* [5.21]; *solid curve* [5.109]

Table 5.10. Fitting parameters for capture cross section (5.3.4) [5.100]

Target	A	B	C	D	Appr. error, %
H	5.967×10^{-17}	5.870×10^{5}	1.913×10^{-3}	1.383×10^{-7}	13
H_2	5.707×10^{-17}	5.283×10^{4}	7.800×10^{-4}	2.721×10^{-8}	15
He	1.818×10^{-17}	1.856×10^{6}	2.753×10^{-4}	1.370×10^{-9}	70

(3) with increasing v up to $v_0 = 1$ a.u., the most populated level l_m increases and tends to the maximum value $l_m = n - 1$. Because of the strong mixing of states in this collision regime, the relative distribution tends to its statistical limit $\sigma_{nl}/\sigma_n \approx (2l+1)/n^2$;

(4) in the intermediate energy region $E = 10$–200 keV/u, one has $l_m \approx n - 1$ for $n \leqslant n_m$ and $l_m \approx n_m$ for $n > n_m$;

(5) with further increase of E, the maximum l_m value decreases rapidly and tends to the s state, i.e., $l_m = 0$.

The behavior of SSEC cross sections is shown in Fig. 5.12 in the case of Fe^{8+} + H collisions.

In general, the properties of EC in collisions with complex atoms are similar to those in collisions with H and He with one important exception. At high energies, the capture of inner-shell target electrons is dominant and the capture of outer electrons is negligible.

An impirical scaling law was found [5.111] for one-electron capture by fast multicharged ions in gas targets $A = H_2$, He, N_2, Ne, Ar, Kr, Xe in the form

$$F(u) = \sigma_c(E) Z^{1.8}/z^{0.5}\,, \quad u = E/Z^{1.25} z^{0.7} \text{ [keV amu]}\,, \tag{5.3.8}$$

where E is the ion projectile energy and Z is the nuclear charge of the atomic target. About 70% of the experimental data lie within a factor of 2 of the curve described by the function $F(u)$ (Fig. 5.13)

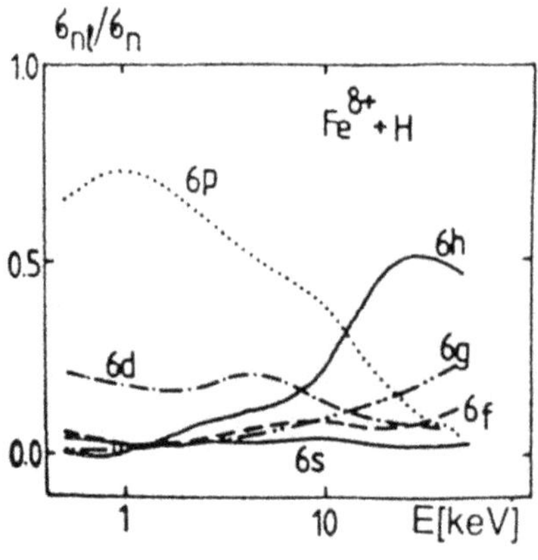

Fig. 5.12. State-selective EC cross sections for the reaction $Fe^{3+} + H \rightarrow Fe^{7+} + H^+$: calculations by the semiclassical close-coupling method on atomic-orbital basis [5.110]

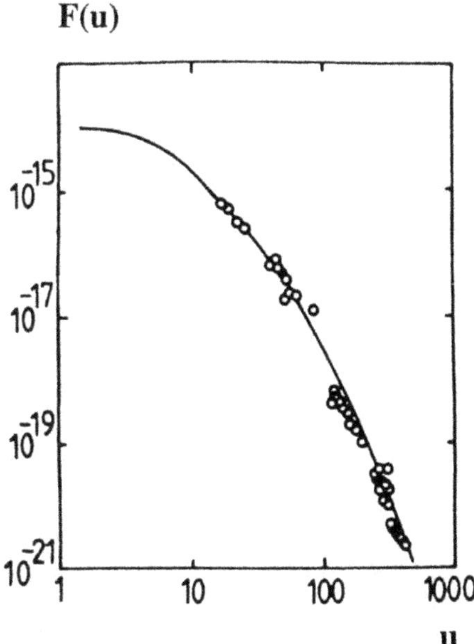

Fig. 5.13. Scaled total EC cross sections of multicharged ions on rare-gas targets. *Solid curve* (5.3.8); *open circles* experiment [5.111]

$$F(u) = \frac{1.1 \times 10^{-8}\,\mathrm{cm}^2}{u^{4.8}}[1 - \exp(-0.037u^{2.2})] \times [1 - \exp(-2.44 \times 10^{-5}u^{2.6})]\,. \quad (5.3.9)$$

Equation (5.3.9) can be used for prediction of the capture cross sections in the reduced energy range $10 < u < 1000$ and ion charges $z > 3$.

The one-electron removal (or loss) cross sections (electron capture plus ionization) are described by the scaling relation [5.36, 111]

$$F(u) = \sigma(E)/z\,, \quad u = E/z\ [\mathrm{keV/amu}]\,. \quad (5.3.10)$$

In [5.36], the expression for one-electron loss cross section was obtained as

$$\sigma_l = N\varkappa^{-3}zF(u)10^{-16}\,\mathrm{cm}^2\,, \quad u = E/\varkappa z\ [\mathrm{keV/amu}]\,, \quad \varkappa = (2I)^{1/2}\,, \quad (5.3.11)$$

where N and I are the number of equivalent electrons in the target shell and its binding energy in atomic units, respectively. The universal function $F(u)$ is given in Table 5.11.

5.4 Collisions Involving H^- Ions

The problems of interaction of H^- ions with positive and negative ions are important from both fundamental and practical points of view, especially with

Table 5.11. The function $F(u)$ in (5.3.11) for electron-loss cross sections [5.36])

u	$F(u)$	u	$F(u)$
4	7.20	20	4.01
5	7.00	22	3.73
6	6.77	24	3.49
7	6.52	26	3.27
8	6.26	28	3.07
9	5.98	30	2.90
10	5.70	32	2.75
12	5.28	34	2.63
14	4.90	36	2.53
16	4.57	38	2.45
18	4.27	40	2.40

respect to conversion efficiencies of H^- to neutral H^0 ion plasma neutralizers proposed for neutral beam heating of future fusion devices [5.112, 113]. A review on experimental and theoretical data of single and multiple electron detachment in collisions of negative ions with neutrals is given in [5.114]. In recent years, the use of advanced crossed-beams or merged-beams techniques made it possible to receive for the first time the results covering collisions of H^- ions with protons, multicharged ions and H^- ions themselves.

5.4.1 $H^+ + H^-$ Collisions

Three fundamental reactions in $H^+ + H^-$ collisions have been investigated:
(*i*) Transfer Ionization (TI)

$$H^+ + H^- \rightarrow H + H^+ + e\,; \tag{5.4.1}$$

(*ii*) Electron Detachment (ED)

$$H^+ + H^- \rightarrow H^+ + H + e\,; \tag{5.4.2}$$

(*iii*) Mutual Neutralization (MN)

$$H^+ + H^- \rightarrow H + H\,. \tag{5.4.3}$$

TI is a two-electron process, when one of the target electrons is captured by a projectile and another one is ejected. The cross section σ_{TI} measured with the use of the merged techniques [5.115] is given in Fig. 5.14. The theoretical treatment of TI is quite complicated because the strong coupling of many discrete states and a whole continuum has to be taken into account. The low-energy behavior of σ_{TI} is well described by the approach given in [5.116], when TI is considered as a mutual neutralization (5.4.3) followed by autoionization.

The ED processes have been investigated in [5.117–121].

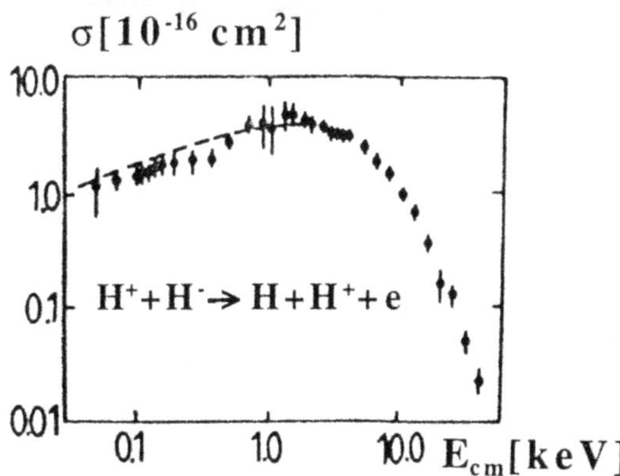

Fig. 5.14. Experimental TI cross section in collision $H^+ + H^- \rightarrow H + H^+ + e$: points – experiment, dahsed curve-model calculations [5.115]

The MN process was investigated extensively experimentally [5.117, 122–126] and theoretically [5.121, 127–128]. First measurements of MN cross section are presented in Fig. 5.15 together with theoretical calculations. It has been pointed out that at energies $E_{cm} < 2$ keV, the contribution of $H^*(n = 2, 3) + H(1s)$ channels to the MN process is very important and at energies $E_{cm} < 5$ keV, the capture of the active $1s'$ electron in H^- plays the main role and gives the largest contribution (80%) to the total cross section.

The molecular treatment for calculations of the total (TI + MN) cross section has been applied in [5.130].

5.4.2 $H^- + H^-$ Collisions

Single- and double-electron detachment as well as the triple ionization, i.e., respectively, the reactions:

$$H^- + H^- \rightarrow H^0 + H^0 + e\,, \tag{5.4.4}$$

$$H^- + H^- \rightarrow H^0 + H^0 + 2e\,, \tag{5.4.5}$$

$$H^- + H^- \rightarrow H^0 + H^0 + 3e\,, \tag{5.4.6}$$

have been studied in [5.131, 132] using the crossed-beam technique. The main theoretical approaches used for description of the processes (5.4.4–6) are the CTM method [5.132, 133], the independent particle model [5.134] and the non-stationary tunneling approach [5.135–137]. According to [5.137], the properties of reactions (5.4.4, 5) are described by promotion of the initial H_2^- quasimolecular term due to the Coulomb interaction: first, into the continuum of the H_2^- quasimolecule at the internuclear distance $R \simeq 36a_0$, and then into the continuum of the H_2 quasimolecule at $R \simeq 18a_0$.

Single-electron detachment is related to the decay of the quasistationary state of the electron bound to the H^0 atom in the presence of an external Coulomb repulsive potential.

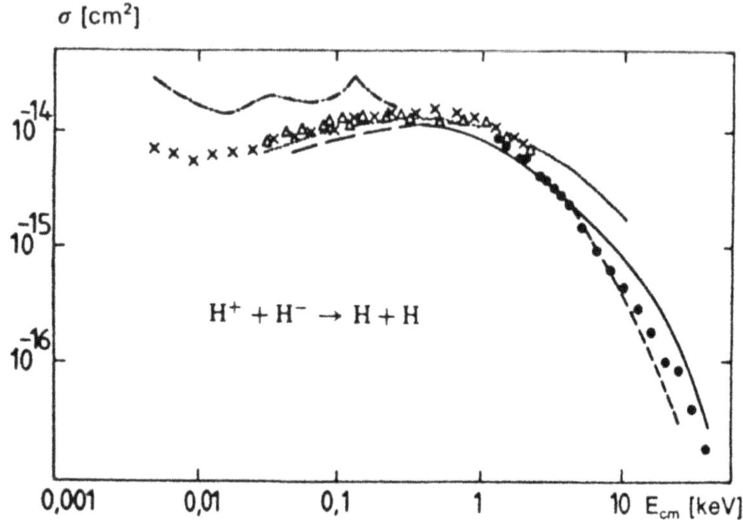

Fig. 5.15. MN cross sections for reaction $H^+ + H^- \to H + H$. Experiment: Δ [5.124]; × [5.125]; • [5.126]; –·– [5.117–118; 122–123]. Theory: *dotted curve* MO calculations [5.127]; *dahsed curve* MO calculations [5.129]; *solid curve* close-coupling calculations [5.121]

Double-electron detachment is defined by two mechanisms: a two-step transition and the direct transition. The two-step mechanism corresponds to the release of one electron and the ionization of another one in subsequent collisions of H^- with H^0. The direct mechanism is a simultaneous transition of two electrons into the continuum with the population of H_2 quasimolecular term.

Experimental data and theoretical calculations of the processes (5.4.4–6) are given in Fig. 5.16.

5.4.3 Collisions of H^- with Multicharged Ions

Reactions arising in collisions of H^- ions with singly and multiply charged ions are investigated in [5.131, 132, 135–137].

Single-electron removal (detachment) from H^- ions in collisions with highly charged ions

$$H^- + X^{z+} \to H^0 + \cdots \tag{5.4.7}$$

has been studied experimentally [5.115, 126, 131] and theoretically [5.135–137]. The reaction (5.4.7) consists of the electron capture and ionization of the target electrons into the continuum. The process (5.4.7) takes place mainly at large internuclear distances where the field of the positive ion is close to the Coulomb field. Single-electron removal cross section $\sigma^{(1)}$ can be described in the framework of the generalized Keldysh theory [5.135, 137] which yields the scaling law

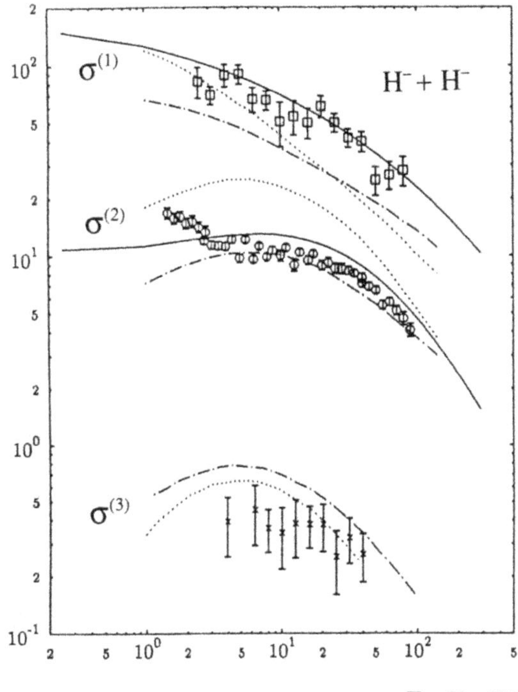

Fig. 5.16. Single $\sigma^{(1)}$ (5.4.4), double $\sigma^{(2)}$ (5.4.5) electron detachment and triple ionization $\sigma^{(3)}$ cross sections. *Symbols*: experimental data [5.131–132]. Theory: *dotted* and *dot-dashed curve* CTMC calculations [5.132] corresponding to the different fitting interaction potentials; *solid curves* non-stationary tunneling approach [5.137]

for the removal cross section $\sigma^{(1)}$:

$$F(u) = \frac{\sigma^{(1)}(v)}{z}, \quad u = \frac{v^2}{z}\ \text{[a.u.]}, \tag{5.4.8}$$

where v is the relative velocity of the colliding particles in atomic units. The universal function $F(u)$ has the asymptotics

$$F(y) \simeq \begin{cases} 0.239\ln(2.5y + 0.8), & y \leqslant 0.3, \\ 0.129\ln(8.12y + 1.01), & y \geqslant 0.3, \end{cases} \tag{5.4.9}$$

$$0.02/z^2 < y < 1.5z, \quad y = 0.25v^2/z = E_{\text{cm}}/100z\ \text{[keV/amu]}. \tag{5.4.10}$$

At high energies $y \gg z$, the function $F(y)$ has the asymptotic

$$F(y) = \frac{1.29}{y}\ln[8.12(1 + 0.577y/z)^{-1/2} + 1.01], \tag{5.4.11}$$

which agrees with the Bethe-Born calculations [5.31] (Fig. 5.17). Figure 5.18 shows a comparison of the scaled cross sections (5.4.8–10) with experimental data. It is seen that the scaling law suggested in [5.135] allows one to make an

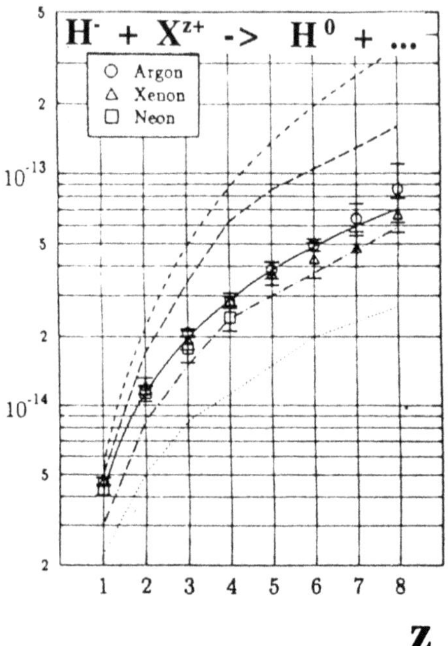

Fig. 5.17. Neutralization of H^- ions in collisions with multichanged X^{z+} ions. *Symbols* experimental data [5.131, 132, 140]. Theory: *dot-dashed curve* Bethe–Born calculations for ionization by bare-ion projectiles [5.141]; *long-dashed* and *dotted curves* CTMC calculation [5.131]; *solid curves* generalization of the Keldysh theory [5.135, 137]

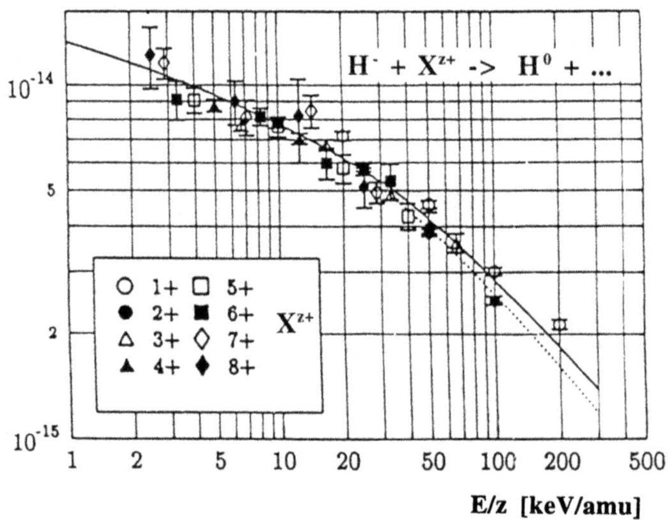

Fig. 5.18. Scaled single electron-removal cross section for the reaction (5.4.7) as a function of the scaled energy. *Symbols*: experimental data [5.131, 132, 140]. Theory: *solid curve* (5.4.8–10); *dotted curve* the same for $z = 1$

accurate prediction of the neutralization cross sections in collisions of H^- ions with multiply charged ions in a wide energy range.

Double electron-removal processes

$$H^- + X^{z+} \to H^+ + \cdots \tag{5.4.12}$$

are investigated experimentally in [5.135] and considered theoretically in [5.135, 137, 142].

According to [5.137], three main physical mechanisms are responsible for the double electron-removal:

(*i*) the shake-off process when one of the atomic electrons is ejected with high velocity after interaction with the projectile and another one due to the correlation if the initial state undergoes a relaxation causing additional ionization;

(*ii*) the double ionization resulting from collision of the primarily ejected electron with the second one;

(*iii*) simultaneous ionization of both electrons due to interaction with the projectile.

The processes (*i*) and (*ii*) play a key role at high relative velocities, while the process (*iii*) is dominant in the low-velocity range. The shake-off process is

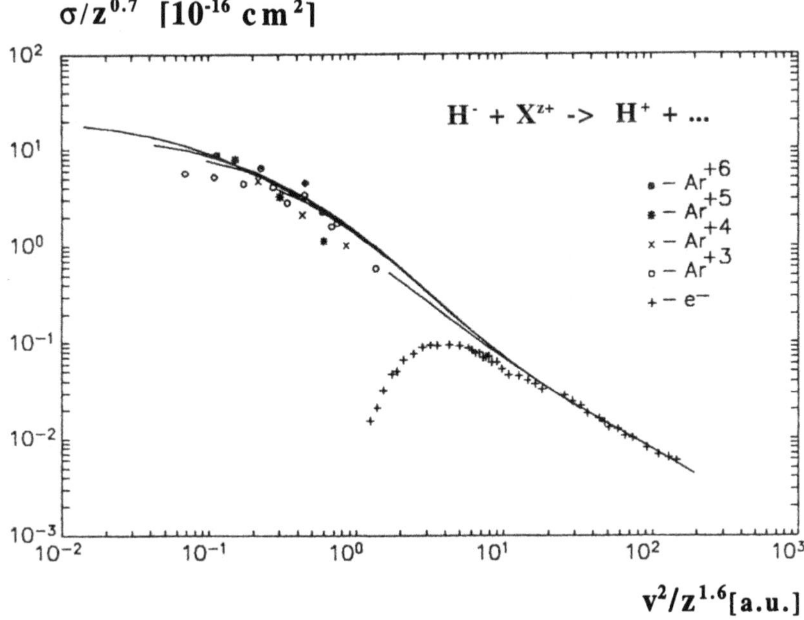

Fig. 5.19. Scaled double electron-removal cross sections. *Symbols*: experiment [5.143]; *crosses* correspond to the double-ionization cross sections by electron impact [5.144]. Theory: *solid curves* generalization of the Keldysh theory [5.135–137]

caused mostly by the distant encounters. Therefore, the corresponding cross section in defined by the behavior of double photoionization of H^- ions. The processes (*ii*) and (*iii*) take place with a large momentum transfer to the target electrons and therefore, they are usually treated using the classical mechanics approaches.

As was found in [5.137], the double electron-removal cross section $\sigma^{(2)}$ is described by the empirical scaling relation

$$F(u) = \sigma(v)/z^{0.7}, \quad u = v^2/z^{1.6}\ [\text{a.u.}], \tag{5.4.13}$$

which describes reasonably well the available experimental data (Fig. 5.19).

References

Chapter 1

1.1 I.I. Sobelman: *Atomic Spectra and Radiative Transitions*, 2nd edn., Springer Ser. Atom. Plas., Vol. 12 (Springer, Berlin, Heidelberg 1992)
1.2 U.I. Safronova and V.S. Senashenko: *Theory for Spectra of Multicharged Ions* (Energoizdat, Moscow 1984) (in Russian)
1.3 W.L. Wieseand, G.A. Martin: In *A Physicist's Desk Reference* (American Institute of Physics, New York 1989, 1994)
1.4 W. Lotz: J. Opt. Soc. Am. **60**, 206 (1970)
1.5 T.A. Carlson, C.W. Nestor, Jr., N. Wasserman, J.D. McDowell: At. Data **2**, 63 (1970)
1.6 A.A. Radzig, B.M. Smirnov: *Reference Data on Atoms, Molecules, and Ions*, Springer Ser. Chem. Phys., Vol. 13 (Springer, Berlin, Heidelberg 1985)
1.7 D.R. Lide (ed): *Handbook of Chemistry and Physics*, 73rd edn. (CRC, Boca Raton 1992–1993)
1.8 Heavy-Ion Spectroscopy and QED Effects in Atomic Systems. Proc. Nobel Symp. 85 ed. by I. Lindgren, I. Martinson and, R. Schuch: Phys. Scr., **T46** (1993)
1.9 W.R. Johnson, G. Soff: At. Data Nucl. Data Tables **33**, 405 (1985)
1.10 H. Persson, S. Salomonson, P. Sunnergren, I. Lindgren: Phys. Rev. Lett. **76**, 204 (1996)
1.11 P. Mohr: In *Atomic, Molecular and Optical Physics Reference Book*, ed. by G.W.F. Drake, (AIP, New York 1996)
1.12 M.I. Eides, V.A. Shelyto: Phys. Rev. A **52**, 954 (1995)
1.13 V.G. Pal'chikov, Yu.L. Sokolov, V.P. Yakovlev: Physica Scr. **55**, 33 (1997)
1.14 H. Beyer, H.-J. Kluge, V.P. Shevelko: *X-Ray Radiation of Highly Charged Ions*, Springer Ser. At. Plasm., (Springer, Berlin, Heidelberg 1997) (in preparation)
1.15 R.D. Cowan: *The Theory of Atomic Structure and Spectra* (Univ. California Press, Berkeley 1981)
1.16 V.P. Shevelko, L.A. Vainshtein: *Atomic Physics for Hot Plasmas* (IOP, Bristol 1993)
1.17 D.R. Bates, A. Damgaard: Philos. Trans. R. Soc. London **242**, 101 (1949)

Chapter 2

2.1 I.I. Sobelman: *Atomic Spectra and Radiative Transitions*, 2nd edn., Springer Ser. At. Plasm., Vol. 12 (Springer, Berlin, Heidelberg 1992)
2.2 D.A. Varshalovich, A.N. Moskalev, V.K.: *Quantum Theory of Angular Momentum* (World Scientific, Singapore 1988)
2.3 V.P. Shevelko, L.A. Vainshtein: *Atomic Physics for Hot Plasmas* (IOP, Bristol 1993)
2.4 R. Marrus, P.J. Mohr: Adv. Atom. Mol. Phys. **14**, 181 (1978)
2.5 H. Bethe, E.E. Salpeter: Quantum Mechanics of One- and Two-Electron Atoms (Plenum Press, New York 1977)
2.6 W.L. Wiese: Nucl. Fusion Suppl. **2**, 7 (1992)
2.7 A. Musgrove, R. Zalubas: Natl. Bur. Stand. (US), Spec. Publ. **363**, Suppl. 3 (1985)
2.8 A. Musgrove, R. Zalubas: Natl. Bur. Stand. (US), Spec. Publ. **363**, Suppl. 4 (1985)
2.9 C.E. Moore: Natl. Bur. Stand. (US), Ref. Data Ser. 3, Sect. 6 (1972)

2.10 W.L. Wiese, M.W. Smith, B.M. Glennon: Atomic Transition Probabilities, NBS (USA) **4**, Vol. 1 (1966) and **22**, Vol. 2 (1969)
2.11 R.L. Kelly: J. Phys. Chem. Ref. Data **16**, Suppl. 1 (1987)
2.12 J. Reader, C.H. Corliss: In *Handbook of Chemistry and Physics*, ed. by D.R. Lide 73st edn. (CRC, Boca Raton, 1992–1993)
2.13 A.R. Striganov, N.S. Sventitskij: *Tables of Spectral Lines of Neutral and Ionized Atoms* (IFI/Plenum, New York 1968)
2.14 W.C. Martin: J. Phys. Chem. Ref. Data **2**, 257 (1973)
2.15 W.C. Martin: Phys. Rev. A **36**, 3575 (1987)
2.16 C.E. Moore: Atomic Energy Levels, Natl. Bur. Stand. (US), Data Ser. 35, Vols 1–3 (1971) (reprint of NBS Circ. 467), US Govt. Printing Office, Washington, DC.
2.17 G.A. Odintzova, A.R. Striganov: J. Phys. Chem. Ref. Data **8**, 63 (1979)
2.18 C.E. Moore: Natl. Bur. Stand. (US), Ref. Data Ser. **3**, Sect. 3 (1970)
2.19 C.E. Moore: Natl. Bur. Stand. (US), Ref. Data Ser. **3**, Sect. 4 (1972) and Sect. 5 (1975)
2.20 C.E. Moore: Natl. Bur. Stand. (US), Ref. Data Ser. **3**, Sect. 7 (1976); Sect. 8 (1979); Sect. 9 (1980); Sect. 10 (1982); Sect. 11 (1985)
2.21 W.C. Martin, R. Zalubas: J. Phys. Chem. Ref. Data **10**, 152 (1981)
2.22 W.L. Wiese, M.W. Smith, B.M. Miles: Natl. Bur. Stand. (US), Ref. Data Ser. **22** (1969)
2.23 W.C. Martin, R. Zalubas: J. Phys. Chem. Ref. Data **9**, 1 (1980)
2.24 W.C. Martin, R. Zalubas: J. Phys. Chem. Ref. Data **8**, 817 (1979)
2.25 W.C. Martin, R. Zalubas: J. Phys. Chem. Ref. Data **12**, 323 (1983)
2.26 W.C. Martin, R. Zalubas, A. Musgrove: J. Phys. Chem. Ref. Data **14**, 751 (1985)
2.27 W.C. Martin, R. Zalubas, A. Musgrove: J. Phys. Chem. Ref. Data **19**, 821 (1990)
2.28 J. Sugar, C. Corliss: J. Phys. Chem. Ref. Data **14**, Suppl. 2 (1985)
2.29 V. Kaufman, J. Sugar: J. Phys. Chem. Ref. Data **17**, 1679 (1988)
2.30 G.A. Martin, J.R. Fuhr, W.L. Wiese: J. Phys. Chem. Ref. Data **17**, Suppl. 3 (1988)
2.31 J.R. Fuhr, G.A. Martin, W.L. Wiese: J. Phys. Chem. Ref. Data **17**, Suppl. 4 (1988)
2.32 J. Sugar, A. Musgrove: J. Phys. Chem. Ref. Data **19**, 527 (1990)
2.33 W.L. Wiese, G.A. Martin: *Handbook of Chemistry and Physics*, 70st edn, (CRC Press, Boca Raton 1989)
2.34 J. Sugar, and A. Musgrove: J. Phys. Chem. Ref. Data **17**, 155 (1988)
2.35 W.C. Martin, R. Zalubas, L. Hagan: Natl. Bur. Stand. (US), Ref. Data Ser. **60**, (1978)
2.36 W.F. Meggers, C.H. Corliss, B.F. Scribuer: Natl. Bur. Stand. (US) Monogr. **145** (1975)
2.37 C.E. Moore: Natl. Bur. Stand. (US), Ref. Data Ser. **40** (1972); Natl. Bur. Stand. (US) Circ. **488** (1968) Sect. 1–5.
2.38 Yu. V. Bogdanov, S.I. Kanorsky, A.A. Papchenko, I.I. Sobelman, V.N. Sorokin, I.I. Struck, E.A. Yukov: Proc. Lebedev Inst., Vol. **218**, 210 (Nova Science, N.Y. 1990)
2.39 M. Nawaz, W.A. Farooq, J.-P. Connerade: J. Phys. B **25**, 5327 (1992)
2.40 V.P. Shevelko: *Radiative transitions and atomic collisions involving inner shells of multielectron atoms.* Thesis, P.N. Lebedev Physics Institute (Moscow 1973) (unpublished); B.N. Chichkov, V.P. Shevelko: Phys. Ser. **23**, 1055 (1981)
2.41 C.E. Theodosiou: Phys. Rev. A **30**, 2881 (1984)
2.42 A.N. Filippov: Sov. Phys. – Zhurn. Teor. Ekper. Fiz. **2**, 24 (1932)
2.43 G.A. Martin, W.L. Wiese: J. Phys. Chem. Ref. Data **5**, 537 (1976)
2.44 G.V. Marr, D.M. Creek: Proc. R. Soc. (London) A **304**, 245 (1968)
2.45 D.W. Norcross: Phys. Rev. A **7**, 606 (1973)
2.46 A. Lingaard, S.E. Nielsen: At. Data Nucl. Data Tables **19**, 533 (1977)
2.47 I. Martin, C. Barrentos: J. Phys. B **64**, 867 (1986)
2.48 A.N. Filippov, V.K. Prokofjew: Z. Phys. **56**, 475 (1929)
2.49 D. Hofsaess: Z. Phys. A **281**, 1 (1977)
2.50 J.C. Weisheit: Phys. Rev. A **5**, 1621 (1972)
2.51 L.N. Shabanova, A.N. Khlyustalov: Opt. Spectrosc. **56**, 128 (1984)
2.52 E. Caliebe, K. Niemax: J. Phys. B **12**, L45 (1979)
2.53 J. Migdalek, W.E. Baylis: Can. J. Phys. **57**, 1708 (1979)

2.54 G.I. Goldberg: Optical Transition Probabilition (Jerusalem Natl. Sci. Foundation, 1962)
2.55 B. Warner: Mon. Not. Astron. Soc. **139**, 115 (1968)
2.56 L.N. Shabanova, Yu. N. Manakov, A.N. Khlyustalov: Opt. Spectrosc. **47**, 1 (1980)
2.57 G. Pichler: J. Quant. Spectrosc. Radiat. Transfer **16**, 146 (1976)
2.58 R.J. Extron: J. Quant. Spectrosc. Radiat. Transfer **16**, 309 (1976)
2.59 G.F. Fulop, H.H. Stroke: *Atomic Physics* (Plenum, N.Y. 1973)
2.60 M.A. Mazing, P.D. Serapinas: Opt. Specrosc. **27**, 143 (1969)
2.61 E.S. Chang: Phys. Rev. A **31**, 465 (1985)
2.62 J. Jin, D.A. Church: Phys. Rev. A **49**, 3463 (1994)
2.63 C. Tanner: In *Atomic Physics*, 14, ed. by D.J. Wineland, C.E. Wieman, C.J. Smith (AIP, New York 1995), p. 130
2.64 Ya. F. Verolainen, A. Ya. Nikolaich: Sov. Phys. – Uspekhi, **25**, 431 (1982)
2.65 B.C. Fawcett, M. Wilson: At. Data Nucl. Data Tables **47**, 241 (1991)
2.66 W. Sheares-Izumi: At. Data Nucl. Data Tables **20**, 531 (1977)
2.67 E.H.S. Burhop: *The Auger Effect and other Radiationless Transitions* (Cambridge Univ. Press, Cambridge 1952)
2.68 M.J. Seaton: Proc. R. Soc. London **77**, 184 (1961)
2.69 P. Feldman, R. Novick: Phys. Rev. **160**, 143 (1967)
2.70 U. Fano: Phys. Rev. **124**, 1866 (1968)
2.71 O. Bely: J. Phys. B **1**, 23 (1968)
2.72 A. Temkin (ed.): *Autoionization* (Plenum, New York 1985)
2.73 Y. Hahn: In *Proc. XIV ICPEAC, AIP Conf. Proc. 205*, p. 550 (eds A. Dalgarno, R.S. Freund, R.M. Koch, M.S. Lubell, T. Lucatorto) (AIP, New York 1990)
2.74 P.F. Gruzdev: Transition *Probabilities and Radiative Lifetimes of Levels Atoms and Ions* (Energoatomizdat, Moscow 1990) (in Russian)
2.75 L.N. Ivanov, V.S. Letokhov: Commun. Atom. Mol. Phys. D **16**, 169 (1985)
2.76 T. Andersen, S. Mannervik: Commun. Atom. Mol. Phys. D **28**, 185 (1985)
2.77 J.O. Gaardsted, T. Andersen: Commun. Atom. Mol. Phys. D **28**, 77 (1992)
2.78 V.A. Davidkin, B.A. Zon: *Correction and Relativistic Effects in Atoms and Ions* (Nauka, Moscow 1978) (in Russian)
2.79 L.A. Bureeva: Sov. Astrophys. J. **45**, 1218 (1968)
2.80 D.H. Menzel: Nature **218**, 756 (1968)
2.81 K. Omidvar, P.T. Guimaraes: Astrophys. J. suppl. **73**, 555 (1990); At. Data Nucl. Data Tables **28**, 1 (1983)
2.82 D.H. Menzel: Astrophys. J. Suppl. **18**, 221 (1969)
2.83 I.I. Sobelman, L.A. Vainshtein, E.A. Yukov: *Excitation of Atoms and Broadening of Spectral Lines*, 2nd edn., Springer Ser. Atom. Plasm., Vol. 15 (Springer, Berlin, Heidelberg 1995)

Chapter 3

3.1 H.A. Bethe, E.E. Salpeter: *Quantum Mechanics of One- and Two-Electron Atoms* (Plenum, New York 1977)
3.2 U. Fano, J.W. Cooper: Rev. Mod. Phys. **40**, 441 (1968)
3.3 G.V. Marr: *Photoionization in Gases* (Academic, New York 1967)
3.4 D.H. Sampson: *Atomic Photoionization* (Springer, Berlin, Heidelberg 1982)
3.5 A.F. Starace: In *Handbuch der Physik*, Vol. 31, ed. by W. Mehlhorn (Springer, Berlin, Heidelberg 1982) pp. 1–121
3.6 I.I. Sobelman: *Atomic Spectra and Radiative Transitions*, 2nd edn., Springer Ser. At. Plasm., Vol. 12 (Springer, Berlin, Heidelberg 1992)
3.7 M. Ya. Amusia: *Atomic Photoeffect* (Pergamon, Oxford 1990)
3.8 H.P. Kelly: In *X-Ray and Inner-Shell Processes*, AIP Conf. Proc. 215, 292 (ed. by T.A. Carlson, M.O. Krause, S.T. Manson (AIP, New York 1990)

3.9 V. Schmidt: Rep. Prog. Phys. **55**, 1483 (1992)
3.10 T.-N. Chang (ed.): *Many-Body Theory of Atomic Structure and Photoionization* (World Scientific, Singapore 1993)
3.11 A. Ron, I.B. Goldberg, J. Stein, S.T. Manson, R.H. Pratt, R.Y. Yin: Phys. Rev. A **50**, 1312 (1994)
3.12 I.B. Borovskii, R.V. Vedrinskii, V.I. Kraizman, V.P. Savchenko: Sov. Phys. – Uspekhi **149**, 275 (1986)
3.13 I. Stöhr (ed.): *X-ray Absorption: Principles, Applications, Techniques of EXAFS, SEXAFS and XANES* (Wiley, New York 1988)
3.14 G.V. Marr, J.B. West: Atom. Data **18**, 497 (1976)
3.15 J.B. West, J. Norton: Atom. Data **22**, 103 (1978)
3.16 D. Hofsaess: Atom. Data **24**, 285 (1979)
3.17 B.L. Henke, P. Lee, T.J. Tanaka, R.L. Shimabukuro, B.K. Fujikawa: At. Data Nucl. Data Tables **27**, 1 (1982)
3.18 J.J. Yeh, I. Lindau: Atom. Data Nucl. Data Tables **32**, 1 (1985)
3.19 R.E.H. Clark, R.D. Cowan, F.W. Bobrowicz: Atom. Data Nucl. Data Tables **34**, 415 (1986)
3.20 E. Jannitti, P. Nicolosi, G. Tondello: In *X-Ray and Inner-Shell Processes*, AIP Conf. Proc. **215**, ed. by T.A. Carlson, M.O. Krause, S.T. Manson (AIP, New York 1990)
3.21 M.J. van der Wiel, G. Wiebes: Physica **53**, 225 (1971)
3.22 D.M.P. Holland, K. Codling, J.B. West, G.V. Marr: J. Phys. B **12**, 2465 (1979)
3.23 E.B. Saloman, J.H. Hubbell, J.H. Scofield: At. Data Nucl. Data Tables **38**, 1 (1988)
3.24 H.B.G. Casimir, P. Polder: Phys. Rev. **73**, 360 (1948)
3.25 L.A. Bureeva: Sov. Astrophys. J. **45**, 1218 (1968)
3.26 I.L. Beigman, L.A. Vainshtein, B.N. Chichkov: Sov. Phys. – JETP **53**, 541 (1981)
3.27 M. Pajek, R. Schuch: Phys. Rev. A **45**, 7894 (1992)
3.28 N.B. Delone, S.P. Goreslavsky, V.P. Krainov: J. Phys. B **27**, 4403 (1994)
3.29 L.N. Shabanova, Yu. N. Manakov, A.N. Khlyustalov: Opt. Spectrosc. **47**, 1 (1980)
3.30 G.V. Marr, D.M. Creek: Proc R. Soc. (London) A **304**, 233 (1968)
3.31 D.H. Norcross: Phys. Rev. A **7**, 606 (1973)
3.32 T.V. Cook, F.B. Dunning, G.V. Foltz, R.F. Stebbings: Phys. Rev. A **15**, 1536 (1977)
3.33 V.P. Shevelko: In *Spectroscopy of Multicharged Ions in Hot Plasmas*, ed. by U.I. Safronova (Nauka, Moscow 1991) p. 147
3.34 Y. Yan, M.J. Seaton: J.Phys. B **20**, 6409 (1987)
3.35 A. Wague: Z. Phys. D **13**, 123 (1989)
3.36 P.M.J. Sawey, K.A. Berrington: J. Phys. B **23**, L817 (1990)
3.37 K. Omidvar, A.M. McAllister: Phys. Rev. A **51**, 1063 (1995)
3.38 W.J. Karzas, R. Latter: Astrophys. J. Suppl. **6**, 167 (1961)
3.39 K. Omidvar: XVIII ICPEAC, Book of Abstracts, ed. by T. Andersen, B. Fastrup, F. Folkman, H. Knudsen, Vol. I (Aahus University, Denmark 1993) p. 56
3.40 N.B. Delone, V.P. Krainov: *Atoms in Strong Light Fields*, Springer Ser. Chem. Phys., Vol. 28 (Springer, Berlin, Heidelberg 1985)
3.41 N.L. Manakov, V.D. Ovsyannikov, L.P. Rapoport: Phys. Rep. **141**, 319 (1986)
3.42 N.B. Delone, V.P. Krainov: *Multiphoton Processes in Atoms*, Springer Ser. At. Plasm., Vol. 13 (Springer, Berlin, Heidelberg 1994)
3.43 C.A. Nicalaides, Th. Mercouris, G. Aspromallis: J. Opt. Soc. Am. B **7**, 494 (1990)
3.44 S.V. Khristenko, S.I. Vetchinkin: Opt. Spectrosc. **26**, 186 (1969)
3.45 N.L. Manakov, V.A. Sviridov, A.G. Fainshtein: Sov. Phys. – JETP **95**, 790 (1989)
3.46 M.S. Adams, M.V. Fedorov, V.P. Krainov, D.D. Meyerhofer: Phys. Rev. A **52**, 125 (1995)
3.47 A.V. Vinogradov, V.P. Shevelko: Phys. Scr. **19**, 275 (1979)
3.48 W.R. Johnson, D. Kolb, K.H. Huang: At. Data **28**, 333 (1983)
3.49 A.A. Radzig, B.M. Smirnov: *Reference Data on Atoms, Molecules and Ions*, Springer Ser. Chem. Phys., Vol. 13 (Springer, Berlin, Heidelberg 1985)
3.50 *Handbook of Chemistry and Physics*, ed. by D.R. Lide, 73rd edn. (CRC, Boca Raton 1992–1993)
3.51 R. Szmytkowski, A.M. Alhasan: Phys. Scr. **52**, 309 (1995)

3.52 V.P. Shevelko, A.D. Ulantsev: Opt. Spectrosc. **65**, 590 (1988)
3.53 P. Gombas: *Die statistische Theorie des Atoms und ihre Anwendunqen* (Springer, Wien 1949)
3.54 J.M. Standard, P.R. Certain: J. Chem. Phys. **83**, 3002 (1985)
3.55 R.H. Pratt, I.J. Fend: In *Atomic Inner-Shell Physics*, ed. by B. Crasemann (Plenum, New York 1985) pp 533–580 (and references therein)
3.56 W. Heitler: *Quantum Theory of Radiation*, 3rd edn. (Univ. Press, Oxford 1954)
3.57 A. Sommerfeld: *Atombau und Spectrallinien* (Vieweg, Braunschweig 1939)
3.58 L.D. Landau, L.M. Lifshitz: *The Classical Theory of Fields*, 3rd edn. (Pergamon, New York 1971)
3.59 J.D. Jackson: *Classical Electrodynamics*, 2nd edn. (Wiley, New York 1975)
3.60 V.I. Kogan, A.B. Kukushkin, V.S. Lisitsa: Phys. Rep. **213**, 1 (1992)
3.61 J.M. Jauch, R. Rohrlich: *The Theory of Photons and Electrons*, 2nd edn. (Springer, Berlin, Heidelberg 1980)
3.62 M.Ya. Amusia: Phys. Rep. **162**, 249 (1988)
3.63 V.N. Tsytovitch, I.M. Oiringel (eds.): *Polarization Bremsstrahlung of Particles and Atoms* (Nauka, Moscow 1987)
3.64 R.H. Pratt, H.K. Tseng, C.M. Lee: At. Data Nucl. Data Tables **20**, 175 (1975); erattum At. Data Nucl. Data Tables **26**, 477 (1981)
3.65 L. Kissel, A. Quarles, R.H. Pratt: At. Data Nucl. Data Tables **28**, 381 (1983).
3.66 H.A. Kramers: Philos. Mag. **46**, 836 (1923)

Chapter 4

4.1 S. Trajmar: Nucl. Fusion Suppl. **2**, 15 (1992)
4.2 S. Trajmar, J.W. McConkey: Adv. At. Mol. Opt. Phys. **33**, 63 (1994)
4.3 A.R. Filippelli, C.C. Lin, L.W. Anderson, J.W. Conkey: Adv. At. Mol. Opt. Phys. **33**, 1 (1994)
4.4 M. Inokuti: Rev. Mod. Phys. **43**, 297 (1971)
4.5 Y. Itikawa: Phys. Rep. **143**, 69 (1986)
4.6 J.W. Henry, A.E. Kingston: Adv. At. Mol. Phys. **25**, 267 (1988)
4.7 J.W. Henry: Rep. Prog. Phys. **56**, 3277 (1993)
4.8 I.I. Sobelman, L.A. Vainshtein, E.A. Yukov: *Excitation of Atoms and Broadening of Spectral Lines*, 2nd edn., Springer Ser. Chem. Phys., Vol. 7 (Springer, Berlin Heidelberg 1995)
4.9 I.L. Beigman, V.S. Lebedev: Phys. Rep. **250**, 95 (1995)
4.10 K. Bartschart: J. Phys. B **26**, 3595 (1993)
4.11 H. Ehrhardt, K. Jung, G. Knoth, M. Rädle: In *Atomic and Molecular Physics*, ed. by K. Rai, D.N. Tripathi (World Scientific, Singapore 1987)
4.12 H. Griem: J. Quant. Spectrosc. Radiat. Transfer **40**, 403 (1988)
4.13 Nucl. Fusion Suppl. 4, ed. by R.K. Janev (IAEA, Vienna 1993)
4.14 A.K. Pradhan, J.W. Gallagher: At. Data Nucl. Data Tables **52**, 227 (1992)
4.15 U. Krishnan, B. Stumpf: At. Data Nucl. Data Tables **51**, 151 (1992)
4.16 W. Fite: In *Atomic and Molecular Processes*, ed. by D.R. Bates (Academic, New York 1962)
4.17 W.E. Kauppila, W.R. Ott, W.L. Fite: Phys. Rev. A **1**, 1099 (1970)
4.18 R.L. Long, Jr., D.M. Cox, S.J. Smith: J. Res. NBST A **72**, 521 (1968)
4.19 A.N. Mahan, A. Gallagher, S.J. Smith: Phys. Rev. A **13**, 156 (1976)
4.20 J.F. Williams: J. Phys. B **14**, 1197 (1981)
4.21 J. Lower, I.E. McCarthy, E. Weigold: J. Phys. B **20**, 4771 (1987)
4.22 J.P. Doering, S.O. Vaughan: J. Geophys. Res. **91**, 3279 (1986)
4.23 T.W. Shyn, S.Y. Cho: Phys. Rev. A **40**, 1315 (1989)
4.24 A. Konovalov, I.E. McCarthy: J. Phys. B **27**, L 741 (1994)
4.25 A.E. Kingston, J.E. Lauer: Proc. R. Phys. Soc. (London) **87**, 399 (1966); Proc. R. Phys. Soc. **88**, 597 (1966)
4.26 K. Omidvar: Phys. Rev. **140**, A 38 (1965)

4.27 M.R. Flannery: J. Phys. B **23**, L 501 (1990)
4.28 V.P. Shevelko: Proc. P.N. Lebedev Phys. Inst. **218**, 143 (Nova Science, New York 1993)
4.29 K.A. Berrington: J. Phys. B **24**, 1385 (1991)
4.30 J. Callaway, K. Unnikrishnan: Phys. Rev. A **48**, 4292 (1993); J. Phys. B **26**, L 419 (1993)
4.31 M.P. Scott, P.G. Burke: J. Phys. B **26**, L 191 (1993)
4.32 M.P. Scott, B.R. Odgers, P.G. Burke: J. Phys. B **26**, L827 (1993)
4.33 F.J. de Heer, R. Hoekstra, A.E. Kingston, H.P. Summers: Nucl. Fusion Suppl. **3**, 19 (IAEA, Vienna 1992)
4.34 T. Kato, R.K. Janev: Nucl. Fusion Suppl. **3**, 33 (IAEA, Vienna 1992)
4.35 K.A. Berrington, P.M.J. Sawey: At. Data Nucl. Data Tables **55**, 81 (1993)
4.36 V.A. Gostev, D.V. Elakhovskii, Yu. V. Yaitsev, L.A. Luisova, A.D. Khakhaev: Opt. Spectrosc. **48**, 251 (1980)
4.37 D.L. Rall, F.A. Sharpton, M.B. Schulman, L.W. Anderson, J.E. Lawler, C.C. Lin: Phys. Rev. Lett. **62**, 2253 (1989)
4.38 F.J. de Heer, H.O. Folkerts, F.W. Bliek, R. Hoekstra, T. Kato, A.E. Kingston, K.A. Berrington, H.P. Summers: FOM-Report 0653 (FOM – Institute, Amsterdam 1995)
4.39 V.P. Shevelko, H. Tawara: Rep. NIFS-DATA-28 (National Institute for Fusion Science, Nagoya 1995)
4.40 M.R. Flannery, K.J. McCann: Phys. Rev. A **12**, 846 (1975)
4.41 W.C. Fon, K.A. Berrington, P.G. Burke, A.E. Kingston: J. Phys. B **14**, 2921 (1981)
4.42 N.R. Badnell: J. Phys. B **17**, 4013 (1984)
4.43 K.A. Berrington, P.G. Burke, L.C.G. Freitas, A.E. Kingston: J. Phys. B **18**, 4135 (1985)
4.44 K.C. Mathur, R.P. McEachran, L.A. Parcell, A.D. Stauffer: J. Phys. B **20**, 1599 (1987)
4.45 A.E. Kingston: Unpublished (1992)
4.46 I. Bray, I.E. McCarthy: Phys. Rev. A **47**, 317 (1993)
4.47 V.P. Shevelko, L.A. Vainshtein: *Atomic Physics for Hot Plasmas* (IOP, Bristol 1993)
4.48 V.P. Shevelko: Physica Scr. **37**, 47 (1991); Phys. Scr. **43**, 266 (1991)
4.49 V.I. Ochkur: Sov. Phys. – JETP **18**, 503 (1964)
4.50 I.C. Percival, D. Richards: Adv. Atom. Mol. Phys. **11**, 1 (1975)
4.51 K. Alder, A. Bohr, T. Huns, M. Mottelson, A. Winther: Rev. Mod. Phys. **28**, 432 (1956)
4.52 I.L. Beigman, S.A. Chernyagin: Short Commun. Phys. 5–6, 65 (Lebedev Physics Inst., Moscow 1994)
4.53 I.L. Beigman, M.I. Syrkin: Sov. Phys. – JETP **62**, 226 (1985)
4.54 V.M. Borodin, A.K. Kazansky: J. Phys. B **26**, 1863 (1993)
4.55 R.G. Rolfes, L.G. Gray, O.P. Makarov, K.B. McAdam: J. Phys. B **26**, 2191 (1993)
4.56 D.R. Herrick: Molec. Phys. **35**, 1211 (1978)
4.57 R.F. Stebbings, F.B. Dunning (eds.): *Rydberg States of Atoms and Molecules* (Cambridge Univ. Press, Cambridge 1983)
4.58 T.F. Gallagher: *Rydberg Atoms* (Cambridge Univ. Press, Cambridge 1994)
4.59 I.L. Beigman, L.A. Vainshtein: Sov. Phys. – JETP **25**, 119 (1967)
4.60 M. Matsuzawa: Phys. Rev. A **9**, 241 (1974)
4.61 I.L. Beigman, M.G. Matusovsky: J. Phys. B **24**, 4117 (1991)
4.62 M.I. Syrkin: Phys. Rev. A **51**, 847 (1995)
4.63 R.K. Peterkop: *Theory of Ionization of Atoms by Electron Impact* (Assoc. Univ. Press, Colorado, OH 1977)
4.64 T.D. Märk, G.H. Dunn: *Electron Impact Ionization* (Springer, Berlin, Heidelberg 1985)
4.65 K. Dolder: Adv. At. Mol. Opt. Phys. **32**, 69 (1994)
4.66 H. Tawara, T. Kato: At. Data Nucl. Data Tables **36**, 167 (1987)
4.67 R.S. Freund, R.C. Wetzel, R.J. Shul, T.R. Hayes: Phys. Rev. A **41**, 3575 (1990)
4.68 Y. Itikawa: At. Data Nucl. Data Tables **49**, 209 (1991); ibid. **63**, 315 (1996)
4.69 K.L. Bell, H.B. Gilbody, J.G. Hughes, A.E. Kingston, F.J. Smith: J. Phys. Chem. Ref. Data **12**, 891 (1983)
4.70 M.A. Lennon, K.L. Bell, H.B. Gilbody, J.G. Hughes, A.E. Kingston, M.J. Murray, F.J. Smith: J. Phys. Chem. Ref Data **17**, 1285 (1988)

4.71 M.J. Higgings, J.G. Hughes, H.B. Gilbody et al: *Atomic and Molecular Data for Fusion*, Part 3, Recommended Cross Sections and Rates for Electron Impact Ionization of Atoms and Ions: Copper to Uranium, Report CLM-R294, UKEA (Culham Lab., Abingdon 1989)
4.72 A. Müller: In *Physics of Ion Impact Phenomena*, ed. by D. Mathur, Springer Ser. Chem. Phys., Vol. 54 (Springer, Berlin, Heidelberg 1991)
4.73 H.Genz: AIP Conf. Proc. **94**, 85 (AIP, New York 1982)
4.74 H. Paul, J. Muhr: Phys. Rep. **135**, 47 (1986)
4.75 R. Mayol, F. Salvat: J. Phys. B **23**, 2117 (1990)
4.76 X. Long, M. Liu, F. Ho, X. Peng: At. Data Nucl. Data Tables **45**, 353 (1990)
4.77 R.E.H. Clark, J. Abdallah, Jr.: Astrophys. J., **381**, Part 1, 597 (1991); Phys. Scr. **T37**, 28 (1991)
4.78 V.P. Shevelko, A.M. Solomon, V.S. Vukstich: Phys. Scr. **43**, 158 (1991)
4.79 W. Lotz: Z. Phys. **232**, 101 (1970)
4.80 W. Lotz: Z. Phys. **216**, 241 (1968)
4.81 W. Lotz: Z. Phys. **220**, 466 (1969)
4.82 L. Vriens: In *Case Studies in Atomic Physics*, Vol. 1, ed. by E.W. McDaniel, M.R.C. McDowell (North Holland, Amsterdam 1969) p. 335
4.83 V.I. Ochkur: J. Phys. B25, 445 (1992)
4.84 I.L. Beigman, V.P. Shevelko: Phys. Scr. **51**, 60 (1995); Z.-Q. Wu, S.-C. Li, V.P. Shevelko: Phys. Scr. (1997, to be published)
4.85 R.C. Stabler: Phys. Rev. A **133**, 1268 (1964)
4.86 A.E. Kingston: Proc. R. Phys. Soc. (London) **87**, 193 (1966); ibid. 393 (1966)
4.87 K. Omidvar: Phys. Rev. A **140**, 26 (1965); Phys. Rev. A **166**, 164 (1966)
4.88 M. Matsuzawa: Phys. Rev. A **9**, 241 (1974)
4.89 A. Müller: Phys. Lett. A **113**, 415 (1980)
4.90 M. Stenke, K. Aichele, D. Hathiramani, G. Hofmann, V.P. Shevelko, M. Steidl, R. Völpel, H. Tawara, E. Salzborn: J. Phys. B J. Phys. B **28**, 4853 (1995)
4.91 P.L. Colestock, K.A. Konnor, R.L. Hickok, R.A. Dandl: Phys. Rev. Lett. **40**, 1717 (1978)
4.92 J.H. McGuire: Adv. At. Mol Opt. Phys. **29**, 217 (1992)
4.93 J.H. McGuire: J. Phys. B **28**, 913 (1995)
4.94 J.H. McGuire: In *Atomic Inner Shell Processes*, ed. by B. Crasemann (Academic, New York 1995), Ch. 7
4.95 J.H. McGuire: In *Atomic, Molecular and Optical Physics Reference Book*, ed. by G.W.F. Drake (AIP, New York 1996), Ch. 40
4.96 J.H. McGuire: *Introduction to Dynamic Correlation: Multiple Electron Transitions in Atomic Collisions* (Tulane University, New Orleans 1996) (in press)
4.97 B.L. Schram: Physica **32**, 197 (1966)
4.98 R.C. Wetzel, F.A. Baiocchi, T.R. Hayes, R.S. Freund: Phys. Rev. A **35**, 559 (1987)
4.99 K. Tinschert, A. Müller, R. Becker, E. Salzborn: J. Phys. B **20**, 1823 (1987)
4.100 E. Krishnakumar, S.K. Srivastava: J. Phys. B **21**, 105 (1988)
4.101 H. Lebius, J. Binder, H.R. Koslowski, K. Wiesemann, B.A. Huber: J. Phys. B **22**, 83 (1989)
4.102 R.S. Freund, R.C. Wetzel, R.J. Shull, T.R. Hayes: Phys. Rev. A **41**, 3575 (1990)
4.103 P. McCallion, M.B. Shah, H.B. Gilbody: J. Phys. B **25** 1051 (1992)
4.104 P. McCallion, M.B. Shah, H.B. Gilbody: J. Phys. B **25** 1061 (1992)
4.105 J.A. Syage: Phys. Rev. A **46**, 5666 (1992)
4.106 M.B. Shah, P. McCallion, K. Okuno, H.B. Gilbody: J. Phys. B **26**, 2393 (1993)
4.107 M.A. Bolorizadeh, C.J. Patton, M.B. Shah, H.B. Gilbody: J. Phys. B **27**, 175 (1994)
4.108 M. Stenke, D. Hathiramani, G. Hofmann, V.P. Shevelko, M. Steidl, R. Völpel, E. Salzborn: Nucl. Instrum. Methods Phys. Res. B **98**, 138 (1995)
4.109 A. Müller, R. Frodl: Phys. Rev. Lett. **44**, 29 (1980)
4.110 D.W. Hughes, R.K. Feeney: Phys. Rev. A **23**, 2241 (1981)
4.111 D.R. Hertling, R.K. Feeney, D.W. Hughes, W.E. Sayle II: J. Appl. Phys. **53**, 5427 (1982)
4.112 A. Müller, W. Groh, U. Kneisel, R. Heil, H. Ströher, E. Salzborn: J. Phys. B **16**, 2039 (1983)
4.113 A. Müller, C. Achenbach, E. Salzborn, R. Becker: J. Phys. B **17**, 1427 (1984)

4.114 M.S. Pindzola, D.C. Griffin, C. Bottcher, D.H. Crandall, R.A. Phaneuf, D.C. Gregory: Phys. Rev. A **29**, 1749 (1984)
4.115 A. Müller, K. Tinschert, C. Achenbach, R. Becker, E. Salzborn: J. Phys. B **18**, 3011 (1985)
4.116 A.M. Howald, D.C. Gregory, R.A. Phaneuf, D.H. Crandall, M.S. Pindzola: Phys. Rev. Lett. **56**, 1675 (1986)
4.117 W. Lotz: J. Opt. Soc. Am. **58**, 915 (1968); ibid. **60**, 206 (1970)
4.118 T.A. Carlson, C.W. Nestor, Jr., N. Wasserman, J.D. McDowell: Atomic Data Tables **2**, 63 (1970)
4.119 M. Gryzinski: Phys. Rev. A **138**, 336 (1965)
4.120 M. Zambra, D. Belic, P. Defrance, D.J. Yu: J. Phys. B **27**, 2383 (1994)
4.121 H. Deutsch, K. Becker, T.D. Märk: J. Phys. B **29**, L497 (1996)
4.122 V. Fisher, Yu. Ralchenko, A. Goldgirsh, D. Fisher, Y. Maron: J. Phys. B **25**, 4593 (1995)
4.123 V.P. Shevelko, H. Tawara: Phys. Scr. **52**, 649 (1995)
4.124 V.P. Shevelko, H. Tawara: J. Phys. B **28**, L589 (1995)
4.125 C. Bélenger, P. Defrance, E. Salzborn, D.B. Uskov, V.P. Shevelko, H. Tawara: J. Phys. B (1996) (submitted)

Chapter 5

5.1 E.W. Thomas: *Excitation in Heavy Particle Collisions* (Wiley, New York 1972)
5.2 E.E. Nikitin, S.Y. Umanskii: *Theory of Slow Atomic Collisions*, Springer Ser. Chem. Phys., Vol. 30 (Springer, Berlin Heidelberg 1984)
5.3 R.K. Janev, H. Winter: Phys. Rep. **117**, 265 (1985)
5.4 R.K. Janev, L.P. Presnyakov, V.P. Shevelko: *Physics of Highly Charged Ions* (Springer, Berlin, Heidelberg 1985)
5.5 H.B. Gilbody: Adv. Atom. Mol. Phys. **22**, 143 (1986)
5.6 R.K. Janev, W.D. Langer, K. Evans, D.E. Post (eds.): *Electron Processes in Hydrogen-Helium Plasmas*, Springer Ser. At. Plasm., Vol. 4 (Springer, Berlin, Heidelberg 1987)
5.7 R. McCarroll: *Recent Studies in Atomic and Molecular Processes*, ed. by A.E. Kingston (Plenum, New York 1987)
5.8 J.S. Briggs, J.H. Macek: Adv. At. Mol. Phys. **28**, 1 (1991)
5.9 W. Fritsch: Phys. Rep. **202**, 1 (1991)
5.10 B.H. Bransden, M.R. McDowell: *Charge Exchange and the Theory of Ion-Atom Collisions* (Clarendon, Oxford 1992)
5.11 D S W. Crothers, L.J. Dube: Adv. At. Mol. Phys. **30**, 287 (1993)
5.12 E.W. McDaniel, J.B.A. Mitchell, M.E. Rudd: *Atomic Collisions (Heavy Particle Projectiles)* (Wiley, New York 1993)
5.13 W. Fritsch: Nucl. Fusion Suppl. **3**, 41 (1992)
5.14 F.J. de Heer, R. Hoekstra, H.P. Summer: Nucl. Fusion Suppl. **3**, 47 (1992)
5.15 K. Reymann, K.-H. Schartner, B. Sommer: Phys. Rev. A **38**, 2290 (1988)
5.16 M. Anton, D. Detleffsen, K.-H. Schartner: Nucl. Fusion Suppl. **3**, 51 (1992)
5.17 M. Anton, D. Detleffsen, K.-H. Schartner, A. Werner: J. Phys. B **26**, 2005 (1993)
5.18 C.O. Reinhold, R.E. Olson, W. Fritsch: Phys. Rev. A **41**, 4837 (1990)
5.19 M. Bailey, R. Bruch, E. Rauscher, S. Bliman: Plasma Phys. Controlled Fusion (1996) (in press)
5.20 G.H. Olivera, C.A. Ramirez, R.D. Rivarola: Phys. Rev. A **47**, 1000 (1993)
5.21 C.F. Barnett: *Atomic Data for Fission*, Vol. 1, Rep. ORNL – 6086, (Oak Ridge Natl. Lab., Oak Ridge 1990)
5.22 N. Shimakura, S. Suzuki, M. Kimura: Phys. Rev. A **47**, 3930 (1993)
5.23 N. Toshima, H. Tawara: NIFS-DATA-26 (National Institute for Fusion Science, Nagoya 1995)
5.24 T.F. Gallagher: Phys. Rep. **210**, 319 (1992)
5.25 I.L. Beigman, V.S. Lebedev: Phys. Rep. **250**, 95 (1995)
5.26 K.B. MacAdam, L.G. Gray, R.G. Rolfes: Phys. Rev. A **42**, 5269 (1990); X. Sun, K.B. MacAdam: Phys. Rev. A **47**, 3913 (1993)
5.27 D. Detleffsen, M. Anton, A. Werner, K.-H.Scharter: J. Phys. B **27**, 4195 (1994)

5.28 A.R. Schlatmann, R. Hoekstra, H.O. Folkerts, R. Morgenstern: J. Phys. B **25**, 3155 (1992) and references therein
5.29 J. Schweinzer, D. Wutte, H.P. Winter: J. Phys. B **27**, 137 (1994)
5.30 G. Horvath, H.P. Winter, F. Aumayer, J. Schweinzer: XIX ICPEAC, Books of Abstracts, ed. by J.B.A. Mitchell, J.W. McConkey, C.E. Brion (Whistler, British Columbia, Canada 1995) p. 61
5.31 R.K. Janev, L.P. Presnyakov: J. Phys. B **13**, 4233 (1980)
5.32 R.K. Janev: Nucl. Fusion Suppl. **6**, 171 (1995); Phys. Rev. A **53**, 219 (1996)
5.33 M.F. Watts, C.J. Hopkins, G.C. Angel, G.H. Dunn, H.B. Gilbody: J. Phys. B **9**, 3739 (1986)
5.34 H.B. Gilbody: Nucl. Fusion Suppl. **3**, 55 (1992)
5.35 T. Tabata, R. Ito, T. Shirai, Y. Nakai, H.T. Hunter, R.A. Phanenf: Nucl. Fusion Suppl. **2**, 91 (1992)
5.36 L.P. Presnyakov, D.B. Uskov: Sov. Phys. – JETP **59**, 515 (1984)
5.37 R.E. Olson: J. Phys. B **13**, 483 (1980)
5.38 A. Igarashi, T. Shirai: Phys. Rev. A **50**, 4945 (1994)
5.39 C.J. Patton, M.B. Shah, M.A. Bolorizadeh, J. Geddes, H.B. Gilbody: J. Phys. B **28**, 1821 (1995); ibid. p. 3889
5.40 M.B. Shah, C.J. Patton, J. Geddes, H.B. Gilbody: Nucl. Instrum. Methods B **98**, 280 (1995)
5.41 L.C. Tribedi, K.G. Prasad, P.N. Tandon, Y. Chen, C.D. Lin: Phys. Rev. A **49**, 1015 (1994)
5.42 M. McCartney, D.S.F. Crothers: J. Phys. B **26**, 4561 (1993)
5.43 M.H. Chen, B. Grasemann: Atom. Data Nucl. Data Tables **41**, 257 (1989)
5.44 E. Brazilewicz, J. Brazilewicz, T. Czyzenski, L. Glowacka, M. Jaskova, Th. Kauer, A.P. Kobzev, M. Pajek, D. Trautmann: J. Phys. B **24**, 1669 (1991)
5.45 M. Pajek, A.P. Kobzev, D. Trautmann, Th. Kauer: Nucl. Instrum. Methods B **52**, 109 (1990)
5.46 W.M. Ariyasinghe, H.T. Awuku, D. Powers: Phys. Rev. A **42**, 3819 (1990)
5.47 N.B. Mahli, T.J. Gray: Phys. Rev. A **44**, 7199 (1991)
5.48 A.G. Kochur, V.L. Sukhorukov, A.I. Didenko, Ph.V. Demekhin: J. Phys. B **28**, 387 (1995)
5.49 Z. Szokefalvi-Nagy, I. Demeter, L.H. Quynh: Nucl. Instrum. Methods B **75**, 54 (1993)
5.50 M.B. Shah, D.S. Elliot, H.B. Gilbody: J. Phys. B **20**, 2481 (1987)
5.51 M.B. Shah, H.B. Gilbody: J. Phys. B **14**, 2361 (1981)
5.52 W.L. Fite, R.F. Stebbings, D.G. Hummer, R.T.Brackmann: Phys. Rev A **119**, 663 (1960)
5.53 D.R. Bates, G. Griffing: Proc. R. Phys. Soc. (London) A **66**, 961 (1953)
5.54 D. Banks, K.S. Barnes, J. Wilson: J. Phys. B **9**, L141 (1976)
5.55 R. Shakeshaft: Phys. Rev. A **18**, 1930 (1978)
5.56 J.H. McGuire: Phys. Rev. A **26**, 143 (1982)
5.57 D.S.F. Crothers, J.F. Cunn: J. Phys. B **16**, 3229 (1983)
5.58 D.J.W. Hardie, R.E. Olson: J. Phys. B **16**, 1983 (1983)
5.59 W. Fritsch, C.D. Lin: Phys. Rev. A **27**, 3361 (1983)
5.60 T.G. Winter, C.D. Lin: Phys. Rev. A **29**, 3071 (1984)
5.61 M.B. Shah, H.B. Gilbody: J. Phys. B **18**, 899 (1985)
5.62 K.L. Bell, A.E. Kingston: J. Phys. B **2**, 635 (1969)
5.63 R.G. Mantague, M.F.A. Harrison, A.C.H. Smith: J. Phys. B **17**, 3295 (1984)
5.64 G.H. Gillespie: J. Phys. B **15**, L 729 (1982); Phys. Lett. **93** A, 327 (1983)
5.65 W. Wu, C.L. Cocke, S. Datz, J.P. Giese, C.R. Vane: In *XIX ICPEAC, Books of Abstracts*, ed. by J.B.A. Mitchell, J.W. McConkey, C.E. Brion (Whistler, British Columbia Canada 1995); Phys. Rev. Lett. (1995) (in press)
5.66 A. Barany, S. Ovchinnikov: Phys. Scr. T **46**, 243 (1993)
5.67 R.K. Janev, G. Ivanovski, E.A. Soloviev: Phys. Rev. A **49**, R645 (1994)
5.68 L.H. Andersen, P. Hvelplund, H. Knudsen, S.P. Møller, A.N. Sørensen, K. Elsner, K.G. Rensfelt, E. Uggerhøj: Phys. Rev. A **36**, 3612 (1987)
5.69 A. Dalgarno, H.R. Sadeghpour: Phys. Rev. A **46**, R 3591 (1992)
5.70 J.H. McGuire: Adv. At. Mol. Opt. Phys. **29**, 217 (1992)
5.71 J.H. McGuire, N. Berrah, R.J. Bartlett, J.A.R. Samson, J.A. Tanis, C.L. Cocke, A.S. Schlachter: J.Phys. B **28**, 913 (1995)
5.72 J. Wang, J.H. McGuire, J. Burgdorfer: Phys. Rev. A **51**, 4687 (1995)

5.73 L.P. Presnyakov, H. Tawara, I. Yu. Tolstikhina, D.B. Uskov: J. Phys. B **28**, 785 (1995)
5.74 J. Ullrich, R. Dörner, H. Berg, C.L. Cocke et at.: Nucl. Instrum. Methods B **87**, 70 (1994); W. Wu, S. Datz, N.L. Jones, H.F. Krause, B. Rosner, K.D. Sorge, C.R. Vane: Phys. Rev. Lett. **76**, 4324 (1996)
5.75 C.L. Cocke, R.E. Olson: Phys. Rep. **205**, 193 (1991)
5.76 M. Horbatsch: Suppl. Z. Phys. D **21**, S 63 (1991)
5.77 M. Horbatsch: J. Phys. B **25**, 3797 (1992)
5.78 I. Ben-Itzhak, T.G. Gray, J.C. Legg, J.H. McGuire: Phys. Rev. A **37**, 3685 (1988)
5.79 R.E. Olson: Phys. Rev. A **40**, 2843 (1989)
5.80 A.S. Schlachter, K.H. Berkner, H.F. Beyer, W.G. Graham, W. Groh, R. Mann, A. Müller, R.E. Olson, R.V. Pyle, J.W. Stearns, J.A. Tanis: Phys. Scr. T **3**, 153 (1983)
5.81 S. Kelbch, J. Ulrich, R. Mann, P. Richard, H. Schmidt-Böcking: J. Phys. B **18**, 323 (1985)
5.82 S. Kelbch, J. Ullrich, W. Rauch, H. Schmidt-Böcking, M. Horbatsch, R.M. Dreizler, S. Hagmann, R. Anholt, A.S. Schlachter, A. Müller, P. Richard, Ch. Stoller, C.L. Cocke, R. Mann, W.E. Meyerhof, J.D. Rasmussen: J. Phys. B **19**, L47 (1986)
5.83 A. Müller, B. Schuch, W. Groh, E. Salzborn: Z. Phys. D **7**, 251 (1987)
5.84 R.D. Debois, S.T. Manson: Phys. Rev. A **35**, 2007 (1987)
5.85 R.E. Olson, J. Ulrich, H. Schmidt-Böcking: Phys. Rev. A **39**, 5572 (1989)
5.86 C.L. Cocke: Phys. Rev. A **20**, 749 (1979);
S. Kelbch, C.L. Cocke, S. Hagmann, M. Horbatsch, C. Kelbch, R. Koch, H. Schmidt-Böcking, J. Ullrich: J. Phys. B **23**, 1277 (1990)
5.87 H. Berg, R. Dörner, C. Kelbch, S. Kelbch, J. Ullrich, S. Hagmann, P. Richard, H. Schmidt-Böcking, A.S. Schlachter, M. Prior, H.J. Crawford, J.M. Engelage, I. Flores, D.H. Lyod, J. Pedersen, R.E. Olson: J. Phys. B **21**, 3929 (1988)
5.88 H. Berg: Report GSI-93-12 (Darmstadt, Germany 1993)
5.89 J. Ullrich, C.L. Cocke, S. Kelbch, R. Mann, P. Richard, H. Schmidt-Böcking: J. Phys. B **17**, L785 (1984)
5.90 T. Matsuo, T. Tonuma, H. Kumagai, H. Tawara: Phys. Rev. A **50**, 1178 (1995)
5.91 O.I. Tolstikhin, D.B. Uskov, V.P. Shevelko: XVIII ICPEAC, Book of Abstracts, ed. by T. Andersen, B. Fastrup, F. Folkman, H. Knudsen, Aahus University, Denmark, July (1993) p. 478
5.92 H. Berg, J. Ullrich, E. Bernstein, M. Unverzage, L. Spielberger, J. Euler, D. Schardt, O. Jagutzki, H. Schmidt-Böcking, R. Mann, P.H. Mokler, S. Hagmann, P.D. Fainstein: J. Phys. B **25**, 3655 (1992)
5.93 C.L. Cocke: In *Semiclassical Descriptions of Atomic and Nuclear Collisions*, ed. by J. Bang, J. De Boez (North-Holland, Amsterdam 1986) p. 205
5.94 W.K. Wu, B.A. Huber, K. Wiesemann: At. Data Nucl. Data Tables **40**, 57 (1988)
5.95 N.V. Fedorenko: Sov. Phys. – JTP **15**, 1947 (1972)
5.96 H.D. Betz: Rev. Mod. Phys. **44**, 465 (1972)
5.97 C.D. Lin: J. Phys. B **23**, 4215 (1990)
5.98 H. Klinger, A. Müller, E. Salzborn: J. Phys. B **8**, 230 (1975)
E. Salzborn: IEEE Trans. NS-**23**, 947 (1976)
A. Müller, Ch. Achenbach, E. Salzborn: Phys. Rev. Lett. A **70**, 410 (1979)
5.99 M. Barat, P. Roncin: J. Phys. B **25**, 2205 (1992)
5.100 R.A. Phaneuf, R.K. Janev, H.T. Hunter: Nuclear Fusion (Special Suppl.) 7 (1987) (IAEA, Vienna 1987)
5.101 R.W. McCullough: Nucl. Instrum. Methods B **98**, 170 (1995)
5.102 R.K. Janev: Phys. Lett. A **160**, 67 (1991)
5.103 S.B. Hansen, L.G. Gray, E. Horsdal-Peterson, K.B. MacAdam: J. Phys. B **24**, L 315 (1991)
5.104 R.E. Olson: J. Phys. B **25**, 4241 (1992)
5.105 Th. Wormann, Z. Roller-Lutz, H.O. Lutz: Phys. Rev. A **47**, R1594 (1993)
5.106 H.-P. Winter: J. Phys. B **27**, 137 (1994)
5.107 M. Gieler, F. Aumayr, M. Weber, H.P. Winter, J. Schweinzer: J. Phys. B **26**, 2153 (1993)
5.108 R.E. Olson: J.Phys. B **26**, L 817 (1993)

5.109 R. Hoekstra, H.P. Summers, F.J. de Heer: Nucl. Fusion Suppl. **3**, 63 (1992)
5.110 W. Fritsch: Phys. Scr. T **37**, 75 (1991)
5.111 A.S. Schlachter, J.W. Stearns, W.G. Graham, K.H. Berkner, R.V. Pyle, J.A. Tanis: Phys. Rev. A **27**, 3372 (1983)
5.112 AIP Conference Proceedings No. 210, 1990 (Brookhaven Natl. Lab. 1989)
5.113 F. Melchert, W. Debus, R. Schulze, R.E. Olson, E. Salzborn: Z. Phys D (Suppl.) **21**, S 249 (1991)
5.114 J. Risley: In *Proc. 11th ICPEAC*, ed. by N. Oda, K. Takayanagi, Invited papers and progress reports (North-Holland, Amsterdam 1980) p. 679
5.115 M. Schön, S. Krüdener, F. Melchert, K. Rinn, M. Wagner, E. Salzborn, M. Karamera, S. Szucs, M. Terao, D. Fussen, R. Janev, X. Urbain, F. Brouillard: Phys. Rev. Lett. **59**, 1565 (1987)
5.116 X. Urbain, A. Giusti-Suzor, D. Fussen, C. Kubah: J. Phys. B **19**, L273 (1986)
5.117 J.T. Moseley, W. Aberth, J. Peterson: Phys. Rev. Lett. **24**, 435 (1970)
5.118 B. Peart, R. Grey, K. Dolder: J. Phys. B **9**, 3047 (1976)
5.119 K.L. Bell, A.E. Kingston, P.J. Madden: J. Phys, B **11**, 3977 (1978)
5.120 D. Fussen, W. Claeys: J. Phys. B **17**, L89 (1984)
5.121 A.M. Ermolaev: J. Phys. B **21**, 81 (1988); J. Phys. B **25**, 3133 (1992)
5.122 R.D. Rundell, K.L. Aitken, M.F.A. Harrison: J. Phys B **2**, 954 (1969)
5.123 T.D. Gaily, M.F.A. Harrison: J. Phys B **3**, L25 (1970)
5.124 B. Peart, M.A. Bennet, K. Dolder: J. Phys. B **18**, L439 (1985); B. Peart, D.A. Hayton: J. Phys. B **25**, S109 (1992)
5.125 S. Szucs, M. Karemera, M. Terao, F. Brouillard: J. Phys. B **17**, 1613 (1984)
5.126 W. Schon, S. Krüdener, F. Melchert, K. Rinn, M. Wagner, E. Salzborn: J. Phys. B **20**, L759 (1987)
5.127 V. Sidis, C. Kubach, D. Fussen: Phys. Rev. A **27**, 2431 (1983)
5.128 F. Borondo, A. Macias, A. Riera: Chem. Phys. Lett. **100**, 63 (1983)
5.129 R. Shingal, B.H. Bransden: J. Phys. B **20**, L533 (1987)
5.130 L.F. Errea, C. Harel, P. Jimeno, H. Jouin, L. Mendez, A.Riera: Nucl. Instrum. Methods B **98**, 335 (1995)
5.131 F. Melchert, W. Debus, M. Liehr, R.E. Olson, E. Salzborn: Europhys. Lett. **9**, 433, (1989)
5.132 R. Schulze, F. Melchert, M. Hagmann, S. Krüdener, J. Kruger, E. Salzborn, C.O. Reinhold, R.E. Olson: J. Phys. B **24**, L 7 (1991)
5.133 R.E. Olson, A. Salop: Phys. Rev. A **16**, 531 (1977)
5.134 J.H. McGuire, L. Weaver: Phys. Rev. A **16**, 41 (1977)
5.135 F. Melchert, M. Benner, S. Krüdener, R. Schulze, S. Meuser, K. Huber, E. Salzborn, D.B. Uskov, L.P. Presnyakov: Phys. Rev. Lett. **74**, 888 (1995)
5.136 F. Melchert, R. Schulze, S. Krüdener, S. Meuser, E. Salzborn, D.B. Uskov, A. Ulantsev, L.P. Presnyakov: J. Phys. B **28**, 3299 (1995)
5.137 D.B. Uskov: AIP Conf. Proc. **360**, 687 (AIP, New York 1995)
5.138 F. Ebel, E. Salzborn: J. Phys B **20**, 4531 (1987)
5.139 B. Peart, S.J. Foster: J. Phys. B **20**, L691 (1987)
5.140 F. Melchert: AIP Conf. Proc. **295**, (AIP, New York 1993) p. 575
5.141 Y.K. Kim, M. Inokuti: Phys. Rev. A **4**, 665 (1971)
5.142 L.P. Presnyakov, H. Tawara, D.B. Uskov: Nucl. Instrum. Methods B **98**, 332 (1995)
5.143 E. Salzborn: *Private communication* (1995) Unpublished
5.144 D.J. Yu, S. Rachafi, J. Jureta, P. Defrance: J. Phys. B **25**, 4593 (1992)

Subject Index

GPSR Compliance
The European Union's (EU) General Product Safety Regulation (GPSR) is a set of rules that requires consumer products to be safe and our obligations to ensure this.

If you have any concerns about our products, you can contact us on

ProductSafety@springernature.com

In case Publisher is established outside the EU, the EU authorized representative is:

Springer Nature Customer Service Center GmbH
Europaplatz 3
69115 Heidelberg, Germany

www.ingramcontent.com/pod-product-compliance
Ingram Content Group UK Ltd.
Pitfield, Milton Keynes, MK11 3LW, UK
UKHW021827190726
13853UKWH00003B/1243
* 9 7 8 3 6 6 2 0 3 4 3 5 4 *